T0184784

A GRADUATE
COURSE IN
ALGEBRA

A GRADUATE

COURSE IN

ALGEBRA

VOLUME 2

Ioannis Farmakis
Department of Mathematics, Brooklyn College
City University of New York, USA

Martin Moskowitz
Ph.D. Program in Mathematics, CUNY Graduate Center
City University of New York, USA

 **World Scientific**

NEW JERSEY · LONDON · SINGAPORE · BEIJING · SHANGHAI · HONG KONG · TAIPEI · CHENNAI · TOKYO

Published by

World Scientific Publishing Co. Pte. Ltd.

5 Toh Tuck Link, Singapore 596224

USA office: 27 Warren Street, Suite 401-402, Hackensack, NJ 07601

UK office: 57 Shelton Street, Covent Garden, London WC2H 9HE

Library of Congress Cataloging-in-Publication Data
Names: Farmakis, Ioannis. | Moskowitz, Martin A.
Title: A graduate course in algebra : in 2 volumes / by Ioannis Farmakis
    (City University of New York, USA), Martin Moskowitz (City University of New York, USA).
Description: New Jersey : World Scientific, 2017– |
    Includes bibliographical references and index.
Identifiers: LCCN 2017001101| ISBN 9789813142626 (hardcover : alk. paper : v. 1) |
    ISBN 9789813142633 (pbk : alk. paper : v. 1) | 9789813142664 (hardcover : alk. paper : v. 2) |
    ISBN 9789813142671 (pbk : alk. paper : v. 2) | ISBN 9789813142602 (set : alk. paper) |
    ISBN 9789813142619 (pbk set : alk. paper)
Subjects: LCSH: Algebra--Textbooks. | Algebra--Study and teaching (Higher)
Classification: LCC QA154.3 .F37 2017 | DDC 512--dc23
LC record available at https://lccn.loc.gov/2017001101

British Library Cataloguing-in-Publication Data
A catalogue record for this book is available from the British Library.

Printed in Singapore

Contents

Preface and Acknowledgments

This book is a two-volume graduate text in Algebra. The first volume which contains basic material is also suitable for self study as well as preparing for the graduate Qualifying Exam in Algebra, consists of Chapters 1-6, while the second volume, designed for second year topics course in algebra, consists of Chapters 7-13. Our book was written over the course of the past five years, the second author having taught the subject many times. In our view Algebra is an integral and organic part of mathematics and not to be placed in a bubble. For this reason the reader will find a number of topics represented here not often found in an algebra text, but which relate to nearby subjects such as Lie groups, Geometry, Topology, Number Theory and even Analysis.

As a convenience to the instructor using our book we have created a pamphlet containing a large number of exercises together with their solutions.

Here we list the topics covered in Volume II.

Chapter 7 (Multilinear Algebra)

Here we study the important notion of *orientation*, proving among other things that real square matrices are continuously deformable into one another if and only if the signs of their determinants are conserved. We then turn to Riesz's theorem for a separable Hilbert space and use it to define the *Hodge star* $(*)$ *operator* which is especially important in differential geometry and physics. We prove *Liouville's formula* making use of the fact that the diagonalizable matrices are *Zariski dense* in $M(n, \mathbb{C})$. Finally, we study Grassmannians and their relationship to the exterior product.

Chapter 8 (Symplectic Geometry)

We begin by showing that the symplectic group $\mathrm{Sp}(V)$ is generated by transvections. Then we prove the Williamson normal form which states that any positive definite symmetric real $2n \times 2n$ matrix is diagonalizable by a symplectic matrix. This leads to the well known *Gromov non-squeezing theorem* which states that the two-dimensional shadow of a symplectic ball of radius r in $\mathbb{R}^{2n}$ has area at least πr^2. Finally, we discuss the curious relationship between Gromov's non-squeezing theorem and the Heisenberg *uncertainty principle* of quantum mechanics.

Chapter 9 (Commutative Rings with Identity)

In this chapter we systematically find all *principal ideal domains* of a certain type which are not *Euclidean rings*. We present Kaplansky's characterization of a Unique Factorization Domain (UFD), the *power series ring $R[[x]]$*, the *prime avoidance lemma* and the *Hopkins-Levitsky theorem*. Then we turn to *Nakayama's lemma*, where we give a number of applications as well as several variants. We then define the *Zariski topology* and derive some of its important properties (as well as giving a comparison with the *Euclidean topology*). Finally, we discuss the *completion of a ring* and prove the completion of $R[x]$ is $R[[x]]$.

Chapter 10 (Valuations and p-adic Numbers)

Here we study the topology of $\mathbb{Q}_p$ and its curiosities concerning isosceles triangles and spheres, (where any point is a center). As an application, we present *Monsky's theorem* that if a square is cut into triangles of equal area, the number of triangles must be even.

Chapter 11 (Galois Theory)

In addition to the standard features of Galois theory, we prove the *theorem of the primitive element* and emphasize the role of characteristic zero. We also give a purely algebraic proof of the *fundamental theorem of algebra*.

Chapter 12 (Group Representations)

First we deal with the basic results of representations and characters of a finite group G over $\mathbb{C}$, including divisibility properties. For example, $\deg \rho$ divides the index of the center of G for every irreducible representation ρ of G. We then prove *Burnside's p, q theorem* which is an important criterion for solvability and Clifford's theorem on the decomposition of the restriction of an irreducible representation to an invariant subgroup. We then turn to infinite groups where we study linear groups over a field of characteristic zero, real representations and their relationship to calculating homotopy groups and preserving quadratic forms, and finally return to Pythagorean triples of Volume I.

Chapter 13 (Representations of Associative Algebras)

In this final chapter we prove Burnside's theorem, the Double Commutant theorem, the *big Wedderburn theorem* concerning the structure of finite dimensional semi-simple algebras over an algebraically closed field of characteristic zero. We then give two proofs of *Frobenius' theorem* (the second one using algebraic topology) which classifies the finite dimensional division algebras over $\mathbb{R}$. Finally, using quaternions, we prove Lagrange's 4-square theorem. This involves the Hürwitz integers and leads very naturally to the study of the geometry of the unique 24-cell polyhedron in $\mathbb{R}^4$ and its exceptional properties, which are of particular interest in physics.

A significant bifurcation in all of mathematics, and not just in algebra, is the distinction between the commutative and the non-commutative. In our book this distinction is ever present. We encounter it first in Volume I in the case of groups in Chapters 1 and 2, then in rings and algebras in Chapter 5 and modules in Chapter 6, where the early sections deal with the commutative, while the later ones address the non-commutative. Chapter 9 is devoted to commutative algebra while Chapters 12 and 13 to non-commutative algebra, both being of great importance.

We hope our enthusiasm for the material is communicated to our readers. We have endeavored to make this book useful to both beginning and advanced graduate students by illustrating important concepts and techniques, offering numerous exercises and providing historical notes that may be of interest to our readers. We also hope the book will be of use as a reference work to professional mathematicians specializing in areas of mathematics other than algebra.

The endgame of our endeavor has turned out to be a family affair! We wish to heartily thank Andre Moskowitz for reading the text and making numerous useful suggestions for its improvement and Oleg Farmakis for creating all the graphics. We also thank our wives, Dina and Anita, for their forbearance during the long and challenging process of bringing this book to completion.

August 2016

Ioannis Farmakis   Martin Moskowitz

Brooklyn College-CUNY   The Graduate Center-CUNY

INTERDEPENCE OF THE CHAPTERS

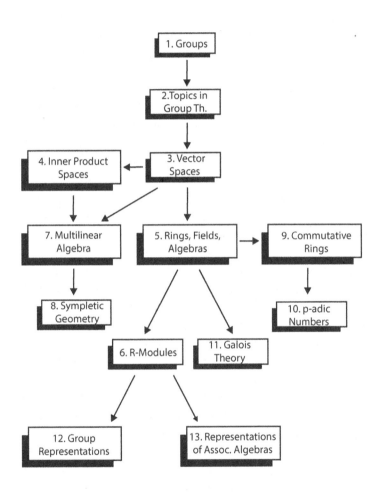

Chapter 7

Multilinear Algebra

In this chapter we will always assume the characteristic of the field k is not 2. Later we will impose additional restrictions on k. We remind the reader that $l.s_k$ means the k-linear span of those vectors.

7.1 Multilinear Functions

Definition 7.1.1. Let $r \geq 2$ be an integer and $V_1, \ldots, V_r$ be finite dimensional vector spaces, all over the same field, k. If W is another such vector space we shall call a function $\omega : V_1 \times \ldots \times V_r \to W$ a *multilinear map* if ω is a k-linear function on each V_i while holding all the other variables fixed. We write $M_k(V_1, \ldots V_r, W)$ for the set of all such multilinear functions. When $r = 1$ this is $\mathrm{Hom}_k(V, W)$. When $r = 2$ we call such functions *bilinear* and write $B(V_1, V_2, W)$.

Proposition 7.1.2. $M_k(V_1, \ldots V_r, W)$ *is a vector space over k and*

$$\dim_k \Big(M_k(V_1, \ldots V_r, W) \Big) = \dim_k(W) \prod_{i=1}^{r} \dim_k(V_i).$$

Proof. This follows because such a function is determined by its values on a basis of $V_1 \times \ldots \times V_r$, which, in turn, are determined by their coordinates in W. $\qquad\square$

Definition 7.1.3. When W has dimension 1, i.e. $W = k$, we shall call ω a *multilinear form*. The most important special case of this is when all $V_i = V$. Then ω is called a *multilinear r-form*. Of course, if $r = 1$, this is just the dual space V^* of V.

An important example here is $\text{Hom}_k(V, W)$, the space of linear operators from V to W. It has dimension $\dim V \dim W$ and consists of $m \times n$ matrices. Let $v_1, \dots v_n$ be a basis of V, $w_1, \dots w_m$ be a basis of W, and $T : V \to W$ be a k-linear transformation. Then for each $i = 1, \dots n$, $T(v_i) = \sum_j t_{i,j} w_j$. Then $(t_{i,j})$ is the matrix of T relative to these bases. It has m rows and n columns. In particular, the matrices $E_{i,j}$, which are of the same form but where there is a 1 in the (i, j) spot and zeros elsewhere, are a basis for such T and $T = \sum_{i,j} t_{i,j} E_{i,j}$. In particular, when $V = W$ we get $\text{End}_k(V)$, the endomorphism algebra of V. In that case the $E_{i,j}$ are the matrix units.

Exercise 7.1.4. Use the matrix above to get an associated bilinear form $V \times W \to k$ by defining $\beta(v_i, w_j) = t_{i,j}$ and extending to $V \times W$ using the bilinearity of β. Show this map is injective and hence by dimension a vector space isomorphism. Thus $B(V, W) = \text{Hom}_k(V, W)$.

7.2   The Symmetric Algebra

The universal property for tensor products (see 3.7.3 of Volume I) says that every bilinear map $U \times V \longrightarrow W$ is essentially the same as a linear map $U \otimes V \longrightarrow W$ and if this linear map is something natural and surjective, we can view W as a quotient space of $U \otimes V$. Many interesting products in algebra can be regarded as quotient spaces of tensor products. One of the most familiar is the *polynomial algebra*. Here is its definition.

Let V be a finite dimensional vector space over a field k and set

$$T^0(V) = k, \quad T^1(V) = V, \quad \text{and} \quad T^n(V) = \underbrace{V \otimes \cdots \otimes V}_{n-\text{times}} = V^{\otimes n}.$$

Let C^n be the subspace of $T^n(V)$ spanned by all vectors of the form

$$(x_1 \otimes x_2 \otimes \cdots \otimes x_i \otimes \cdots \otimes x_j \otimes \cdots \otimes x_n) - (x_1 \otimes x_2 \otimes \cdots \otimes x_j \otimes \cdots \otimes x_i \otimes \cdots \otimes x_n)$$

where $x_i \in V$ for any $i, j = 1, ..., n$.

The tensor product is not commutative, but when we quotient $T^n(V)$ by the subspace C^n, we will obtain a commutative product. This is because the thing we are dividing out by annihilating all the non-commutative elements. Similarly, the tensor product is a big infinite dimensional object, but when we quotient $T^n(V)$ by the subspace C^n, we obtain something of finite dimension because the thing we are dividing out is also a big infinite dimensional object.

Definition 7.2.1. The n-th *symmetric power* of V is the quotient space

$$\mathsf{Sym}^n(V) = T^n(V)/C^n.$$

We will write

$$v_1 \vee \cdots \vee v_n := v_1 \otimes \cdots \otimes v_n + C^n, \quad \text{or simply } v_1 \ldots v_n$$

and observe that this product is symmetric in the sense that

$$v_1 \vee \cdots \vee v_n = v_{\sigma(1)} \vee \cdots \vee v_{\sigma(n)}.$$

Theorem 7.2.2. (*Universal property of symmetric powers*) *Let V and W be vector spaces and $f : \mathsf{Sym}^n(V) \longrightarrow W$ be a symmetric n-linear map, i.e., for each permutation $\sigma \in S_n$,*

$$f(v_1, ..., v_n) = f(v_{\sigma(1)}, ..., v_{\sigma(n)}),$$

for $v_i \in V$, $i = 1, \ldots, n$. Then, there exists a unique linear map

$$\widetilde{f} : \mathsf{Sym}^n(V) \longrightarrow W, \quad \text{such that} \quad \widetilde{f}(v_1 \vee \cdots \vee v_n) = f(v_1, ..., v_n),$$

with $v_i \in V$, $i = 1, \ldots, n$.

Proof. Using the universal property of the n-fold tensor product $V^{\otimes n}$, we obtain a unique linear map

$$f_1 : V^{\otimes n} \longrightarrow W, \quad \text{such that} \quad f_1(v_1 \otimes \cdots \otimes v_n) = f(v_1, \ldots, v_n),$$

for all $v_1, ..., v_n \in V$. In view of the symmetry of f, the linear map f_1 vanishes on C^n. Hence it factors through a linear map $\widetilde{f} : \mathsf{Sym}^n(V) \longrightarrow W$, with the desired property. $\qquad\square$

Lemma 7.2.3. *The number of n-tuples of non-negative integers with sum k is*

$$\binom{n+k-1}{k}.$$

Proof. The claim about the number of monomials of degree k follows immediately from this lemma, which should be contrasted with the fact that the number of n-tuples of zeros and ones with sum k is $\binom{n}{k}$.

We can describe any such n-tuple in the following way. Take a line of $n+k-1$ boxes. Then choose $n-1$ boxes, and place barriers in these boxes. Let

1. a_1 be the number of empty boxes before the first barrier.

2. a_2 be the number of empty boxes between the first and second barriers.

3. ...

4. a_n be the number of empty boxes after the last barrier.

Then a_1, ..., a_n are non-negative integers with sum k. Conversely, given n non-negative integers with sum k, we can represent it with $n-1$ barriers in $n+k-1$ boxes: place the first barrier after a_1 empty boxes, the second after a_2 further empty boxes, and so on. Thus the required number of n-tuples is equal to the number of ways to position $n-1$ barriers in $n+k-1$ boxes, which is

$$\binom{n+k-1}{n-1} = \binom{n+k-1}{k}.$$

$\square$

Proposition 7.2.4. *Let* $\dim V = n$. *Then the dimension of* $Sym^k(V)$ *is*

$$\dim\left(Sym^n(V)\right) = \binom{n+k-1}{k}.$$

Proof. Suppose that $\dim V = n$.   Let $e_1, \ldots, e_n$ be a basis and $k[x_1, \ldots, x_n]$ be the polynomial ring over k in indeterminates $x_1, \ldots, x_n$. Consider the map $e_i \mapsto x_i$. This map extends to a unique k-linear map $V \longrightarrow k[x_1, \ldots, x_n]$. Since the polynomial ring is commutative, the universal property of symmetric powers gives us a k-linear homomorphism $\mathsf{Sym}^n(V) \longrightarrow k[x_1, \ldots, x_n]$. Now, $\mathsf{Sym}^n(V)$, as a quotient space, is spanned by the images of pure tensors of the form e_{i_1}, $e_{i_2}, \ldots, e_{i_n}$. Moreover, any such can be reordered in $\mathsf{Sym}^n(V)$ using the symmetric property to assume $i_1 < \ldots < i_n$. Hence, $\mathsf{Sym}^n(V)$ is spanned by the ordered monomials of the form $e_{i_1} \cdots e_{i_j}$, for all $i_1 \leq i_2 \leq i_j$ and all $j \geq 1$. Clearly, such a monomial maps to $x_{i_1} \cdots x_{i_j}$ in the polynomial ring $k[x_1, \ldots, x_n]$. Since by definition the ordered monomials give a basis for the k-vector space $k[x_1, \ldots, x_n]$, they must in fact be linearly independent in $\mathsf{Sym}^n(V)$ as well. Thus we have constructed an isomorphism. Thus, $S(V)$ is isomorphic to $k[x_1, \ldots, x_n]$, the isomorphism mapping a basis element e_i of V to the indeterminate x_i in the polynomial ring. In particular, $S_n(V)$ has basis given by all ordered monomials of the form $e_{i_1} \cdots e_{i_k}$ with $1 \leq i_1 \leq \cdots \leq i_j \leq n$. In these terms, the universal property of the symmetric algebra tells us the polynomial algebra $k[x_1, \ldots, x_n]$ is the free commutative k-algebra on the generators $x_1, \ldots, x_n$. Note, in particular, that if $\det V = n$ the basis theorem implies the result. $\square$

We leave to the reader the proof of the following facts:

1. The map $V^n \longrightarrow \mathsf{Sym}^n(V)$, defined by $(x_1 \ldots x_n) \mapsto x_1 \vee \cdots \vee x_n$ is k-linear.

2. $\mathsf{Sym}^n(V)$ is commutative.

3. If $\{e_1, \ldots, e_l\}$ is a basis for V, then any element of $\mathsf{Sym}^n(V)$ can be written as a polynomial of degree n in $e_1, \ldots e_l$.

4. There is a bilinear map

$$\mathsf{Sym}^m(V) \times \mathsf{Sym}^n(V) \longrightarrow \mathsf{Sym}^{m+n}(V)$$

such that

$$(x_1 \vee \cdots \vee x_m , y_1 \vee \cdots \vee y_n) \mapsto x_1 \vee \cdots \vee x_m \vee y_1 \vee \cdots \vee y_n.$$

When expressed in terms of a basis, this map is just polynomial multiplication. In other words, $\mathsf{Sym}^n(V)$ can be thought of as the space of polynomials of degree n in the elements of V, and the operation "$\vee$" is just the ordinary commutative polynomial multiplication. When the elements are expressed in terms of a basis, these look exactly like polynomials.

Example 7.2.5. Consider the product

$$(1,1,1) \vee (-1,1,1) \vee (0,1,-1) \in \mathsf{Sym}^3(\mathbb{R}^3).$$

Let $x = (1,0,0)$, $y = (0,1,0)$, $z = (0,0,1)$. When we expand this product in terms of x, y, z, we obtain:

$$(1,1,1) \vee (-1,1,1) \vee (0,1,-1) = (x+y+z) \vee (-x+y+z) \vee (y-z)$$
$$= x^2 z - x^2 y + y^3 + y^2 z - yz^2 - z^3.$$

To consider polynomials of mixed degrees, we look at the direct sum of all the symmetric products of V:

$$\mathsf{Sym}^\bullet(V) := \mathsf{Sym}^0(V) \oplus \mathsf{Sym}^1(V) \oplus \mathsf{Sym}^2(V) \oplus \mathsf{Sym}^3(V) \oplus \cdots$$

The reader can check that the multiplication

$$\mathsf{Sym}^m(V) \times \mathsf{Sym}^n(V) \longrightarrow \mathsf{Sym}^{m+n}(V)$$

makes $\mathsf{Sym}^\bullet(V)$ a commutative ring.

7.3   The Exterior Algebra

Let V be a finite dimensional vector space over k of dimension n and $1 \leq r \leq n$. We denote by S_r the permutation group on r letters. Its order, $|S_r| = r!$. Let sgn stand for the signum of a permutation; that is whether it is composed of an even or odd number of transpositions. Thus $\text{sgn}(\sigma) = \pm 1$ according to whether this number is even or odd. The reader should verify that sgn is well defined (meaning that if a permutation is a product of an even or odd number of transpositions, any other way of writing it will give respectively an even or odd number of transpositions) and a homomorphism $\text{sgn} : S_r \to \{\pm 1\}$.

Definition 7.3.1. A multilinear form ω is called an *alternating form* if for each $(x_1, \ldots, x_r) \in V \times \ldots \times V$ and each $\sigma \in S_r$, $\omega(x_{\sigma(1)}, \ldots, x_{\sigma(r)}) = \text{sgn}(\sigma)\omega(x_1, \ldots, x_r)$.

The name "alternating" because ω is merely invariant under the alternating group, A_r, rather than the full symmetric group, S_r.

Trivial cases: Of course if ω is identically zero this is an alternating form. Also, if $r = 1$ all linear forms are alternating.

We first give some features of alternating linear forms.

Lemma 7.3.2. *Let ω be an r-form. Then ω is alternating if and only if for each $x \in V$ we have $\omega(\ldots, x, \ldots, x, \ldots) = 0$.*

Proof. If ω is alternating and $x \in V$ we have $2\omega(\ldots, x, \ldots, x, \ldots) = 0$. Hence $\omega(\ldots, x, \ldots, x, \ldots) = 0$. Conversely, suppose this condition holds for all $x \in V$. Let $x, y \in V$ and $\beta(x, y) = \omega(\ldots, x, \ldots, y, \ldots)$. Then β is a bilinear form and

$$\beta(x + y, x + y) = \beta(x, x) + \beta(x, y) + \beta(y, x) + \beta(y, y).$$

Since $\beta(x + y, x + y) = \beta(x, x) = \beta(y, y) = 0$, $\beta(x, y) + \beta(y, x) = 0$ and so $\beta(x, y) = -\beta(y, x)$. This proves ω is alternating whenever σ is a transposition. Since sgn is a homomorphism and the transpositions generate S_r, ω is alternating. □

Instead of just having two of the x's equal, the following corollary is a more general statement.

Corollary 7.3.3. *Let ω be an alternating r-linear form and $x_1, \ldots, x_r$ be r elements of V. If these are linearly dependent, then $\omega(x_1, \ldots, x_r) = 0$.*

Proof. If $x_1, \ldots, x_r$ is linearly dependent then one of them can be written as a linear combination of the others. Let us say $x_k = \sum_{i \neq k} c_i x_i$. Then by linearity in the k-th argument,

$$\omega(x_1, \ldots, x_r) = \omega(\ldots x_i, \ldots, \sum_{i \neq k} c_i x_i, \ldots)$$

$$= \sum_{i \neq k} c_i \omega(\ldots x_i, \ldots, x_i \ldots) = 0.$$

$\square$

We denote the set of alternating multilinear forms by $\bigwedge^r(V)$. Some authors refer to this as the r-th exterior power of V. It is a subspace of $M_k(V, \ldots V, k)$ and therefore of finite dimension. We will calculate its dimension shortly. Now assume k has characteristic zero. There is a "natural" k-linear projection $P^r : \bigotimes^r V \to \bigwedge^r V$ given by

$$x_1 \otimes \ldots \otimes x_r \mapsto \frac{1}{|S_r|} \sum_{\sigma \in S_r} \operatorname{sgn}(\sigma)(x_{\sigma(1)} \otimes \ldots \otimes x_{\sigma(r)}),$$

and extending by linearity. As we shall see, its properties depend on invariant integration on the group S_r.

Proposition 7.3.4. $P^r(\bigotimes^r V) \subseteq \bigwedge^r V$ *and* $(P^r)^2 = P^r$.

Proof. We first prove the second statement.

$$(P^r)^2(x_1 \otimes \ldots \otimes x_r) = P^r\left(\frac{1}{|S_r|} \sum_{\sigma \in S_r} \operatorname{sgn}(\sigma)(x_{\sigma(1)} \otimes \ldots \otimes x_{\sigma(r)})\right).$$

But this is

$$\frac{1}{|S_r|} \sum_{\sigma \in S_r} \operatorname{sgn}(\sigma) P^r(x_{\sigma(1)} \otimes \ldots \otimes x_{\sigma(r)}) =$$

$$= \frac{1}{|S_r|} \sum_{\sigma \in S_r} \operatorname{sgn}(\sigma) \frac{1}{|S_r|} \sum_{\tau \in S_r} \operatorname{sgn}(\tau)(x_{\tau\sigma(1)} \otimes \ldots \otimes x_{\tau\sigma(r)}).$$

Because sgn is a homomorphism with values in an Abelian group this latter term is just

$$\frac{1}{|S_r|} \sum_{\sigma \in S_r} \frac{1}{|S_r|} \sum_{\tau \in S_r} \text{sgn}(\tau\sigma)(x_{\tau\sigma(1)} \otimes \ldots \otimes x_{\tau\sigma(r)}).$$

On the other hand, S_r is a finite group and translation just permutes the group elements. Therefore, it follows that this last term is

$$\frac{1}{|S_r|} \sum_{\sigma \in S_r} \frac{1}{|S_r|} \sum_{\mu \in S_r} \text{sgn}(\mu)(x_{\mu(1)} \otimes \ldots \otimes x_{\mu(r)}) = P^r(x_{\sigma(1)} \otimes \ldots \otimes x_{\sigma(r)}).$$

Because the $x_1 \otimes \ldots \otimes x_r$ generate $\bigotimes^r(V)$, $(P^r)^2 = P^r$.

We now show $P^r(\otimes^r(V)) \subseteq \bigwedge^r(V)$. Let $\tau \in S_r$. Then,

$$\tau(P^r)(x_1 \otimes \cdots \otimes x_r) = \left(\frac{1}{|S_r|} \sum_{\sigma \in S_r} \text{sgn}(\sigma)\tau(x_{\sigma(1)} \otimes \cdots \otimes x_{\sigma(r)})\right).$$

But this is

$$\left(\frac{1}{|S_r|} \sum_{\sigma \in S_r} \text{sgn}(\sigma)x_{\tau\sigma(1)} \otimes \cdots \otimes x_{\tau\sigma(r)}\right).$$

Replace $\tau\sigma$ by μ. That is, translate $\sigma \mapsto \tau^{-1}\sigma$. Then,

$$\left(\frac{1}{|S_r|} \sum_{\mu \in S_r} \text{sgn}(\tau^{-1}\mu)x_{\tau\sigma(1)} \otimes \cdots \otimes x_{\tau\sigma(r)}\right).$$

But because sgn is a group homomorphism this is

$$\text{sgn}(\tau^{-1})\left(\frac{1}{|S_r|} \sum_{\mu \in S_r} \text{sgn}(\mu)x_{\mu(1)} \otimes \cdots \otimes x_{\mu(r)}\right)$$

$$= \text{sgn}(\tau^{-1})P^r(x_1 \otimes \cdots \otimes x_r).$$

However, it is also true that $\text{sgn}(\tau^{-1}) = \text{sgn}(\tau)$. Therefore

$$\tau(P^r)(x_1 \otimes \cdots \otimes x_r) = \text{sgn}(\tau)P^r(x_1 \otimes \cdots \otimes x_r).$$

$\square$

Hence $P^r(\bigotimes^r V) = \bigwedge^r V$. In this way $\bigwedge^r V$ is a quotient space of $\bigotimes^r V$. It is the space of all skew-symmetric tensors in $\bigotimes^r V$. We will now show $\bigwedge^r V$ comes equipped with an r-linear map

$$V \times \cdots \times V \to \overset{r}{\bigwedge} V$$

given by $(x_1, \ldots, x_r) \mapsto x_1 \wedge \cdots \wedge x_r$, which factors through $T^r(V)$. For this reason one might say that this is *the mother of all alternating r-linear maps*.

Let $x_1, \ldots, x_n$ be a basis of V. Then the $x_{i_1} \otimes \cdots \otimes x_{i_r}$, where the x_i's are chosen arbitrarily from this basis, is itself a basis of $\bigotimes^r V$. Hence the images $P^r(x_{i_1} \otimes \cdots \otimes x_{i_r}) = x_{i_1} \wedge \cdots \wedge x_{i_r}$ generate $\bigwedge^r(V)$. Because P^r is linear, it follows that $x_1 \wedge \cdots \wedge x_r$ is an r-linear function on $V \times \cdots \times V$.

We know the effect of a permutation $\tau \in S_r$ (after applying the linear projection P^r) on $x_1 \wedge \cdots \wedge x_r$. For, as we saw in the proof just above,

$$S_\tau(x_1 \wedge \cdots \wedge x_r) = x_{\tau(1)} \wedge \cdots \wedge x_{\tau(r)} = \text{sgn}(\tau)(x_1 \wedge \cdots \wedge x_r).$$

Which of these $x_1 \wedge \cdots \wedge x_r$ are linearly independent? As we know, we just get a scalar multiple of such a term when we unpermute. Also we get zero if any of the i_s's are equal. Thus we may assume we have only *distinct indices in increasing order*. Suppose there was a relation among them,

$$\sum_{i_1 < \ldots < i_r} c_{i_1, \ldots, i_r} x_{i_1} \wedge \cdots \wedge x_{i_r} = 0.$$

We can unpermute these by absorbing the scalar multiple ± 1 and the constants and get $x_{i_1} \wedge \cdots \wedge x_{i_r} = 0$ for some fixed $i_1 < \ldots < i_r$. Since the $x_{i_1}, \ldots, x_{i_r}$ are linearly independent in V this contradicts the result above. Thus $\{x_{i_1} \wedge \cdots \wedge x_{i_r} : i_1 < \ldots < i_r\}$ is a basis of $\bigwedge^r V$ and $\dim \bigwedge^r(V) = \binom{n}{r}$.

As a consequence of this we get two conclusions. The first tells us that $x_1, \ldots, x_r$ of V are linearly independent if and only if $x_1 \wedge \ldots \wedge x_r \neq 0$. While the second tells us that if one has two different bases for a subspace W of V (of $\dim r$), then each of their wedge products is some non-zero multiple of the other. We formalize this in the next two lemmas.

Lemma 7.3.5. *If $x_1, \ldots, x_r$ of V are linearly independent, then $x_1 \wedge \cdots \wedge x_r \neq 0$.*

Proof. Suppose $x_1, \ldots, x_r$ are linearly independent. Extend this to a basis $x_1, \ldots, x_r, \ldots x_n$ of V. By the very definition of the wedge, $x_1 \wedge \ldots \wedge x_r$ is a basis vector of $\bigwedge^r V$. Hence it is non-zero. $\square$

Lemma 7.3.6. *Let W be a subspace of V with two bases, $v_1, \ldots v_r$ and $x_1, \ldots x_r$. Then, there is some $\lambda \neq 0 \in k$ so that $v_1 \wedge \ldots \wedge v_r = \lambda \cdot x_1 \wedge \ldots \wedge x_r$.*

Proof. Since $x_1, \ldots x_r$ is a basis of V each of the v_i can be expressed as a linear combination of the $x_1, \ldots x_r$. From this it follows that $v_1 \wedge \cdots \wedge v_r$ can be expressed as a linear combination of the various $x_{i_1}, \wedge \cdots \wedge x_{i_r}$. These in turn are each $\pm(x_1 \wedge \cdots \wedge x_r)$. Thus $v_1 \wedge \cdots \wedge v_r = \lambda \cdot x_1 \wedge \ldots \wedge x_r$. The only question that remains is: does $\lambda = 0$? But this cannot be since if $\lambda = 0$, then $v_1 \wedge \ldots \wedge v_r = 0$, violating Lemma 7.3.5 since $v_1, \ldots, v_r$ are linearly independent. $\square$

The following corollary enables us to calculate dimensions.

Corollary 7.3.7. *If $r > n$, $\dim \bigwedge^r V = 0$. For $1 \leq r \leq n$,*

$$\dim \bigwedge^r V = \binom{n}{r}.$$

In particular,

Corollary 7.3.8. *For $1 \leq r \leq n$, $\bigwedge^r V$ and $\bigwedge^{n-r} V$ are isomorphic since they have the same dimension. Also, $\dim \bigwedge^n V = 1$.*

Remark: Relation to cross product in $\mathbb{R}^3$

The wedge product generalizes the cross product in the following sense. If $V = \mathbb{R}^3$, then

$$\dim \bigwedge^1 \mathbb{R}^3 = \dim \bigwedge^2 \mathbb{R}^3 = 3.$$

So we have the following identifications (which are isomorphisms):

$$\bigwedge^1 \mathbb{R}^3 \longrightarrow \mathbb{R}^3, \quad \alpha = \alpha_1 e_1 + \alpha_2 e_2 + \alpha_3 e_3 \mapsto (\alpha_1, \alpha_2, \alpha_3),$$

$$\bigwedge^2 \mathbb{R}^3 \longrightarrow \mathbb{R}^3, \quad \beta = \beta_1 e_2 \wedge e_3 + \beta_2 e_3 \wedge e_1 + \beta_3 e_1 \wedge e_2 \mapsto (\beta_1, \beta_2, \beta_3).$$

Now, let $\eta, \xi \in \bigwedge^1 \mathbb{R}^3$, with $\eta = \sum_1^3 \eta_i e_i$ and $\xi = \sum_1^3 \xi_i e_i$. Then,

$$\eta \wedge \xi = (\eta_1 \xi_2 - \eta_2 \xi_1) e_1 \wedge e_2 + (\eta_1 \xi_3 - \eta_3 \xi_1) e_1 \wedge e_3 + (\eta_2 \xi_3 - \eta_3 \xi_2) e_2 e_3.$$

Therefore, if we identify η and ξ with the vectors (η_1, η_2, η_3), and (ξ_1, ξ_2, ξ_3), then the isomorphisms above make $\eta \wedge \xi$ correspond as above to the cross product of these vectors.

Exercise 7.3.9. We define the alternating algebra $\bigwedge V$ as follows. As a k-space, it is given by

$$\bigwedge V = \bigoplus_{r=1,\dots,n} \bigwedge^r V.$$

To multiply elements in $\bigwedge V$, let $u \in \bigwedge^r(V)$ and $w \in \bigwedge^s(V)$. Then u is a linear combination of $x_1 \wedge \dots \wedge x_r$ and w is a linear combination of $x_{r+1} \wedge \dots \wedge x_s$. Consider partitions of the set $(i_1, \dots, i_{r+s})$ into two parts $(i_1, \dots, i_r | i_{r+1}, \dots i_{r+s})$ consisting of r and s elements respectively. Then, $u \wedge w$ is given by

$$(u \wedge w)(x_1, \dots, x_{r+s})$$

$$= \frac{r! s!}{(r+s)!} \sum_{I_r \mid I_{r+s}} \mathrm{sgn}(i_1, \dots, i_{r+s}) u(x_{i_1}, \dots, x_{i_r}) w(x_{i_{s+1}}, \dots, x_{i_{r+s}}),$$

where $I_r = \{i_1, \dots i_r\}$, and $I_{r+s} = \{i_{r+1}, \dots i_{r+s}\}$.

1. Show $\dim \bigwedge V = 2^n$.

2. Prove $\bigwedge V$ is an associative algebra.

3. Prove $\bigwedge V$ is not commutative. In fact, if $u \in \bigwedge^r(V)$ and $w \in \bigwedge^s(V)$, then $u \wedge w = (-1)^{rs} w \wedge u$.

7.3.1   Determinants

As we know from Corollary 7.3.8, $\dim \bigwedge^n V = 1$. Moreover, if $x_1, \ldots, x_n$ are linearly independent and ω is a non-zero n-linear alternating form, then $\omega(x_1, \ldots, x_n) \neq 0$. This is because the x's form a basis of V. Hence, if $y_1, \ldots, y_n$ is any set of n vectors, we can write them as linear combinations of the x's. Expanding by multilinearity we get a big linear combination of ω always involving the x's. If in such a term two of the x's coincide, this term is zero and may be neglected. Thus, in each term all the x's are distinct. If $\omega(x_1, \ldots, x_n) = 0$, then all these terms would also be $\pm$ zero. Hence $\omega(y_1, \ldots, y_n) = 0$, a contradiction. In particular, as we shall see, any n-linear alternating form is a multiple of det.

We seek a function $\delta : M(n, k) \to k$, but instead of writing $\delta(a_{i,j})$ we write $\delta(a_i)$, where the a_i are the n column vectors of the matrix $a_{i,j}$. Here $V = k^n$ and $\delta : V \times \cdots \times V \to k$ is a non-trivial alternating n-linear function. Since, as we know, δ is essentially unique; it will be uniquely determined by normalizing it: $\delta(e_i) = 1$, where (e_i) are the standard basis of V. Thus, we will also assume $\delta(I) = 1$. Now let $T \in \mathrm{End}_k(V)$ and for each alternating n-multilinear form ω define $T(\omega)(x_1, \ldots, x_n) = \omega(Tx_1, \ldots, Tx_n)$. This makes the left hand side again an alternating n-multilinear form and gives an action

$$\mathrm{End}_k(V) \times \bigwedge^n V \to \bigwedge^n V.$$

In particular,

$$\delta(Tx_1, \ldots, Tx_n) = \delta(T)\delta(x_1, \ldots, x_n), \quad (*)$$

where $\delta(T) \in k$ and depends on T alone. Now we write $\det(T)$ for $\delta(T)$. A special case of this is $Tx = cx$ for $x \in V$. Then $\det(cI) = c^n$. This is because

$$\det(cx_1, \ldots, cx_n) = c^n \det(x_1, \ldots, x_n) = \det(T) \det(x_1, \ldots, x_n),$$

and det is non-trivial. In particular, $\det(0) = 0$.

Corollary 7.3.10. *For T and $S \in \mathrm{End}_k(V)$, $\det(TS) = \det(T)\det(S)$.*

Proof.

$$\begin{aligned}
\det(TS)\det(x_1, \ldots, x_n) &= \det(TSx_1, \ldots, TSx_n) \\
&= \det(T)\det(Sx_1, \ldots, Sx_n) \\
&= \det(T)\det(S)\det(x_1, \ldots, x_n).
\end{aligned}$$

Since det is non-trivial, $\det(TS) = \det(T)\det(S)$. □

Corollary 7.3.11. *Let $\sigma \in S_n$, $x_1, \ldots, x_n$ be a basis of V and P_σ be the permutation matrix of this basis corresponding to σ. Then $\det(P_\sigma) = \mathrm{sgn}(\sigma)$.*

This follows immediately from (*) and the multilinearity of det.

Corollary 7.3.12. *For $T \in \mathrm{End}_k(V)$, T is invertible if and only if $\det(T) \neq 0$. Moreover, $\det(T^{-1}) = \frac{1}{\det(T)}$.*

Proof. If T is invertible then $TT^{-1} = I$. Hence $\det(T)\det(T^{-1}) = 1$. Hence $\det(T) \neq 0$ and $\det(T^{-1}) = \frac{1}{\det(T)}$. Now suppose $\det(T) \neq 0$. Then by (*) we get $\det(Tx_1, \ldots, Tx_n) = \det(T)\det(x_1, \ldots, x_n)$. Since det is a non-trivial n-form we see that $\det(Tx_1, \ldots, Tx_n)$ is also a non-trivial n-form. Therefore, $(Tx_1, \ldots, Tx_n)$ is a basis of V. Hence T is invertible. □

Corollary 7.3.13. $\det(T)$ *is independent of the choice of basis.*

Proof. If T is the matrix of a linear transformation with respect to some basis, then PTP^{-1} is the matrix with respect to a different basis. Hence $\det T = \det(PTP^{-1})$. □

Corollary 7.3.14. *Let $A = (a_{i,j})$ be the matrix of a linear transformation and A^t be its transpose. Then $\det(A) = \det(A^t)$.*

This follows immediately from the uniqueness of det (and the normalization $\det(I) = 1$), together with the fact that $A \mapsto A^t$ is k-linear and $I^t = I$.

Our next corollary is called the *determinant expansion theorem.*

Corollary 7.3.15. *Let $A = (a_{i,j})$ be the matrix of a linear transformation, then $\det(A) = \sum_{\sigma \in S_n} \mathrm{sgn}(\sigma) a_{\sigma(1)1} \cdots a_{\sigma(n)n}$.*

In particular, for $n = 1$, $\det(a) = a$ and for 2×2, $\det A = ad - bc$. In general, $\det(A)$ is a *homogeneous polynomial of degree n* in n^2 variables with coefficients ± 1. In particular, the coefficients are in $\mathbb{Q}$.

Proof. Let $(x_1, \ldots, x_n)$ be a basis for V and $Ax_i = \sum_j a_{i,j} x_j$. Then since *det* is multilinear and alternating

$$\det(Ax_1, \ldots, Ax_n) = \sum_{\sigma \in S_n} \mathrm{sgn}(\sigma) a_{\sigma(1)1} \cdots a_{\sigma(n)n} \det(x_1, \ldots, x_n).$$

But by (*) the left hand side of this equation is $\det(A) \det(x_1, \ldots, x_n)$.
□

We leave the next result to the reader. It can be proved using the determinant expansion theorem, just above, by induction on j. An important special case is when each of the A_i is 1×1, that is, A is triangular.

Corollary 7.3.16. *Let A be a block triangular matrix with diagonal blocks $A_1, \ldots, A_j$. Then, $\det(A) = \det(A_1) \cdots \det(A_j)$.*

In particular, A is invertible if and only if each A_i is invertible.

The following corollary which follows from the one just above is also left to the reader.

Corollary 7.3.17. *Let $(0) \subseteq W_1 \subseteq \cdots \subseteq W_j = V$ be a sequence of T-invariant subspaces of V and consider the induced maps $\widetilde{T}_i$ on the various quotients*

$$W_1/(0) \xrightarrow{\widetilde{T}_1} W_2/W_1 \xrightarrow{\widetilde{T}_2} \cdots \xrightarrow{\widetilde{T}_{j-1}} V/W_{j-1}.$$

Then,

$$\det(T) = \prod_{i=1}^{j-1} \det(\widetilde{T}_i).$$

Exercise 7.3.18. Show that

$$\det \begin{pmatrix} 1 & 1 & \cdots & 1 \\ x_1 & x_2 & \cdots & x_n \\ x_1^2 & x_2^2 & \cdots & x_n^2 \\ x_1^3 & x_2^3 & \cdots & x_n^3 \\ \vdots & \vdots & \cdots & \vdots \\ x_1^{n-1} & x_2^{n-1} & \cdots & x_n^{n-1} \end{pmatrix} = (-1)^{n(n-1)/2} \prod_{i<j}(x_i - x_j).$$

This matrix is known as the *Vandermonde matrix.*

7.3.2   Cramer's Rule

We now get a formula for the inverse of an invertible matrix A of order n. For $(i,j) = 1, \ldots, n$, let $c_{i,j} = (-1)^{i+j}\det(A_{i,j})$. This number is called the (i,j)-th *co-factor*. Here $A_{i,j}$ is the $(n-1) \times (n-1)$ order matrix gotten by striking out the i-th row and j-th column. By the expansion theorem we get

$$\det(A) = (-1)^{1+j}\det(A_{1,j}) + \ldots + (-1)^{n+j}\det(A_{n,j})$$
$$= \sum_{k=1}^{n} a_{k,j}c_{k,j}.$$

But this is $\sum_{k=1}^{n}(c_{j,k})^t a_{k,j}$. Hence

$$1 = \sum_{k=1}^{n} \frac{(c_{j,k})^t}{\det(A)} a_{k,j}.$$

Now consider the matrix $\widetilde{A}$ gotten by replacing the j-th column of A by the l-th column of A, where $j \neq l$. Then $\det(\widetilde{A}) = 0$ and so the expansion theorem yields $\sum_{k=1}^{n} a_{k,j}c_{k,l} = 0$, $j \neq l$. That is

$\sum_{k=1}^{n}(c_{j,k})^{t}a_{k,j} = 0.$ Combining these tells us $\frac{C}{\det(A)}A = I,$ where $C = (c_{ij}).$ Thus

$$A_{i,j}^{-1} = \frac{C_{i,j}}{\det(A)} = \frac{(-1)^{i+j}\det(A_{j,i})}{\det(A)}.$$

For example, the inverse of the 2×2 matrix

$$A = \begin{pmatrix} a & b \\ c & d \end{pmatrix}, \quad \text{with } ad - bc \neq 0, \quad \text{is } A^{-1} = \frac{1}{ad - bc}\begin{pmatrix} d & -b \\ -c & a \end{pmatrix}.$$

In particular, all its coefficients of A^{-1} are rational functions in the entries of A with nowhere vanishing denominators. The polynomials involved all have coefficients in $\mathbb{Q}$.

Now suppose we have a non-homogeneous system of n equations in n unknowns,

$$a_{11}x_1 + \ldots + a_{1n}x_n = b_1$$

$$\ldots\ldots\ldots\ldots\ldots\ldots\ldots\ldots$$

$$\ldots\ldots\ldots\ldots\ldots\ldots\ldots\ldots$$

$$a_{m1}x_1 + \ldots + a_{mn}x_n = b_n.$$

To solve this system we have *Cramer's rule.*

Corollary 7.3.19. *Given the system,*

$$A\boldsymbol{x} = \boldsymbol{b},$$

where A is a non-singular $n \times n$ matrix, its unique solution is

$$\boldsymbol{x} = A^{-1}\boldsymbol{b}.$$

Proof. Applying 7.3.2 we get

$$\mathbf{x} = \frac{1}{\det(A)}C\mathbf{b}.$$

Thus there is a unique solution given by

$$x_1 = \tfrac{1}{\det(A)} \left(\begin{array}{cc} & \text{last } n-1 \\ \mathbf{b} & \text{columns} \\ & \text{of } A \end{array} \right)$$

$$\cdots\cdots\cdots\cdots\cdots\cdots\cdots\cdots\cdots\cdots\cdots$$

$$\cdots\cdots\cdots\cdots\cdots\cdots\cdots\cdots\cdots\cdots\cdots$$

$$x_n = \tfrac{1}{\det(A)} \left(\begin{array}{cc} \text{first } n-1 & \\ \text{columns} & \mathbf{b} \\ \text{of } A & \end{array} \right)$$

Explicitly,

$$x_i = \frac{\det(a_1, \ldots a_{i-1}, b_i, a_{i+1}, \ldots a_n)}{\det(A)}$$

On the other hand, if $\det(A) = 0$, then there are no solutions because A is not surjective. Then there are $\mathbf{b} \in V$ so that $A\mathbf{x} \neq \mathbf{b}$ for any $\mathbf{x} \in V$.

When $k = \mathbb{R}$, or $\mathbb{C}$, the solution $\mathbf{x}$ depends continuously on A.

7.3.3   Orientation

An important notion in algebraic topology, and differential geometry (especially integration on manifolds) is that of *orientation*. Here we will define orientation only for a vector space which, as we shall see, perforce *must be real*.

Consider a circle contained in a plane in $\mathbb{R}^3$. What does it mean to move along this circle counterclockwise? Whether a circle is traversed clockwise or counterclockwise depends on whether it is viewed from below or above this plane. *To fix an orientation of this plane means to choose one interpretation over the other.* In a sense, this is rather like placing a mark to indicate which is the appropriate side of the plane. Mathematically, this is done by fixing a basis of the plane (a two-dimensional vector space) and identifying this basis, via some linear isomorphism, with the canonical basis of $\mathbb{R}^2$, or equivalently, fixing an *ordered basis*.

More precisely, let V be a real vector space with $\dim V = n$, $0 < n < \infty$, and $\{v_1, \ldots, v_n\}$, $\{w_1, \ldots, w_n\}$ be two bases of V. We know that there is a unique invertible $n \times n$-matrix $A = (a_{ij})$ such that

$$w_i = \sum_1^n a_{ij} v_j.$$

Definition 7.3.20. If $\det A > 0$, the two bases $(v_i)_{i=1,...,n}$ and $(w_j)_{j=1,...,n}$ are said to have the *same orientation* and we write $(v_i)_{i=1,...,n} \sim (w_j)_{j=1,...,n}$. Otherwise they are said to have the *opposite orientation*.

Example 7.3.21. The standard basis e_1, e_2, e_3 of $\mathbb{R}^3$ and the basis e_3, e_1, e_2, define the same (positive) orientation, since

$$\det(A) = \det \begin{pmatrix} 0 & 1 & 0 \\ 0 & 0 & 1 \\ 1 & 0 & 0 \end{pmatrix} = 1 > 0,$$

while the two bases e_1, e_2, e_3 and e_2, e_1, e_3, define different orientations, since

$$\det(A) = \det \begin{pmatrix} 0 & 1 & 0 \\ 1 & 0 & 0 \\ 0 & 0 & 1 \end{pmatrix} = -1 < 0.$$

We leave to the reader the proof of the following theorem.

Theorem 7.3.22. *The property of having the same orientation induces an equivalence relation on the set of all bases of the vector space V, and there are precisely two equivalence classes for this relation[1].*

Definition 7.3.23. An *orientation* on V is a choice of one of these two equivalence classes. Once an orientation has been chosen, the bases lying in the chosen class are said to be *positively oriented*, while those in the other class are called *negatively oriented*.

---

[1]This shows that the choice of an orientation of a vector space is an arbitrary choice: it would have been equally possible to have called the positively oriented bases negatively oriented, and vice versa. It is not a random fact that the actual choice of orientation is frequently based on an appeal to the structure of the human body (left-right), or the right (or left) hand rule of Physics, or the motion of the Sun in the heavens (clockwise or counterclockwise).

Exercise 7.3.24. Why is the complex case excluded?

An alternative and useful definition of the orientation of a real vector space V is the following. If V is a 1-dimensional real vector space, each non-singular 1×1-matrix is of the form (x), $x \neq 0$ with $\det(x) = x$. In this case, each of the two equivalence classes of 1×1-matrices with non-zero determinant coincides with one of the two connected components of $V - \{0\}$ (one of the two half-lines). An orientation on V is a choice of one of these two components. This shows that one can think of an orientation of an n-dimensional real vector space V as a *"labeling"* rule which, for every $v \in V$, labels one of the two components of the set $\bigwedge^n V - \{0\}$, by $\bigwedge^n V_+$, which henceforth we will call the *"positive"* part of $\bigwedge^n V$, and the other component by $\bigwedge^n V_-$, which we will call the *"negative"* part of $\bigwedge^n V$. This can be summarized in the following theorem.

Theorem 7.3.25. (Characterization of Orientations) *Let $n \geq 1$ and $0 \neq \omega \in \bigwedge^n V$. Then the set*

$$\mathcal{O}_\omega := \big\{ B = \{b_1, ..., b_n\} \mid \omega(b_1, \ldots, b_n) > 0 \big\}$$

is an orientation on V. Conversely, to any given orientation $[B]$ there exists an $\omega \in \bigwedge^n V$, $\omega \neq 0$, such that $\mathcal{O}_\omega = [B]$.

Proof. ($\Longrightarrow$). Let $B = \{b_1, \ldots, b_n\}$ and $C = \{c_1, \ldots, c_n\}$ be two bases of V and let $T : V \longrightarrow V$ be the linear map sending B to C. Then,

$$\omega(c_1, \ldots, c_n) = \omega(T(b_1), \ldots, T(b_n)) = \det(T)\omega(b_1, \ldots, b_n). \qquad (1)$$

Using this, we will show that

$$B,\ C \in \mathcal{O}_\omega \iff B \sim C.$$

Indeed, if $B,\ C \in \mathcal{O}_\omega$, then

$$\det(T) = \frac{\omega(b_1, \ldots, b_n)}{\omega(c_1, \ldots, c_n)}$$

and so, $B \sim C$.

Conversely, if $\det(T) > 0$, then by (1) we see that both $\omega(b_1, \ldots, b_n)$, and $\omega(c_1, \ldots, c_n)$ must have the same sign.

($\Longleftarrow$). Now, if $(b_1, \ldots, b_n)$ is any basis lying in the positive oriented class $[B]$, we define $\omega : V^n \longrightarrow \mathbb{R}$ as the alternating multilinear map uniquely determined by $\omega(b_1, \ldots, b_n) = 1$. By construction $\omega \neq 0$ and the first part of the proof shows that any other basis $C \in [B]$ also satisfies $\omega(c_1, \ldots c_n) > 0$.

Conversely, if $(c_1, \ldots c_n)$ is any other basis such that $\omega(c_1, \ldots c_n) > 0$ and T is the transition matrix sending B to C, the first part of the proof shows again that

$$\det(T) = \frac{\omega(b_1, \ldots, b_n)}{\omega(c_1, \ldots, c_n)} = \frac{1}{\omega(c_1, \ldots, c_n)}.$$

Thus $C \in [B]$. $\qquad\qquad\qquad\qquad\qquad\qquad\qquad\qquad\qquad\qquad\quad\square$

The fact that a non-zero finite dimensional real vector space has exactly two orientations, is not true in the general case when we deal with a topological space, or a manifold. Some smooth manifolds, like the Möbius strip, admit no orientation. On the other hand, a disconnected smooth manifold may have more than two different orientations.

Definition 7.3.26. Let V and W be oriented real vector spaces. A linear isomorphism

$$T : V \longrightarrow W$$

is said to be *orientation-preserving* if for some positively oriented basis $(v_i)_{i=1,\ldots,n}$ of V, the basis $(T(v_i))_{i=1,\ldots,n}$ of W is also positively oriented. Otherwise T is said to be *orientation-reversing*.

There is a connection between orientation and certain topological concepts. Let us examine the question of whether two bases of the vector space V can be continuously deformed into each other. This question reduces to whether there is a continuous deformation between two non-singular matrices A and B corresponding to these bases under the selection of some auxiliary basis $e_1, \ldots, e_n$. (Just as with other topological concepts, continuous deformability does not depend on the choice of the auxiliary basis.) The condition of non-singularity of the matrices A and B plays an essential role here.

We shall now formulate the notion of continuous deformability for matrices in $\mathrm{GL}(n, \mathbb{R})$.

Definition 7.3.27. A $n \times n$ matrix A is said to be *continuously deformable* into a matrix B if there exists a family of matrices $F(t)$ in $\mathrm{GL}(n, \mathbb{R})$ whose elements depend continuously on a parameter $t \in [0, 1]$ such that $F(0) = A$ and $F(1) = B$.

We leave it to the reader to check that this property of matrices being continuously deformable into each other is an equivalence relation on the set $\mathrm{GL}(n, \mathbb{R})$.

Theorem 7.3.28. *Two non-singular real square matrices of the same order are continuously deformable into each other if and only if the signs of their determinants are the same. Of course, this also applies to linear transformations.*

Proof. Let $g \in \mathrm{GL}(n, \mathbb{R}) = \mathrm{GL}^+(n, \mathbb{R}) \cup \mathrm{GL}^-(n, \mathbb{R})$ (disjoint union), where the $+$ and $-$ indicate the signs of the determinants. These are actually the two connected components of $\mathrm{GL}(n, \mathbb{R})$. By the Iwasawa decomposition theorem (see Volume I, Theorem 4.7.13) $g = kan$. Since $\det an > 0$, $\det g$ is positive or negative if and only if $\det k$ is 1 or -1. This proves that both $\mathrm{GL}^+(n, \mathbb{R})$ and $\mathrm{GL}^-(n, \mathbb{R})$ are connected. Let A and B be two matrices as in the statement. Then, either they are both in the same component or they are in different components. In the later case they can not be continuously deform since these components are disjoint and open. Now, if A and B belong to the same component since this is arcwise connected we are done. $\qquad\square$

Exercise 7.3.29. What would *continuous deformation* mean in $\mathrm{GL}(n, \mathbb{C})$, and why?

Corollary 7.3.30. *Two bases of a real vector space V are continuously deformable into one another if and only if they have the same orientation.*

Now, we return to the relationship of det to volume and recapture our results in Volume I, Chapter 4. Again, let $k = \mathbb{R}$ and P be a parallelepiped in $\mathbb{R}^n$. By translation we may assume P is determined by vectors $a_1, \ldots a_n$ emanating from the origin. We may also assume these vectors are linearly independent, for otherwise $\text{vol}(P) = 0$. Now, what do we demand of vol?

1. $\text{vol}(a_1, \ldots a_n) \geq 0$.

2. For $i \neq j$,

$$\text{vol}(a_1, a_i, \ldots, a_j, \ldots a_n) = \text{vol}(a_1, a_i + a_j, \ldots, a_j, \ldots, a_n).$$

3. $\text{vol}(a_1, \ldots, \lambda a_i, \ldots, a_n) = |\lambda| \, \text{vol}(a_1, \ldots, a_i, \ldots a_n)$.

4. $\text{vol}(e_1, \ldots, e_n) = 1$.

Suppose vol is such a function. Let

$$\mu = \begin{cases} \dfrac{\text{vol}(a_1, \ldots a_n) \det(a_1, \ldots, a_n)}{|\det(a_1, \ldots, a_n)|} & \text{if } a_1, \ldots, a_n \text{ are lin. independent.} \\ 0 & \text{otherwise} \end{cases}$$

A direct calculation which we leave to the reader shows that $\mu(a_1, \ldots, a_n)$ is an alternating n-form with $\mu(e_1, \ldots, e_n) = 1$. Therefore $\mu = \det$. Hence for any P, $\text{vol}(P) = |\det(P)|$.

When does a $T \in \text{End}_{\mathbb{R}}(V)$ preserve volumes of all parallelepipedes? Applying (*) to μ, taking absolute values, and using $\text{vol}(P) = |\det(P)|$, we get the following corrolary.

Corollary 7.3.31. *$T \in \text{End}_{\mathbb{R}}(V)$ preserves volumes if and only if* $|\det(T)| = 1$.

Our next result is known as *Hadamard's inequality*.

Proposition 7.3.32. *For any $A = (a_{i,j}) \in M_n(\mathbb{C})$,*

$$|\det A|^2 \leq \prod_{j=1}^{n} \sum_{i=1}^{n} |a_{i,j}|^2.$$

In particular, if $a = max(|a_{i,j}|)$, then $|\det A|^2 \leq a^n n^{n/2}$.

Proof. The rows $x_1, \ldots, x_n$ of the matrix A can be regarded as n vectors in $\mathbb{C}^n$ which is a Hermitian inner product space under $\langle, \rangle$. In it these rows give rise to a parallelepiped P. Of course, if they are linearly dependent i.e. $\text{vol}(P) = 0$ (since it is of dimension $< n$), then $\det A = \text{vol}(P)$ is also 0 and so the inequality is trivially satisfied. Therefore, we may assume they are linearly independent. Then again, $\det A = \text{vol}(P)$. Now consider the parallelepiped P_0 formed by these same vectors, but which are now pairwise orthogonal. Evidently this has larger volume. What is its volume? Each of these vectors has length $(\sum_{i=1}^n |a_{i,j}|^2)^{\frac{1}{2}}$ and the volume is the product of the lengths. Thus

$$\det(A) = \text{vol}(P_0) = \prod_{j=1}^n (\sum_{i=1}^n |a_{i,j}|^2)^{\frac{1}{2}}.$$

Squaring yields Hadamard's inequality. Obviously equality occurs if and only if these vectors are orthogonal. $\square$

Exercise 7.3.33. Let k be any field and denote by $\mathcal{S}$ the set $\left\{ \lambda I_{n \times n} \ \ \lambda \neq 0, \ \ \lambda \in k \right\}$. Prove that when n is odd $\text{GL}(n, \mathbb{R}) = \mathcal{S} \cdot \text{SL}(n, \mathbb{R})$ and when n is even this is false. Also, show that $\text{GL}(n, \mathbb{C}) = \mathcal{S} \cdot \text{SL}(n, \mathbb{C})$.

7.3.4   The Volume Form on a real vector space

Let V be a real vector space of dimension $n \geq 1$ with a chosen orientation, and $\langle \cdot, \cdot \rangle$ be a positive-definite inner product on V.

Definition 7.3.34. The *volume form* on V corresponding to this choice of an orientation and to $\langle \cdot, \cdot \rangle$, is the unique n-form dV on V such that for any positively oriented basis $v_1, \ldots, v_n$ of V we have

$$dV = \sqrt{\det g} \ v_1^\star \wedge \cdots \wedge v_n^\star,$$

where g is the $n \times n$ symmetric matrix $g = (g_{ij})$ with $g_{ij} = \langle v_i, v_j \rangle$.

Let $e = \{e_1, \ldots, e_n\}$ be an orthonormal basis of V, $v = \{v_1, \ldots, v_n\}$ be any other basis, and $A = (a_{ij})$ be the invertible transition matrix between the two. Then,

$$v_i = \sum_{j=1}^{n} a_{ij} e_j, \quad i = 1, \ldots, n.$$

We claim that

$$dV(v_1, \ldots, v_n) = |\det(A)|.$$

Indeed,

$$\langle v_i, v_j \rangle = \sum_{r,s} a_{ir} a_{js} \langle e_r, e_s \rangle.$$

Alternatively,

$$[g]_v = A^t [g]_e A,$$

where $[g]_v = (\langle v_i, v_j \rangle)$ and $[g]_e = (\langle e_r, e_s \rangle)$. From this we see

$$\det(\langle v_i, v_j \rangle) = (\det(A))^2 \det(\langle e_r, e_s \rangle)$$

and since $\{e_1, \ldots, e_n\}$ is an orthonormal basis,

$$g(e_r, e_s) = \delta_{rs}.$$

Hence,

$$\det(g(e_r, e_s)) = 1,$$

which gives us $\det(g(e_i, e_j))^{1/2} = \det(A)$. Thus

$$dV(v_1, \ldots, v_n) = |\det(A)|.$$

Using this, we leave the proof of the following proposition to the reader.

Proposition 7.3.35. *The volume form dV satisfies the following properties:*

1. *dV does not depend on the choice of the positively oriented basis $v_1, \ldots, v_n$ of V.*

2. *If $e_1, \ldots, e_n$ is a positively oriented orthonormal basis of V, then*

$$dV = e^1 \wedge \cdots \wedge e^n.$$

3. *Let $e_1, \ldots, e_n$ be a positively oriented orthonormal basis of V. If $v_1, \ldots, v_n$ is any other basis of V, there is a unique matrix $A = (a_{ij})$ such that $v_i = \sum_j a_{ij} e_j$. Then*

$$dV(v_1, \ldots, v_n) = |\det(A)|.$$

Thus $dV(v_1, \ldots, v_n)$ is equal to the volume of the n-dimensional box with sides $v_1, \ldots, v_n$ in V.

Example 7.3.36. Let $\dim(V) = 1$. Let $v \in V$, $v \neq 0$, so that $\{v\}$ is a basis of V. Then, g is a 1×1-matrix with the entry $g_{11} = \langle v, v \rangle = \| v \|^2$. Hence, the volume form is $dV = \| v \| v^\star$ if v is positive, relative to the orientation on V, and $\omega = - \| v \| v^\star$ if v is negative. In particular if $e \in V$ is a positive unit vector, then $dV = ce$. Note that if $w \in V$ is any other vector, then $w = ce$, where $c = \varepsilon \| w \|$, where $\varepsilon = 1$ if $w \neq 0$ is a positive vector, $\varepsilon = -1$ if $w \neq 0$ is negative and $\varepsilon = 0$ if $w = 0$. Then $dV(w) = e^\star(ce) = c = \varepsilon \| w \|$. Thus the volume form $dV(w)$ is the *signed length* of w.

7.3.5   The Hodge star ($*$) Operator

The Hodge star operator or Hodge dual is an important operator in differentiable manifolds and is defined through an operation on the exterior algebra of the tangent space $T_p(M)$ at each point $p \in M$ of a manifold M. Here, we will define this $*$ operator on the exterior algebra of a general real, finite dimensional, oriented, inner product space. This operator sets up a duality between $\bigwedge^k V$ and $\bigwedge^{n-k} V$. It is one of Grassmann's most fundamental contributions because it algebraizes the principle of duality in projective geometry and elsewhere.

Let H be a real or complex separable Hilbert space. This means that H has a countable orthonormal basis. Obviously all finite dimensional real or complex vector spaces are separable Hilbert spaces. Here we define the dual space $H^\star$ to H by:

$$H^\star = \{\varphi : H \to \mathbb{C} \ (or \ \mathbb{R}), \ \varphi \text{ continuous}\}$$

and prove the following result establishing the connection between H and H^*.

Theorem 7.3.37. *(**Riesz's Theorem**)*[2] *Let H be a separable real or complex Hilbert space, and $H^\star$ be its dual. Then, for any $\varphi \in H^\star$ there is a unique vector $v \in H$ such that*

$$\varphi(x) = \langle x, v \rangle$$

for any $x \in H$. In addition, $\|\varphi\| = \|v\|$.

Actually, the result holds for Hilbert spaces, separable or not. For a proof of this see e.g. M.A. Naimark, [93].

Proof. H being separable, there is an orthonormal basis e_1, e_2,..... Set $\varphi(e_i) = \lambda_i$, consider the vector $v \in H$, let $\langle v, e_i \rangle = \eta_i$ and set

$$v_n = \sum_{i=1}^{n} \eta_i e_i.$$

The fact that e_1, e_2,... is an orthonormal basis implies

$$\lim_{n \to \infty} \|v - v_n\| = 0 \tag{1}$$

and, since φ is a linear and continuous, and therefore bounded,

$$\varphi(v_n) = \sum_{i=1}^{n} \lambda_i \eta_i,$$

and

$$\| \varphi(v) - \varphi(v_n) \| \leq \| \varphi \| \| v - v_n \| .$$

Using (1)

$$\varphi(v) = \lim_{n \to \infty} \varphi(v_n) = \sum_{i=1}^{\infty} \lambda_i \eta_i.$$

---

[2]This theorem, named in honor of Frigyes Riesz, is attributed simultaneously to Riesz and Fréchet-Riesz in 1907. It is a justification of P. Dirac's *bra-ket* notation (1939) of Quantum Mechanics, where every *ket* $|\psi\rangle$ has a corresponding *bra* $\langle\phi|$.

The sequence (λ_i) is square summable. Indeed, since

$$|\varphi(x)| \leq \| \varphi \| \| x \|,$$

we see

$$\left| \sum_{i=1}^{\infty} \lambda_i \eta_i \right| \leq \| \varphi \| \left(\sum_{i=1}^{\infty} \eta_i \right)^{1/2}. \tag{2}$$

Equation (2) holds for every square summable sequence η_i. Now, pick a positive integer n and define the sequence

$$\eta_i = \begin{cases} 0 & \text{if } i > n, \\ \lambda_i & \text{if } i \leq n. \end{cases}$$

This sequence is square summable and using equation (2) one gets

$$\left| \sum_{i=1}^{n} \lambda_i^2 \right| \leq \| \varphi \| \left(\sum_{i=1}^{n} \lambda_i^2 \right).$$

Therefore,

$$\left(\sum_{i=1}^{n} \lambda_i^2 \right)^{1/2} \leq \| \varphi \| \tag{3}$$

and since the sequence of partial sums is bounded from above, the sequence λ_i is square summable. This implies the vector

$$v = \sum_i \lambda_i e_i$$

is well defined and $\varphi(x) = \sum_i \lambda_i \eta_i = \langle x, v \rangle$.

Finally, equation (3) shows that $\| v \| \leq \| \varphi \|$. Using the Cauchy-Schwarz inequality,

$$|\varphi(x)| = |\langle x, v \rangle| \leq \| x \| \| v \|$$

or,

$$\frac{|\varphi(x)|}{\| x \|} \leq \| v \|,$$

which in turn implies that $\| \varphi \| \leq \| v \|$. So, $\| \varphi \| = \| v \|$. $\qquad\square$

This theorem applies in particular to the case of a finite dimensional real or complex inner product vector space V and its dual space V^*. There is a canonical isomorphism

$$\psi : V \to V^\star \;\; : \;\; v \mapsto \psi(v) = \langle -, v \rangle \text{ such that } V^\star \ni \varphi \mapsto \langle \varphi, v \rangle = \varphi(v).$$

This shows $V^\star$ is also an inner product space with inner product defined by

$$\langle \omega, \eta \rangle_{V^*} = \langle \psi^{-1}(\omega) \, , \, \psi^{-1}(\eta) \rangle_V.$$

Henceforth, we will always assume that $V^\star$ is equipped with this inner product, which we will also denote by $\langle \, , \, \rangle$.

The inner product of V induces a unique inner product on $\bigwedge^k V$, $0 \le k \le n$, as the following theorem shows.

Theorem 7.3.38. *Let V be a real vector space ($\dim V = n$), with an inner product $\langle \cdot, \cdot \rangle$. Then, if we define for any $0 \le k \le n$*

$$\langle v_1 \wedge v_2 \wedge \cdots \wedge v_k, \, w_1 \wedge w_2 \wedge \cdots \wedge w_k \rangle_{\bigwedge^k} := \det((a_{lm})_{l,m=1,\dots,k}),$$

where $a_{lm} = \langle v_k, w_l \rangle$ and we extend it by linearity, we get an inner product on $\bigwedge^k V$. (This is the same as in the Hermitian case.)

Proof. We know that there exists an orthonormal basis $B = \{e_1, \dots, e_n\}$ of V, and $\bigwedge^k B$ is a basis of $\bigwedge^k V$. So we define

$$\langle e_1 \wedge e_2 \wedge \cdots \wedge e_k, \, e_1 \wedge e_2 \wedge \cdots \wedge e_k \rangle_{\bigwedge^k} := \det(\langle e_l, e_m \rangle)$$

and we extend it linearly onto $\bigwedge^k V$. Hence $\langle \cdot, \cdot \rangle_{\bigwedge^k}$ is bilinear and symmetric since $\langle \cdot, \cdot \rangle$ is. Also,

$$\langle e_1 \wedge e_2 \wedge \cdots \wedge e_k \, , \, e_1 \wedge e_2 \wedge \cdots \wedge e_k \rangle_{\bigwedge^k} := \det(\langle e_l, e_m \rangle) = \det(\delta_{lm}) = 1 > 0.$$

Thus, $\langle \cdot, \cdot \rangle$ is positive definite. In addition, if $C = \{\varepsilon_1, \cdots, \varepsilon_n\}$ is another orthonormal basis of V, then there is an orthogonal transformation $T \in O(n, \mathbb{R})$ such that $C = T(B)$. In this case we get

$$\det(\langle c_l, c_m \rangle) = \det(\langle T(e_l), T(e_m) \rangle) = \det(\langle e_l, T^t T(e_m) \rangle) = \det(\langle e_l, e_m \rangle),$$

which shows that $\bigwedge^k C$ is also an orthonormal basis for $\bigwedge^k V$. Therefore the induced inner product on $\bigwedge^k V$ is unique.   □

From now on we will write simply $\langle \cdot, \cdot \rangle$, instead of $\langle \cdot, \cdot \rangle_{\wedge^k}$.

Now, assume that $(V, \langle \cdot, \cdot \rangle)$ is an oriented inner product vector space, and let $e_1,..., e_n$ be an orthonormal basis. We will denote by $e^1,...,e^n$ the dual basis (in V^*).

The *Hodge $*$ operator* is the unique isomorphism (for each k)

$$* : \overset{k}{\bigwedge} V \longrightarrow \overset{n-k}{\bigwedge} V$$

such that for any positive orthonormal basis, $e_1,...,e_n$ of V

$$*(e^1 \wedge \cdots \wedge e^n) = e^{k+1} \wedge \cdots \wedge e^n. \tag{1}$$

Intuitively, since k-forms $e^1 \wedge \cdots \wedge e^k$ correspond to k-dimensional subspaces $W = \text{lin.sp.}\{e^1, \ldots, e^k\}$ of V^*, $*(e^1 \wedge \cdots \wedge e^n)$ corresponds to the orthogonal complement of W.

Theorem 7.3.39. *(Existence of the Hodge operator) Let V be an oriented inner product space and let dV be the associated volume element. Then, for any $0 \leq k \leq n$, there exists a unique isomorphism*

$$* = *_k : \overset{k}{\bigwedge} V \longrightarrow \overset{n-k}{\bigwedge} V, \ \xi \mapsto *\xi$$

such that, for all $\xi, \eta \in \bigwedge^k V$

$$\xi \wedge *\eta = \langle \xi, \eta \rangle dV$$

or equivalently,

$$\langle \xi \wedge *\eta, dV \rangle = \langle \xi, \eta \rangle.$$

Proof. Since $\dim \bigwedge^n V = 1$, dV is a basis of $\bigwedge^n V$. We consider the map

$$P : \overset{k}{\bigwedge} V \times \overset{n-k}{\bigwedge} V \longrightarrow \mathbb{R}, \ : (\xi, \zeta) \mapsto \lambda$$

where λ is the coefficient of $\xi \wedge \zeta = \lambda dV$. The map P is easily seen to be bilinear and (e.g. using a basis) non-degenerate. Therefore, by Riesz' Theorem 7.3.37, for any linear form $q : \bigwedge^k V \longrightarrow \mathbb{R}$ there is a unique element $\zeta \in \bigwedge^{n-k} V$ so that $P(\xi, \zeta) = q(\xi)$ for all $\xi \in \bigwedge^k V$. Now, given $\eta \in \bigwedge^k V$, apply the Riesz Theorem to the form $q(\xi) = \langle \xi, \eta \rangle$. This determines an element $\zeta \in \bigwedge^{n-k} V$. Define

$$*\eta := \zeta.$$

Then, the equation

$$\xi \wedge *\eta = \langle \xi, \eta \rangle dV, \text{ for all } \xi, \eta \in \bigwedge^k V$$

holds by definition, and from this, the equation (1) follows.

Now, we shall prove the equivalence of the two conditions. Let ξ, $\eta \in \bigwedge^k V$. Since $\dim \bigwedge^n V = 1$ and $dV \neq 0$ there is exactly one $\lambda \in \mathbb{R}$, such that $\xi \wedge *\eta = \lambda dV$. Then, by Proposition 7.3.35 and Theorem 7.3.38, we get

$$\lambda = \lambda \langle dV, dV \rangle = \langle \lambda dV, dV \rangle = \langle \xi \wedge *\eta, dV \rangle = \langle \xi, \eta \rangle.$$

Thus the second condition implies the first one. Conversely the first condition implies the second since

$$\xi \wedge *\eta = \langle \xi, \eta \rangle dV \implies \langle \xi \wedge *\eta, dV \rangle = \langle \xi, \eta \rangle \langle dV, dV \rangle = \langle \xi, \eta \rangle.$$

$$\square$$

The Hodge operator has the properties listed in the following proposition.

Proposition 7.3.40. *For all $\xi, \eta \in \bigwedge^k V$, the Hodge operator satisfies:*

1. $*\eta \wedge \xi = \langle \eta, \xi \rangle dV = \langle \xi, \eta \rangle dV = \xi \wedge *\eta.$

2. *For each permutation $\sigma \in S_n$ such that $\sigma(1) < \cdots < \sigma(k)$ and $\sigma(k+1) < \cdots < \sigma(n)$ we have*

$$*(e^{\sigma(1)}, \ldots, e^{\sigma(k)}) = \text{sgn}(\sigma) e^{\sigma(k+1)} \wedge \cdots \wedge e^{\sigma(n)},$$

where $e_1, \ldots, e_n$ is a positive orthonormal basis of V. In particular

$$*(e^1, \ldots, e^k) = \text{sgn } e^{k+1} \wedge \cdots \wedge e^n.$$

3. $*1 = dV$, and $*dV = 1$.

4. $*_{n-k} *_k \xi = (-1)^{k(n-k)} \xi$.

5. $\langle \xi, \eta \rangle = \langle *\xi, *\eta \rangle$.

6. $\langle \xi, \eta \rangle = *(\xi \wedge *\eta) = *(\eta \wedge *\xi)$.

Proof.

1. Follows directly from Theorem 7.3.39.

2. Set $\zeta = e^{\sigma(1)} \wedge \cdots \wedge e^{\sigma(k)} \in \bigwedge^k V$ and $\eta = \mathrm{sgn}(\sigma) e^{\sigma(k+1)} \wedge \cdots e^{\sigma(n)} \in \bigwedge^{n-k} V$. To show that $\eta = *\zeta$, we have to prove that for each $\xi \in \bigwedge^k V$

$$\xi \wedge \eta = \langle \xi, \eta \rangle dV.$$

Since both sides are linear in ξ we will check this on a basis. So let $\xi = e^{i_1} \wedge \cdots \wedge e^{i_k}$ with $i_1 < \cdots < i_k$. We need to show that

$$\mathrm{sgn}(\sigma) e^{i_1} \wedge \cdots \wedge e^{i_k} \wedge e^{\sigma(k+1)} \wedge \cdots \wedge e^{\sigma(n)}$$

$$= \langle e^{i_1} \wedge \cdots \wedge e^{i_k}, e^{\sigma(k+1)} \wedge \cdots \wedge e^{\sigma(n)} \rangle e^1 \wedge \cdots \wedge e^n.$$

We distinguish the following two cases:

Case 1. If $\{i_1, \ldots, i_k, \sigma(k+1), \ldots, \sigma(n)\} \neq \{1, \ldots, n\}$, the permutation must have a fixed point, say $\sigma(j) = i_j$, and so the left side is zero. The right side is also zero because $(i_1, \ldots, i_k)$ can no longer be a permutation of $(\sigma(1), \ldots, \sigma(k))$, which implies that

$$e^{i_1} \wedge \cdots \wedge e^{i_k} \perp e^{\sigma(1)} \wedge \cdots \wedge e^{\sigma(k)}.$$

Case 2. If $\{i_1, \ldots, i_k, \sigma(k+1), \ldots, \sigma(n)\} = \{1, \ldots, n\}$, it follows that $\{\sigma(1), \ldots, \sigma(k)\} = \{i_1, \ldots, i_k\}$, and so there is a permutation $\tau \in S_k$ such that $(\sigma(1), \ldots, \sigma(k)) = (\tau(i_1), \ldots, \tau(i_k))$.

Then,

$$\text{sgn}(\sigma)e^{i_1} \wedge \cdots \wedge e^{i_k} \wedge e^{\sigma(k+1)} \wedge \cdots \wedge e^{\sigma(n)}$$
$$= \text{sgn}(\sigma)e^{\tau^{-1}(\sigma(1))} \wedge \cdots \wedge e^{\tau^{-1}(\sigma(k))} \wedge e^{\sigma(k+1)} \wedge \cdots e^{\sigma(n)}$$
$$= \text{sgn}(\sigma)\,\text{sgn}(\tau)e^{\sigma(1)} \wedge \cdots \wedge e^{\sigma(k)} \wedge e^{\sigma(k+1)} \wedge \cdots \wedge e^{\sigma(n)}$$
$$= \text{sgn}(\tau)e^1 \wedge \cdots \wedge e^k \wedge e^{k+1} \wedge \cdots \wedge e^n$$
$$= \text{sgn}(\tau)\langle e^{\sigma(1)} \wedge \cdots e^{\sigma(k)} ,\ e^{\sigma(1)} \wedge \cdots e^{\sigma(k)}\rangle e^1 \wedge \cdots \wedge e^n$$
$$= \langle e^{i_1} \wedge \cdots e^{i_k} ,\ e^{\sigma(1)} \wedge \cdots e^{\sigma(k)}\rangle e^1 \wedge \cdots e^n.$$

3. Set $\sigma = \text{id}$ in the above equation and consider the cases $k = 0$ and $k = n$.

4. Again we will work on a basis. Let $\sigma \in S_n$ such that

$$\sigma(1) < \cdots < \sigma(k) \quad \text{and} \quad \sigma(k+1) < \cdots < \sigma(n)$$

and consider $\xi = e^{\sigma(1)} \wedge \cdots \wedge e^{\sigma(k)}$. Using (2), we obtain

$$* * \xi = *\Big(\text{sgn}(\sigma)(e^{\sigma(k+1)} \wedge \cdots \wedge e^{\sigma(n)})\Big)$$
$$= \text{sgn}(\sigma) * (e^{\sigma(k+1)} \wedge \cdots \wedge e^{\sigma(n)})$$
$$= \text{sgn}(\sigma)\,\text{sgn}(\tau)e^{\tau(n-k+1)} \wedge \cdots \wedge e^{\tau(n)},$$

where $\tau \in S_n$ is such that

$$(\tau(1),\ \ldots,\ \tau(n-k)) = (\sigma(k+1),\ \ldots,\ \sigma(n))$$
$$\text{and} \quad \tau(n-k+1) < \cdots < \tau(n).$$

From this we get

$$(\tau(n-k+1),\ \ldots,\ \tau(n)) = (\sigma(1),\ \ldots,\ \sigma(k)).$$

Hence,

$$* * \xi = \text{sgn}(\sigma)\,\text{sgn}(\tau)\xi.$$

Now, we must calculate the signs. To do this, we consider the permutation $\varrho \in S_n$ which sends

$$(1,\ \ldots,\ k,\ k+1,\ \ldots,\ n) \mapsto (k+1,\ \ldots,\ n,\ 1,\ \ldots,\ k).$$

One easily checks that $\tau = \sigma \circ \varrho$, which implies

$$\text{sgn}(\sigma)\,\text{sgn}(\tau) = \text{sgn}(\sigma)\,\text{sgn}(\sigma)\,\text{sgn}(\varrho) = \text{sgn}(\varrho) = (-1)^{k(n-k)}.$$

5. We have

$$\langle *\xi, *\eta \rangle dV = \text{by 1} = *\xi \wedge (* * \eta) = \text{by 4} = (-1)^{k(n-k)} * \xi \wedge \eta$$
$$= \eta \wedge *\xi = \langle \eta, \xi \rangle dV = \langle \xi, \eta \rangle dV.$$

6. We have

$$\langle \xi, \eta \rangle = \langle \eta, \xi \rangle 1 = \text{by 3} = *(\langle \xi, \eta \rangle dV) = *(\xi \wedge *\eta) = *(\eta \wedge *\xi).$$

$$\square$$

To see how all this works, let us first consider the case of $V = \mathbb{R}^2$ equipped with a positive-definite inner product. We will write the basis as $\{dx, dy\}$. The inner product can be expressed in terms of the metric or line element

$$ds^2 = dx^2 + dy^2$$

which also shows that our basis is orthonormal. Since all basis vectors have norm $+1$, the g's can be omitted. Choosing

$$\omega = dx \wedge dy = dy \wedge (-dx),$$

we obtain:

$$*1 = dx \wedge dy$$
$$*dx = dy$$
$$*dy = dx$$
$$*(dx \wedge dy) = 1$$

As another example take $V = \mathbb{R}^3$ equipped with the Lorentzian metric of signature -1 (i.e. the Minkowski 3-space). The metric is

$$ds^2 = dx^2 + dy^2 - dt^2$$

corresponding to the orthonormal basis $\{dx, dy, dt\}$. Notice that $g(dt, dt) = -1$. We take the volume element to be

$$dV = dx \wedge dy \wedge dt = dy \wedge dt \wedge dx = dt \wedge dx \wedge dy.$$

Using skew-symmetry we obtain

$$*1 = dx \wedge dy \wedge dt$$
$$*dx = dy \wedge dt$$
$$*dy = dt \wedge dx$$
$$*dt = -dx \wedge dy$$
$$*(dx \wedge dy) = dt$$
$$*(dy \wedge dt) = -dx$$
$$*(dt \wedge dx) = -dy$$
$$*(dx \wedge dy \wedge dt) = -1.$$

Here the minus signs are due to the presence of dt, whose squared norm is negative.

7.4   Liouville's Formula

Here we prove the following formula due to Liouville.

Theorem 7.4.1. (*Liouville's Formula*) *Let $A \in M(n, \mathbb{C})$. Then*

$$\frac{d}{dt} \det \left(\exp(At) \right)|_{t=0} = \mathrm{tr}(A).$$

Let $\bigwedge^k A$ be the induced linear map on the exterior power $\bigwedge^k \mathbb{C}^n$. In terms of the standard basis of $\mathbb{C}^n$, $\bigwedge^k A$ is represented by the matrix of all minors of A of order k. Then,

$$\overset{0}{\bigwedge} A = 1, \quad \overset{1}{\bigwedge} A = A, \quad \overset{n}{\bigwedge} A = \det(A).$$

Proposition 7.4.2. *For any matrix $A \in M(n, \mathbb{C})$,*

$$\det(\lambda I - A) = \sum_{k=1}^{n} (-1)^k \, \mathrm{tr} \left(\overset{k}{\bigwedge} A \right).$$

Proof. First suppose A is diagonalizable. Since the polynomial $\det(\lambda I - A)$ is independent of the matrix representation, the equation in the statement does not change if we replace A by a conjugate, nor does $\operatorname{tr}\left(\bigwedge^k A\right)$. Hence in this case we can assume that A is diagonal with entries $\lambda_1, \lambda_2, ..., \lambda_n$. If $e_1, ..., e_n$ is the standard base of $\mathbb{C}^n$, then $Ae_i = \lambda_i e_i$ for any $i = 1, ..., n$ and

$$\det(\lambda I - A) = \chi_A(\lambda) = \prod_{i=1}^{n}(\lambda - \lambda_i).$$

Now, from the right side of this equation we see the coefficient, c_k, of λ^{n-k} is

$$c_k = (-1)^k \sum_{1 \leq i_1 < \cdots < i_k \leq n} \lambda_{i_1} \cdots \lambda_{i_k}.$$

On the other hand,

$$\bigwedge^k A(e_{i_1} \wedge \cdots \wedge e_{i_k}) = Ae_{i_1} \wedge \cdots \wedge Ae_{i_k} = \lambda_{i_1} e_{i_1} \wedge \cdots \wedge \lambda_{i_k} e_{i_k}$$
$$= \lambda_{i_1} \lambda \cdots \lambda_{i_k} (e_{i_1} \wedge \cdots \wedge e_{i_k}),$$

which means the vectors,

$$\{e_{i_1} \wedge e_{i_2} \wedge \cdots \wedge e_{i_k} \mid 1 \leq i_1 < \cdots < i_k \leq n\}$$

are eigenvectors of $\bigwedge^k A$ acting on $\bigwedge^k \mathbb{C}^n$. Therefore

$$\operatorname{tr}\left(\bigwedge^k A\right) = \sum_{1 \leq i_1 < \cdots < i_k \leq n} \lambda_{i_1} \cdots \lambda_{i_k}.$$

Hence, $c_k = (-1)^k \operatorname{tr}\left(\bigwedge^k A\right)$.

Now let A be arbitrary. Since both the left and right sides of this equation are polynomials in A, they are each Zariski continuous. Because the diagonalizable matrices are Zariski dense in $M(n, \mathbb{C})$ (see Theorem 9.15.9 of Chapter 9), these agree on A as well.   □

As a result we see $\det(\operatorname{Exp} A) = \exp(\operatorname{tr}(A))$, which is also called Liouville's formula. Replacing A by tA and differentiating at $t = 0$ yields

$$\frac{d}{dt}\det\big(\exp(At)\big)|_{t=0} = \operatorname{tr}(A).$$

Exercise 7.4.3. Prove Liouville's formula by calculating $\frac{d}{dt}\det\big(\operatorname{Exp}(At)\big)|_{t=0}$ using the determinant expansion theorem.

Exercise 7.4.4. Prove $\det(\operatorname{Exp} A) = \exp(\operatorname{tr}(A))$ by using the third Jordan form.

7.5   Grassmannians

We now look at the effect of the action of $\operatorname{GL}(V)$, the general linear group of V on the Grassmann space. Here k is a field of characteristic zero, $P(V)$ denotes the projective space of V and $\pi : V - \{0\} \to P(V)$ the canonical map $v \mapsto \bar{v}$. For an integer r, $1 \le r \le n$, $\mathcal{G}^r(V)$ is the Grassmann space of all r-dimensional subspaces of V. (Of course, $P(V)$ is $\mathcal{G}^1(V)$. If $k = \mathbb{R}$, or $\mathbb{C}$, then each $\mathcal{G}^r(V)$ is a compact connected manifold. For all this see subsubsection, "The Grassmann and Flag Varieties" of Volume I.)

For $g \in \operatorname{GL}(V)$ define $\bar{g} : P(V) \to P(V)$ by $\bar{g}(\bar{v}) = \overline{g(v)}$. Routine calculations prove that $\bar{g}$ is well defined, $\bar{g}\pi = \pi\bar{g}$, $\overline{gh} = \bar{g}\bar{h}$ and if $\lambda \ne 0$, then $\overline{(\lambda g)} = \bar{g}$. Now, suppose one has a linear representation $G \times V \to V$ of G on V. Then this induces a compatible action of G on $P(V)$, making the diagram below commutative.

$$
\begin{array}{ccc}
G \times (V - \{0\}) & \longrightarrow & V - \{0\} \\
\downarrow{\scriptstyle (id,\pi)} & & \downarrow{\scriptstyle \pi} \\
G \times P(V) & \longrightarrow & P(V)
\end{array}
$$

We now use the exterior algebra to embed the Grassmann[3] space in the projective space of something larger, namely $\wedge^r(V)$. This can be

---

[3]Hermann Grassmann (1809-1877) German mathematician tried to get approval from the authorities to teach high school, but was rejected and only allowed to teach in the lower grades. One of the many examinations for which Grassmann

done using the following canonical map,

$$\phi : \mathcal{G}^r(V) \to P(\wedge^r V),$$

which is defined as follows: For an r-dimensional subspace W of V, choose a basis $\{w_1, \ldots, w_r\}$ of it. Then $w_1 \wedge \ldots \wedge w_r$ is a non-zero element of $\wedge^r V$ and so the line through it gives a point in $P(\wedge^r V)$.

We require the following lemma.

Lemma 7.5.1. *Let* $w_1, \ldots, w_r$ *be linearly independent vectors in* V *and*

$$S = \{v \in V : v \wedge w_1 \wedge \ldots \wedge w_r = 0\}.$$

Then S *is a subspace of* V *of dimension less than or equal to* r.

---

sat to try to rectify this situation, required that he submit an essay on the theory of the tides. He did so, taking the basic theory from Laplace's Mécanique céleste and from Lagrange's Mécanique analytique, but using vector methods of his own divising. This essay contains the first known appearance of linear algebra and the notion of a vector space. In 1844, Grassmann published, "The Theory of Linear Extension, a New Branch of Mathematics", which proposed a new foundation for what is now called linear algebra which he develops in a way which is remarkably similar to what one finds in modern literature: subspace, linear independence, span, dimension and projections. He also defines what is now called the exterior algebra and the Grassmann space. This revolutionary text was too far ahead of its time to be appreciated. When Grassmann submitted it to apply for a professorship the Education ministry asked Kummer for a report. Kummer said that there were good ideas in it, but found the exposition deficient and advised against giving Grassmann a university position. Over the next 10-odd years, Grassmann wrote several essays applying his ideas to electrodynamics and algebraic curves and surfaces in the hope that these applications would cause others to take his theory seriously. Some years later he published a thoroughly rewritten second edition, hoping to earn belated recognition, and containing the definitive exposition of his linear algebra and, even though the exposition anticipates the textbooks of the 20th century, this fared no better. Disappointed by the reception of his work in mathematical circles, Grassmann lost his contacts with mathematicians as well as his interest in geometry. In his later years he turned to linguistics, collected folk songs, and learned Sanskrit. He wrote a 2,000-page dictionary and a translation of the Rigveda (more than 1,000 pages) which earned him a membership of the American Orientalists' Society. Grassmann discovered a sound law of Indo-European languages, which was named in his honor. For these philological accomplishments he received an honorary doctorate from the University of Tübingen.

Proof. Clearly S is a subspace of V. Let $s_1, \ldots, s_q$ be a basis of S and extend this to a basis $s_1, \ldots, s_q, s_{q+1}, \ldots, s_n$ of V. A basis of $\wedge^r V$ is then given by $\{s_{i_1} \wedge \ldots \wedge s_{i_r} : s_{i_1}, \ldots, s_{i_r} \in \sigma\}$, where $\sigma = \{i_1 < \ldots < i_r : 1 \leq i_j \leq n\}$. Since $w_1 \wedge \ldots \wedge w_r \in \wedge^r V$,

$$w_1 \wedge \ldots \wedge w_r = \sum a_{i_1}, \ldots, a_{i_r} s_{i_1} \wedge \ldots \wedge s_{i_r},$$

where $a_{i_1}, \ldots, a_{i_r} \in k$. Now, not all these $a_{i_1}, \ldots, a_{i_r} = 0$. For if they were all zero, then $w_1 \wedge \ldots \wedge w_r = 0$, a contradiction, since $w_1, \ldots, w_r$ are linearly independent. Now, for $s_1, \ldots, s_q \in S$,

$$0 = s_j \wedge w_1 \wedge \ldots \wedge w_r = \sum a_{i_1}, \ldots, a_{i_r} s_{i_1} \wedge \ldots \wedge s_{i_r}. \tag{7.1}$$

Let j be fixed. If $j \in \{i_1, \ldots, i_r\}$ for any $i_1, \ldots, i_r \in \sigma$, then $s_j \wedge w_1 \wedge \ldots \wedge w_r = 0$ for that $i_1, \ldots, i_r$. Now, for each $i_1, \ldots, i_r \in \sigma$ with j not in $\{i_1, \ldots, i_r\}$, $s_j \wedge s_{i_1} \wedge \ldots \wedge s_{i_r} \neq 0$. Since $s_j, s_{i_1}, \ldots, s_{i_r}$ are linearly independent, this means by equation (7.1) we have a dependence relation in $\wedge^{r+1} V$ among the linearly independent elements

$$\{s_j \wedge s_{i_1} \wedge \ldots \wedge s_{i_r}, : i_1, \ldots, i_r \in \sigma, \ j \neq i_1, \ldots, i_r\}.$$

Hence, all these $a_{i_1}, \ldots, a_{i_r} = 0$. We have just proven that if $a_{i_1}, \ldots, a_{i_r} \neq 0$, where $i_1, \ldots, i_r \in \sigma$, and $j = 1, \ldots, q$, then $j \in \{i_1, \ldots, i_r\}$. Since there exists $i_1, \ldots, i_r$ with $a_{i_1}, \ldots, a_{i_r} \neq 0$ and $i_1, \ldots, i_r$ are r distinct integers containing $\{1, \ldots, q\}$, we must have $q \leq r$. $\qquad\square$

We are now in a position to prove the following theorem concerning ϕ.

Theorem 7.5.2. *The map* $\phi : \mathcal{G}^r(V) \to P(\wedge^r V)$ *is well-defined and injective. Moreover as we shall see below, when* $k = \mathbb{R}$ *(resp.* $\mathbb{C}$*),* ϕ *is smooth (resp. holomorphic).*

Proof. Define $\phi : \mathcal{G}^r(V) \to P(\wedge^r V)$ as above. ϕ is a well-defined map because if $u_1, \ldots, u_r$ is another basis for W, then $u_1 \wedge \ldots \wedge u_r$ is a non-zero multiple of $w_1 \wedge \ldots \wedge w_r$ by Lemma 7.3.6. Hence ϕ based on the $u_1, \ldots u_r$ gives the same point in $P(\wedge^r V)$ as that based on the $w_1, \ldots w_r$.

To see ϕ is also injective, suppose U and W are r-dimensional subspaces of V with $u_1, \ldots u_r$ a basis of U and $w_1, \ldots w_r$ a basis of W. Since we know ϕ is well defined, if $\phi(W) = \phi(U)$, then

$$w_1 \wedge \ldots \wedge w_r = \lambda\, u_1 \wedge \ldots \wedge u_r \qquad (7.2)$$

where $\lambda \neq 0$. Consider the subspace

$$T_W = \{v \in V \,|\, v \wedge w_1 \wedge \cdots \wedge w_r = 0\}.$$

Clearly $T_W \supseteq W$. Hence, by Lemma 7.5.1, $T_W = W$ and similarly $T_U = U$. Since by equation (7.2), $T_U = T_W$, it follows that $U = W$. $\square$

This fact gives rise to certain useful commutative diagrams. As explained above, $GL(V)$ acts transitively on $\mathcal{G}^r(V)$, so the latter is a quotient space $GL(V)/\operatorname{Stab}_{GL(V)}(W)$, where W is a fixed r-dimensional subspace of V. If $\gamma : GL(V) \to \mathcal{G}^r(V)$ denotes the corresponding projection and $\{w_1, \ldots, w_r\}$ is a basis of W, then since $\{gw_1, \ldots, gw_r\}$ are linearly independent for each $g \in G$, $g \mapsto gw_1 \wedge \cdots \wedge gw_r$ is a map $\psi : GL(V) \to \bigwedge^r V - \{0\}$. Then ϕ factors as $\phi_1 \pi$, where $\phi_1 : \mathcal{G}^r(V) \to \bigwedge^r V - \{0\}$ and $\pi : \wedge^r V - \{0\} \to P(\bigwedge^r V)$ is the natural map. The diagram below, involving the map ϕ, is commutative, since $\phi\gamma(g) = \phi(gW) = gw_1 \wedge \ldots \wedge gw_r = \psi(g)$.

$$
\begin{array}{ccc}
GL(V) & \xrightarrow{\ \ \psi\ \ } & \bigwedge^r V - \{0\} \\
\downarrow{\scriptstyle \gamma} & & \downarrow{\scriptstyle \pi} \\
\mathcal{G}^r(V) & \xrightarrow{\ \ \phi\ \ } & P(\bigwedge^r V)
\end{array}
$$

Moreover, for each $g \in G$ we also get a commutative diagram as follows:

$$
\begin{array}{ccc}
\mathcal{G}^r(V) & \xrightarrow{\ \ \phi\ \ } & P(\bigwedge^r V) \\
\downarrow{\scriptstyle g} & & \downarrow{\scriptstyle \overline{(\wedge^r g)}} \\
\mathcal{G}^r(V) & \xrightarrow{\ \ \phi\ \ } & P(\bigwedge^r V)
\end{array}
$$

where $g : \mathcal{G}^r(V) \to \mathcal{G}^r(V)$ is the map induced on $\mathcal{G}^r(V)$ by $g \in \mathrm{GL}(V)$. For, let $W = \mathrm{lin.sp}_k\{w_i\}$ be any point of $\mathcal{G}^r(V)$ and $g \in G$. Then $\phi(g(W)) = \overline{(gw_1 \wedge \cdots \wedge w_r)}$, while

$$
\begin{aligned}
\overline{(g \wedge \cdots \wedge g)}(\phi(W)) &= \overline{(g \wedge \cdots \wedge g)}\,\overline{(w_1 \wedge \cdots \wedge w_r)} \\
&= \overline{[(g \wedge \cdots \wedge g)(w_1 \wedge \cdots \wedge w_r)]} \\
&= \overline{(gw_1 \wedge \cdots \wedge gw_r)}.
\end{aligned}
$$

In particular if $h \in \mathrm{GL}(V)$ applying the above to $h(W)$, we get

$$
\phi(gh(W)) = \overline{ghw_1 \wedge, \ldots, \wedge ghw_r}.
$$

This leads us immediately to:

Theorem 7.5.3. *Let ρ be a smooth (respectively holomorphic) representation of G on V. Then ϕ is a smooth (respectively holomorphic) equivariant map for the induced action of G on $\mathcal{G}(V)$.*

Proof. Since $\mathcal{G}(V) = \bigcup_r \mathcal{G}^r(V)$ a disjoint union of open G-invariant sets, it clearly suffices, by restricting the action to $\mathcal{G}^r(V)$, to prove the theorem for that case. Now, since, as explained above, $\mathrm{GL}(V)$ acts transitively and continuously on $\mathcal{G}^r(V)$, the latter is a quotient space $\mathrm{GL}(V)/\mathrm{Stab}_{\mathrm{GL}(V)}(W)$ where W is some fixed r-dimensional subspace of V. If $\gamma : \mathrm{GL}(V) \to \mathcal{G}^r(V)$ denotes the corresponding projection and $\{w_1, \ldots, w_r\}$ is a basis of W, then since $\{gw_1, \ldots, gw_r\}$ are linearly independent for each $g \in G$, $g \mapsto gw_1 \wedge, \ldots, \wedge gw_r$ is a map $\psi : \mathrm{GL}(V) \to \wedge^r V - (0)$. Here ϕ factors as $\phi_1 \pi$ where $\phi_1 : \mathcal{G}^r(V) \to \wedge^r V - (0)$ and $\pi : \wedge^r V - (0) \to P(\wedge^r V)$ is the natural map. The diagram below, including the map ϕ_1 is commutative, since $\phi_1\gamma(g) = \phi_1(gW) = gw_1 \wedge, \ldots, \wedge gw_r = \psi(g)$.

$$
\begin{array}{ccc}
\mathrm{GL}(V) & \xrightarrow{\ \ \psi\ \ } & \wedge^r V - \{0\} \\
\downarrow{\scriptstyle\gamma} & & \downarrow{\scriptstyle\pi} \\
\mathcal{G}^r(V) & \xrightarrow{\ \ \phi\ \ } & P(\wedge^r V)
\end{array}
$$

To see that ϕ is smooth (respectively holomorphic), it is enough to show ϕ_1 is and hence that ψ is because γ and π are both smooth (respectively holomorphic). But ψ is smooth (respectively holomorphic) since $g \mapsto gw_1 \otimes, \ldots, \otimes gw_r$ is. Moreover, ϕ intertwines the actions.

$$
\begin{array}{ccc}
\mathcal{G}^r(V) & \xrightarrow{\phi} & P(\wedge^r V) \\
{\scriptstyle g}\downarrow & & \downarrow{\scriptstyle \overline{\wedge^r g}} \\
\mathcal{G}^r(V) & \xrightarrow{\phi} & P(\wedge^r V)
\end{array}
$$

$\square$

We remark that ϕ is not surjective in general. For if it were, then by compactness it would be a homeomorphism of connected manifolds, which would therefore have the same dimension. But

$$
\dim P(\overset{r}{\bigwedge} V) = \frac{n!}{r!(n-r)!} - 1.
$$

Whereas, since $\mathrm{GL}(V)$ acts transitively on $\mathcal{G}^r(V)$ with the stabilizer of W the set of operators of the form

$$
\begin{pmatrix} * & 0 \\ ** & ***, \end{pmatrix}
$$

we see that

$$
\dim \mathrm{Stab}_{\mathrm{GL}(V)}(W) = n^2 - (n-r)n.
$$

Hence

$$
\dim \mathcal{G}^r(V) = \dim \mathrm{GL}(V)/\mathrm{Stab}_{\mathrm{GL}(V)}(W) = (n-r)r.
$$

This shows that $\mathcal{G}^r(V)$ is diffeomorphic with $P(\wedge^r V)$, which is true if $r = n - 1$, but not in general.

7.5.1 Exercises

Exercise 7.5.4. Use the matrix above to get an associated bilinear form $V \times W \to k$ by defining $\beta(v_i, w_j) = t_{i,j}$ and extending to $V \times W$ using the bilinearity of β. Show this map is injective and hence by dimension a vector space isomorphism. Thus $B(V, W) = \operatorname{Hom}_k(V, W)$.

Exercise 7.5.5.

1. Let $S \in \operatorname{Hom}_k(V, W)$, $T \in \operatorname{Hom}_k(W, X)$, with $v_1, \ldots v_n$ a basis of V, $w_1, \ldots w_m$ a basis of W and $x_1, \ldots x_p$ a basis of X. If we take the matrices for M_S and M_T as above, show $M_{ST} = M_S M_T$ (matrix multiplication).

2. Prove $\operatorname{Hom}_k(V, W_1 \oplus \ldots \oplus W_k)$ and $\operatorname{Hom}_k(V, W_1) \oplus \ldots \oplus \operatorname{Hom}_k(V, W_k)$ are isomorphic.

Exercise 7.5.6.

1. Let $\beta : V \times W \to U$ be a bilinear function. Show $\beta(V \times W)$ may not be a subspace of U.

2. Let V, W and U be vector spaces. Prove that

$$B(V, W; U) = \operatorname{Hom}_k(V, \operatorname{Hom}_k(W, U)) = \operatorname{Hom}_k(W, \operatorname{Hom}_k(V, U)).$$

3. Let V, W and U be vector spaces. Prove that the map

$$\operatorname{Hom}_k(W, U) \times \operatorname{Hom}_k(V, W)) \to \operatorname{Hom}_k(V, U)),$$

given by $(S, T) \mapsto ST$ is bilinear.

Exercise 7.5.7. Let $\beta \in B(V, W; U)$, the U-valued bilinear functions and $\{v_1, \ldots v_n\}$ be a basis of V and $\{w_1, \ldots w_m\}$ be a basis of W. Prove the following statements are equivalent.

1. The vectors $\beta(v_i, w_j)$, where $1 \le i \le n$ and $1 \le j \le m$ are a basis of U.

2. Every vector $u \in U$ can be written uniquely as $u = \sum_i \beta(v_i, y_i)$, where $y_i \in W$.

3. Every vector $u \in U$ can be written uniquely as $u = \sum_j \beta(x_j, w_j)$, where $x_j \in V$.

Exercise 7.5.8. Let A and B be algebras over a field k.

1. If $A = S(V)$ and $B = S(W)$ for vector spaces V and W, prove that
$$S(V) \otimes S(W) \cong S(V \oplus W)$$
where $S(V)$ and $S(W)$ are the symmetric algebras.

2. Using (1) prove that
$$k[x_1, ..., x_n] \cong k[x] \otimes \cdots \otimes k[x] \quad (n \text{ times}).$$

Exercise 7.5.9. Let $f : V \longrightarrow V$ be an endomorphism of an n-dimensional vector space over a field F. Let $x^n - c_1 x^{n-1} + c_2 x^{n-2} - \cdots + (-1)^n c_n$ be the characteristic polynomial of the linear map f. For each $k = 1, \ldots, n$, prove that

$$c_k = \operatorname{tr}\left(\bigwedge^k f\right)$$

where

$$\bigwedge^k f : \bigwedge^k V \longrightarrow \bigwedge^k V : \bigwedge^k f(v_1 \wedge \cdots \wedge v_k) = f(v_1) \wedge \cdots \wedge f(v_k)$$

for all $v_1, \ldots, v_k \in V$.

Chapter 8

Symplectic Geometry

As we have seen in inner-product spaces the Euclidean geometry of a vector space V is given by a positive definite symmetric bilinear form. Here, we will study another kind of geometry, the symplectic geometry, given by a certain skew-symmetric bilinear form on V. Although symplectic and Euclidean geometries have some similarities, they differ significantly.

8.1 Symplectic Vector Spaces

Symplectic[1] geometry was invented by W. Hamilton in the early nineteenth century, as a mathematical framework for both classical mechanics and geometrical optics. Physical states in both settings are described by points in an appropriate *phase space* (the space of coordinates for position and momenta). Hamilton's equations associate to any energy function (Hamiltonian) on the phase space, a dynamical system. Hamil-

---

[1]The name "symplectic" was introduced in 1939 by Hermann Weyl as the Greek adjective $\sigma\upsilon\mu\pi\lambda\varepsilon\kappa\tau\iota\kappa\varnothing\varsigma$ corresponding to the word "complex". On p. 165 of his book ([131]), he writes: The name "complex group" formerly advocated by me in allusion to line complexes, as they are defined by the vanishing of antisymmetric forms, has become more and more embarrassing through collision with the word "complex" in the connotation of complex numbers. I therefore propose to replace it by the corresponding Greek adjective "symplectic".

ton realized that his equations are invariant under a very large group of symmetries, called canonical transformations, or, in modern terminology, symplectomorphisms.

Definition 8.1.1. We call a bilinear form $\omega : V \times V \to \mathbb{R}$ on a finite dimensional vector space V

- *skew-symmetric* if $\omega(v, w) = -\omega(w, v)$.

- *non-degenerate* if $\omega(v, w) = 0$ for all $w \in V$ implies $v = 0$.

- *symplectic* if it is skew-symmetric and non-degenerate.

If ω is skew-symmetric then (V, ω) is called a *pre-symplectic vector space*. If it is symplectic, (V, ω) is called a *symplectic vector space*. We define the map

$$\_^{\flat\omega} : V \to V^* \qquad v \mapsto v^{\flat\omega} := \omega(v, \cdot).$$

As we shall see, such a V must have even dimension. Here is an example.

Example 8.1.2. $\mathbb{R}^{2n}$ becomes a symplectic space if it is equipped with the *standard linear symplectic form* ω_0 given by

$$\omega_0(v, w) = \sum_{i=1}^{n} (v_{2i-1} w_{2i} - v_{2i} w_{2i-1}).$$

The following is a more general example (after identifying V^* with V).

Example 8.1.3. Let V be a vector space. The *canonical linear symplectic form on $V \oplus V^*$* is defined by

$$\omega_V \left(\begin{pmatrix} v_1 \\ \varphi_1 \end{pmatrix}, \begin{pmatrix} v_2 \\ \varphi_2 \end{pmatrix} \right) = \varphi_2(v_1) - \varphi_1(v_2).$$

Let us take a look at $\mathbb{R}^2$ equipped with the symplectic form

$$\mathbb{J}_0 = \begin{pmatrix} 0 & -1 \\ 1 & 0 \end{pmatrix}.$$

Consider two vectors $v = (a,b) \begin{pmatrix} a \\ b \end{pmatrix}$ and $w = (c,d) \begin{pmatrix} c \\ d \end{pmatrix}$. Then,

$$w^t \mathbb{J}_0 v = (c,d) \begin{pmatrix} 0 & -1 \\ 1 & 0 \end{pmatrix} \begin{pmatrix} a \\ b \end{pmatrix} = (c,d) \begin{pmatrix} -b \\ a \end{pmatrix}$$

$$= -cb + ad = \det \begin{pmatrix} a & c \\ b & d \end{pmatrix} = \omega_0(v,w).$$

Thus in dimension 2, $\omega_0(v,w)$ is the *signed* area of the parallelogram spanned by the vectors v and w, which implies that a linear symplectic map of $(\mathbb{R}^2, \omega_0)$ preserves both area and orientation.

Given a symplectic vector space (V, ω), the bilinearity of ω shows it can be viewed as the map,

$$\omega : V \otimes V \to \mathbb{R}.$$

Now, using properties of the tensor product, skew-symmetry of ω can be rewritten as

$$\omega(v \otimes w + w \otimes v) = 0,$$

for any $v, w \in V$. This means ω descends to a linear map (also denoted by ω)

$$\omega : V \wedge V \to \mathbb{R},$$

where $\bigwedge^2 V = V \wedge V$ is the quotient of $V \otimes V$ by the relations $v \otimes w + w \otimes v$, for any $v, w \in V$. In other words,

$$\omega \in \left(\overset{2}{\bigwedge} V \right)^{\star} \cong \overset{2}{\bigwedge} V^{\star}.$$

Theorem 8.1.4. *Let (V, ω) be a symplectic vector space. Then ω induces a homomorphism*

$$\omega^{\flat} : V \to V^{\star}, \quad v \mapsto \omega(v, \cdot)$$

which is, in fact, an isomorphism of vector spaces.

By picking a basis for V, any bilinear form ω on V gives rise to a matrix $\mathbb{J}$ such that for all $v, w \in V$

$$\omega(v,w) = v^t \, \mathbb{J} \, w,$$

where v^t denotes the transpose of the vector[2] v. If ω is a symplectic form, its skew-symmetry is equivalent to the fact that $\mathbb{J}$ is anti-symmetric. In other words

$$\mathbb{J}^t = -\mathbb{J},$$

while its non-degeneracy is equivalent to $\mathbb{J}$ being invertible[3].

Proof. Since ω is a non-degenerate form, $\omega^\flat$ is an isomorphism of vector spaces. □

Definition 8.1.5. Let $V = \mathbb{R}^{2n}$, where $n \geq 1$, with standard basis $e_1, ..., e_{2n}$. The bilinear form corresponding to the matrix

$$\mathbb{J}_0 = \begin{pmatrix} 0 & I \\ -I & 0 \end{pmatrix},$$

---

[2]Indeed, if $e_1, ..., e_n$ is a basis of V, then

$$\omega(u, w) = \omega\left(\sum_{i=1}^{n} u_i e_i, \sum_{j=1}^{n} w_j e_j\right) = \sum_{i,j=1}^{n} u_i w_j \omega(e_i, e_j) = \sum_{i,j=1}^{n} u_i \mathbb{J}_{ij} w_j = v^t \, \mathbb{J} \, w$$

where $\mathbb{J}$ is the matrix $(\mathbb{J}_{ij})$.

[3]As usual, we identify a bilinear form ω on V with the linear mapping

$$u \mapsto (v \mapsto \omega(u, v))$$

from V to the dual space V^* of V, this linear map will also be denoted by ω. The mapping which assigns to $v \in V$ the linear form $\varphi \mapsto \varphi(v)$ on V^* induces a linear isomorphism from V onto $(V^*)^*$, which is used to identify $(V^*)^*$ with V. Then the dual (= transposed) mapping ω^* of the linear map $\omega : V \to V^*$ is a linear mapping from $(V^*)^* = V$ to V^* and the skew-symmetry of ω is equivalent to the condition that $\omega^* = -\omega$. The non-degeneracy of ω means that the linear mapping $\omega : V \to V^*$ has zero kernel (= null space), and because $\dim V^* = \dim V$, this is equivalent to the condition that the linear mapping $\omega : V \to V^*$ is bijective. More generally, any linear mapping from V to V^* corresponds in the above fashion to a unique bilinear form on V, and the linear mapping $V \to V^*$ is bijective (= an isomorphism) if and only if the bilinear form is non-degenerate. When the bilinear form is an inner product, i.e. symmetric and positive definite, then it is automatically non-degenerate and we obtain the usual identification of V with V^* by means of the inner product. In this way we may think of a symplectic form as an skew-symmetric analogue of an inner product.

where 0 and I are the zero and identity $n \times n$-matrices, respectively, give rise to a symplectic form ω_0 on $\mathbb{R}^{2n}$, which is henceforth referred to as the *canonical symplectic form* on $\mathbb{R}^{2n}$.

Of course, every finite dimensional real vector space has an inner product, but not every vector space has a symplectic form. Indeed, let V be a finite dimensional vector space over a field k of characteristic $\neq 2$.

Proposition 8.1.6. *V has a symplectic form ω if and only if* $\dim V$ *is even.*

Proof. Let A be the matrix representation for the bilinear form ω. As we have seen the fact that ω is skew-symmetric is equivalent to $A = -A^t$ by Theorem 8.1.4. Taking determinants, we see that

$$\det A = (-1)^{\dim V} \det A.$$

Since ω is non-degenerate, A is invertible, and so we can divide both sides by $\det A$ and get

$$1 = (-1)^{\dim V}.$$

Since the characteristic is not 2, $\dim V$ is even. $\qquad\qquad\square$

(Actually $\dim V$ is even in all characteristics.)

Just as for inner product spaces (we assume the inner product is positive definite), all symplectic vector spaces are really the same (i.e. they are isomorphic). In the case of an inner product space we can always find an orthonormal basis. In the case of a symplectic vector space we can always find a symplectic basis, i.e. a basis $e_1, f_1, \ldots, e_n, f_n$ such that $\omega(e_i, f_j) = \delta_{ij}$, $\omega(e_i, e_j) = 0 = \omega(f_i, f_j)$ where ω is the symplectic form.

To see this, we first pick some $0 \neq e_1 \in V$. Then by non-degeneracy one can find f_1 such that $\omega(e_1, f_1) = 1$. Then we consider the symplectic-orthogonal complement of the span of $\{e_1, f_1\}$ and repeat the process. We remark that this implies any symplectic vector space is even dimensional and up to an isomorphism all symplectic spaces look like $V \oplus V^\star$, where the symplectic form is

$$\omega((v, \varphi), (w, \psi)) = \psi(v) - \varphi(w).$$

Then if $e_1, \ldots, e_n$ is a basis for V and $f_1, \ldots, f_n$ the dual basis for $V^\star$ (i.e. $f_i(e_j) = \delta_{ij}$), the set $\{e_1, f_1, \ldots, e_n, f_n\}$ is a *symplectic basis* for $V \oplus V^\star$.

Definition 8.1.7. A *hyperbolic pair* in (V, ω) is an ordered pair (u, v) of vectors (which are necessarily linearly independent) with $\omega(u, v) = 1$.

Definition 8.1.8. Let (V_1, ω_1) and (V_2, ω_2) be two symplectic spaces and let $f : V_1 \to V_2$ be a linear map. We say that f is a *symplectic map* if and only if

$$\omega_2(f(v), f(w)) = \omega_1(v, w), \quad \forall\, v,\, w \in V_1.$$

A symplectic automorphism $f : V \to V$ of a symplectic space (V, ω) is called a *symplectomorphism*.

Proposition 8.1.9. *A symplectic map f is injective.*

Proof. Let $f : V_1 \to V_2$ be a symplectic map, and let $\mathrm{Ker}(f)$ be its kernel. Suppose $v \in \mathrm{Ker}(f)$ and $w \in V_1$. Then,

$$f(v) = 0 \;\Rightarrow\; \omega_2(f(v), f(w)) = 0 = \omega_1(v, w),$$

and this for any $w \in V_1$. By the non-degeneracy of ω_1, $v = 0$, i.e. f is injective. $\square$

Corollary 8.1.10. *A symplectic map between two symplectic real vector spaces of the same dimension is bijective and is therefore an isomorphism.*

Definition 8.1.11. The symplectomorphisms of a symplectic vector space V form a group called the symplectic group $\mathrm{Sp}(V) = \mathrm{Sp}(n, \mathbb{R})$.

Explicitly, if $v = (v_1, \ldots, v_{2n})$ and $w = (w_1, \ldots, w_{2n})$ are vectors in $\mathbb{R}^{2n}$ (respectively $\mathbb{C}^{2n}$). Then the linear maps leaving invariant the bilinear form,

$$\omega(v, w) = \left(\sum_{i=1}^{n} v_i w_{i+n} - v_{i+n} w_i \right)$$

form the real (respectively complex) symplectic group.

Evidently,

Proposition 8.1.12. *Let ω be a non-degenerate bilinear form on a finite dimensional vector space V, over a field k. Then*

$$G_\omega = \{ g \in M_{2n}(k) : \omega(gv, gw) = \beta(v, w) \; where \, v, w \in k^{2n} \}$$

is always a group.

8.1.1 Symplectic Subspaces

Here we introduce the symplectic analogue of the orthogonal complement of a subspace E of a symplectic vector space V called the symplectic orthogonal complement E^ω of a E in V.

Definition 8.1.13. We say that the subspace W of the symplectic space (V, ω) is a *symplectic subspace* if the restriction of ω to W is non-degenerate.

Definition 8.1.14. If (V, ω) is a symplectic vector space and E a subspace of V, we define its *symplectic orthogonal complement* as

$$E^\omega = \{ v \in V \mid \omega(u, v) = 0 \; \forall \, u \in V \}.$$

Remark 8.1.15. It is critical to note that in general, E and E^ω are not complementary subspaces of V. For, consider the case where $E = \text{lin.sp}(v)$ for some non-zero vector $v \in V$. Since $\omega(\lambda v, \mu v) = \lambda \mu \, \omega(v, v) = 0$, with $\lambda, \mu \in \mathbb{R}$, we see $E \subset E^\omega$. Notice that in this example when restricted to the subspace E, ω is degenerate, which means that E itself is *not a symplectic* subspace.

Proposition 8.1.16. *Let (V, ω) be a pre-symplectic vector space and $W \subset V$ a linear subspace. The following equality holds:*

$$\dim W + \dim W^\omega = \dim V + \dim(W \cap V^\omega).$$

Furthermore, if ω is non-degenerate and W is symplectic, W^ω is also symplectic.

The proof of 8.1.16 uses the following:
Let V be a vector space and $W \subset V$ a subspace. We denote by

$$W^\perp := \{\varphi \in V^*\colon \varphi(w) = 0, \text{ for all } w \in W\}$$

the *annihilator* of W (in V^*).

Lemma 8.1.17. *We have $\dim W + \dim W^\perp = \dim V$.*

For a proof see 3.5.8 of Volume I.

Remark 8.1.18. Let V and W be vector spaces and $T : V \to W$ a linear map. Then we have $\operatorname{Ker}(T^*) = (\operatorname{Im}(T))^\perp \subset W^*$. This follows immediately from the definitions.

Proof of 8.1.16. We prove the first assertion. By 8.1.17 we have the formula $\dim W^\omega + \dim(W^\omega)^\perp = \dim V$. Recall $\flat_\omega : V \to V^*$, $\flat_\omega(v) = \omega(v, \cdot)$. We denote by $\iota_W : W \to V$ the inclusion map, $T := \flat_\omega \iota_W : W \to V^*$, and by $\iota : V \to V^{**}$, $\iota(v)(\varphi) = \varphi(v)$ the canonical map.

We first show that $\flat_\omega = -\flat_\omega^* \iota$. To prove this, let v, $v' \in V$. Since ω is skew-symmetric, we have

$$(\flat_\omega^* \iota)(v)(v') = \flat_\omega^*(\iota(v))(v') = \iota(v)(\flat_\omega v') = (\flat_\omega v')(v)$$
$$= \omega(v', v) = -\omega(v, v') = -(\flat_\omega v)(v').$$

From this we see that $\iota_W^* \flat_\omega = -\iota_W^* \flat_\omega^* \iota = -T^* \iota$ and therefore that $W^\omega = \operatorname{Ker}(\iota_W^* \flat_\omega) = \operatorname{Ker}(T^* \iota)$. Since $\dim V < \infty$, the map ι is an isomorphism of vector spaces. Hence $W^\omega = \operatorname{Ker}(T^* \iota)$ and the equation $\operatorname{Ker} T^* = (\operatorname{Im} T)^\perp$ implies $\dim W^\omega = \dim(\operatorname{Ker} T^*) = \dim(\operatorname{Im} T)^\perp$. The rank-nullity theorem for the map T tells us that $\dim W = \dim(\mathrm{Im}T) + \dim(\mathrm{Ker}\,T)$. Combining this and using 8.1.17 we obtain

$$
\begin{aligned}
\dim W + \dim W^\omega &= \dim \mathrm{Im}T + \dim(\mathrm{Im}T)^\perp + \dim \mathrm{Ker}\,T \\
&= \dim V^* + \dim \mathrm{Ker}\,T \\
&= \dim V + \dim(W \cap V^\omega).
\end{aligned}
$$

This proves the first statement. To prove the second statement let W be a symplectic subspace of a symplectic vector space V. Then we have $V^\omega = 0$ and $W \cap W^\omega = 0$. Using the formula from the first statement it follows that

$$
W + W^\omega = V. \qquad (*)
$$

Let $u \in W^\omega$ be such that $\omega(u,u') = 0$ for all $u' \in W^\omega$. We show that $u = 0$. Let $v' \in V$. By $(*)$ there exist vectors $w' \in W$ and $u' \in W^\omega$ such that $v' = w' + u'$. It follows that $\omega(u,v') = \omega(u,w') + \omega(u,u') = 0 + 0 = 0$. Since ω is non-degenerate it follows that $u = 0$. Hence $\omega|_{W^\omega}$ is non-degenerate and therefore W^ω is a symplectic subspace of V. $\square$

Lemma 8.1.19. *(Splitting Lemma) Let (V,ω) be a pre-symplectic vector space and $W \subset V$ a symplectic subspace. Then the map*

$$
W \times W^\omega \longrightarrow V, \quad (w,w') \mapsto w + w'
$$

is an isomorphism with respect to $(\omega|_W) \oplus (\omega|_{W^\omega})$ and ω.

Proof. The map $W \times W^\omega \to V$ is linear and by the definition of the pre-symplectic forms $\omega|_W$, $\omega|_{W^\omega}$ and W^ω, it pulls ω back to $\omega|_W \oplus \omega|_{W^\omega}$. Since W is symplectic, we have $W \cap W^\omega = 0$. It follows that the map is injective. To see that it is surjective, note that by 8.1.16 we have

$$
\dim W + \dim W^\omega = \dim V + \dim(W \cap V^\omega).
$$

Furthermore, we have $V^\omega \subset W^\omega$ and thus $W \cap V^\omega \subset W \cap W^\omega = 0$. So $\dim(W + W^\omega) = \dim V$ and the map $W \times W^\omega \to V$ is surjective by the rank-nullity theorem. $\square$

Proposition 8.1.20. *E is a symplectic subspace if and only if $E \cap E^\omega = \varnothing$, which by dimension is equivalent to saying that E and E^ω are complementary subspaces.*

Lemma 8.1.21. *Let (V, ω) be a symplectic vector space and let W be the complement of V^ω in V, i.e. $W \oplus V^\omega = V$. Then $\omega|_{W \times W}$ is non-degenerate.*

Proof. Let $w \in W$ such that $\omega(w, z) = 0$ for all $z \in W$. Since $V = W \oplus V^\omega$, it follows that w is skew-orthogonal to V. Hence $w \in V^\omega \cap W = \{0\}$. Therefore $w = 0$, which implies the statement. $\square$

Definition 8.1.22. A subspace W of a pre-symplectic vector space (V, ω) is called

1. *isotropic* if $W \subset W^\omega$,

2. *co-isotropic* if $W^\omega \subset W$,

3. *Lagrangian*[4] if $W = W^\omega$,

4. *symplectic* if $W \cap W^\omega = 0$.

---

[4] Joseph-Louis Lagrange (1736-1813) Italian-French mathematician was self taught. By the time he was 19 Lagrange had made important discoveries which would contribute substantially to the new subject of the calculus of variations, particularly as it applied to mechanics. These results generalized those of Euler and were based on the *principle of least action*. Lagrange also made progress on the propagation of sound, vibrating strings and differential equations. He developed various applications to topics such as fluid mechanics where he introduced what we now call the *Lagrangian*. He shared a prize for work on the three-body problem with Euler. He also worked in number theory proving the 4 square theorem. Then, he proved Wilson's theorem that n is prime if and only if $(n-1)! + 1$ is divisible by n. He also presented his important work, Reflexions sur la résolution algèbrique des Équations, which investigated why equations of degrees ≤ 4 could be solved by radicals, giving new proofs of the formulas. He then moved to Paris. During the Reign of Terror a law was passed ordering the arrest of all foreigners born in enemy countries and their property confiscated. His friend, the chemist Lavoisier, intervened on behalf of Lagrange and he was granted an exception. After a trial that lasted less than a day, a revolutionary tribunal condemned Lavoisier (who had saved Lagrange from arrest) and 27 others to death. They were guillotined that afternoon.

The set of Lagrangian subspaces of V is called the *Lagrangian Grassmannian* and is denoted by $\mathrm{Lag}(V)$.

Let (V, ω) be a pre-symplectic vector space with a linear subspace $W \subset V$. This leads to the following proposition.

Proposition 8.1.23.

- *If W is isotropic then* $\dim W \leq \frac{1}{2}(\dim V + \dim V^\omega)$.

- *If W is co-isotropic then* $\dim W \geq \frac{1}{2}(\dim V + \dim V^\omega)$.

- *If W is Lagrangian then* $\dim W = \frac{1}{2}(\dim V + \dim V^\omega)$.

Proof. By 8.1.16 we have $\dim W + \dim W^\omega = \dim V + \dim(W \cap V^\omega)$. If $W \subset W^\omega$ then $2\dim W \leq \dim W + \dim W^\omega$. If on the other hand $W^\omega \subset W$ then $2\dim W \geq \dim W + \dim W^\omega$ and $V^\omega \subset W^\omega \subset W$, and so $W \cap V^\omega = V^\omega$. $\qquad\square$

Because of the finite-dimensionality of V, any strictly increasing sequence of isotropic subspaces terminates at a maximal one, which shows that every isotropic subspace is contained in at least one Lagrange subspace. Indeed,

Lemma 8.1.24. *Let $W \subset V$ be an isotropic subspace. Then W is contained in a Lagrangian subspace.*

Proof. If $W = W^\omega$, there is nothing to prove. Hence assume that $v \in W^\omega - W$. Then $W_1 = W + \mathbb{R}v \subset V$ has dimension $\dim W + 1$ and is isotropic since $W \subset W^\omega$ and $v \in W^\omega$. Induction on the codimension gives us an index k for which $W_k = W_k^\omega$. $\qquad\square$

Proposition 8.1.25. *The following statements are equivalent:*

1. *$W \subset V$ is Lagrangian.*

2. *W is a maximal isotropic subspace.*

3. *W is a minimal co-isotropic subspace.*

Proof. Let W be a Lagrangian and $U \supset W$ an isotropic subspace. Then

$$U^\omega \subset W^\omega = W \subset U \subset U^\omega.$$

Hence $W = U$. If on the other hand W is maximal isotropic, W is Lagrangian by 8.1.24. $\qquad\square$

Remark 8.1.26. Lemma 8.1.24 implies that Lagrangian subspaces always exist. In fact, there exist isotropic subspaces of dimensions $0, 1, \ldots, \frac{1}{2}(\dim V + \dim V^\omega)$.

The following linear version of Weinstein's theorem classifies Lagrangian subspaces of symplectic vector spaces. To state and prove Weinstein's theorem we need the following lemma.

Lemma 8.1.27. *Let (V, ω) be a symplectic vector space and $W \subset V$ a Lagrangian subspace. Then there exists a Lagrangian subspace $W' \subset V$ that is complementary to W, i.e. that satisfies $V = W \oplus W'$.*

Proof. We choose a subspace $U \subset V$ complementary to W. The map $T : W \to U^*$, such that $Tw := \omega(w, \_)|_U$ is an isomorphism (check it!). We also define $T' : U \to U^*$, defined by $T'u := -\frac{1}{2}\omega(u, \_)|_U$. The subspace $W' := \{u + T^{-1}T'u \mid u \in U\}$ has dimension equal to $\dim W' = \dim U = \dim V - \dim W = \frac{1}{2}\dim V$.

We now prove that W' is isotropic. For every pair of vectors $u, u' \in U$, we have

$$\omega(u+T^{-1}T'u, u'+T^{-1}T'u') = \omega(u, u') + \omega(u, T^{-1}T'u') + \omega(T^{-1}T'u, u')$$
$$= \omega(u, u') - (T'u')(u) + (T'u)(u')$$
$$= \omega(u, u') - \frac{1}{2}\omega(u, u') - \frac{1}{2}\omega(u, u') = 0,$$

and therefore W' is isotropic and thus Lagrangian. Since U intersects W trivially, the same holds for W', and so W' has the required properties. $\qquad\square$

Theorem 8.1.28. *(**The Linear Weinstein Theorem**) Let (V, ω) be a symplectic vector space with a Lagrangian subspace. Then there exists an isomorphism*

$$(W \times W^*, \omega_W) \longrightarrow (V, \omega)$$

that carries $W \times 0$ to W.

Proof. We choose a Lagrangian subspace $W' \subset W$ as in Theorem 8.1.27. We have $\dim V = \dim W + \dim W'$. We define the map

$$\Phi : V \longrightarrow W \times W^*, \quad v \mapsto (w, \omega(\_, w')|_W),$$

where (w, w') is the unique pair satisfying $v = w + w'$. Since $V = W \oplus W'$, this map is well defined. It is linear and satisfies $\Phi^* \omega_W = \omega$. It is injective, since $W \cap W' = (0)$. Because $\dim V = \dim(W \times W^*)$, the map is also surjective. Hence it is an isomorphism of symplectic vector spaces. Furthermore, we have $\Phi(W) = W \times 0$. This proves the theorem. $\qquad\square$

Corollary 8.1.29. *Let (V, ω) be a symplectic vector space and $W \subset V$ and $W' \subset V$ be co-isotropic subspaces. Then there exists an automorphism $\Phi : (V, \omega) \to (V, \omega)$ satisfying $\Phi(W) = W'$ if and only if $\dim W = \dim W'$.*

Proof. The "only if" part is clear. Let $W \subset V$ be a co-isotropic subspace. Choose a linear complement $U \subset W$ of W^ω: $U \oplus W^\omega = W$. Then U is symplectic. Therefore, by the Splitting Lemma, we have $(V, \omega) \cong (U \oplus U^\omega, \omega|_U \oplus \omega|_{U^\omega})$. Define $k := \dim W^\omega$ and $2m = \dim U$. Then $\dim W = 2m + k$ and using Proposition 8.1.16 we have $\dim U^\omega = \dim V - 2m = \dim V - \dim W + k = 2k$. It follows that W^ω is Lagrangian in U^ω, since

$$W^\omega = W^\omega \cap U^\omega \subset W \cap U^\omega \subset (W^\omega)^\omega \cap U^\omega = (W^\omega)^{\omega|_{U^\omega}}.$$

So W^ω is maximal isotropic in U^ω (Lagrangian subspaces of symplectic vector spaces have half the dimension). Hence, by Theorem 8.1.28 there exists an isomorphism

$$\psi : (\mathbb{R}^{2k}, \omega_0) \longrightarrow (U^\omega, \omega|_{U^\omega})$$

satisfying $\psi(\mathbb{R}^k \times 0) = W^\omega$. On the other hand, by 8.1.10, there exists an isomorphism

$$(\mathbb{R}^{2m}, \omega_0) \longrightarrow (U, \omega|U).$$

The Cartesian product of this map and ψ is an isomorphism

$$\Phi : (\mathbb{R}^{2m+2k}, \omega_0 \oplus \omega_0)l \longrightarrow (U \times U^\omega, \omega|_U \oplus \omega|_{U^\omega}) \cong (V, \omega)$$

satisfying $\Phi(\mathbb{R}^{2m+k} \times 0) = W$. The statement follows. $\square$

8.2   Linear Complex Structures and $\mathrm{Sp}(n, \mathbb{R})$

Let V be a finite dimensional real vector space.

Definition 8.2.1. A *linear complex structure* on V is an endomorphism of V, i.e. a linear map $\mathbb{J} : V \to V$ such that $\mathbb{J}^2 = -\mathrm{Id}_V$.

Remark 8.2.2. If $\mathbb{J}$ is a complex structure on V, then V is a complex vector space via the scalar product

$$\mathbb{C} \times V \longrightarrow V, \quad (a + ib, v) \mapsto (a + ib)v = av + b\mathbb{J}v.$$

Hence the real dimension of V is twice the complex dimension of V. This is an even integer.

Complex structures are classified as follows:

Proposition 8.2.3. *(The Classification of Complex Structures)* *Let V be a vector space. Then for any pair of complex structures $\mathbb{J}$ and $\mathbb{J}'$ on V there exists an automorphism Φ of V satisfying $\mathbb{J}' = \Phi^{-1}\mathbb{J}\Phi$.*

Proof. We define complex scalar multiplications

$$\begin{aligned} m &: \mathbb{C} \times V \longrightarrow V, & m(a + ib, v) &= a + b\mathbb{J}v \\ m' &: \mathbb{C} \times V \longrightarrow V, & m'(a + ib, v) &= a + b\mathbb{J}'v \end{aligned}$$

Denoting by $\cdot$ the real scalar product $\mathbb{R} \times V \longrightarrow V$ on V, we get

$$2\dim_{\mathbb{C}}(V, +, m) = \dim_{\mathbb{R}}(V, +, \cdot) = 2\dim_{\mathbb{C}}(V, +, m').$$

Therefore, there exists an isomorphism of complex vector spaces

$$\Phi : (V, +, m) \longrightarrow (V, +, m'),$$

which satisfies $\Phi \mathbb{J} v = \Phi m(i, v) = m'(i, \Phi v) = \mathbb{J}' \Phi v$ for all $v \in V$. This implies that $\mathbb{J}' = \Phi^{-1} \mathbb{J} \Phi$. $\qquad \square$

In the following definition ω denotes a symplectic form, $\mathbb{J}$ a complex structure and g an inner product (by definition, g is always positive definite).

Definition 8.2.4.

1. We call the pair $(\omega, \mathbb{J})$ *compatible* if the bilinear form $\omega(\cdot, \mathbb{J} \cdot)$ is an inner product.

2. We call the pair $(\mathbb{J}, g)$ *compatible* if the bilinear form $-g(\cdot, \mathbb{J} \cdot)$ is symplectic.

3. We call the pair (g, ω) *compatible* if $-\flat_\omega^{-1} \flat_g : V \to V$ is a complex structure.

4. We call the triple $(\omega, \mathbb{J}, g)$ *compatible* if $g = \omega(\cdot, \mathbb{J} \cdot)$.

Example 8.2.5. The standard triple on $V = \mathbb{R}^{2n}$ is given by $(\omega_0, \mathbb{J}_0, g_0)$ where g_0 denotes the standard inner product on $\mathbb{R}^{2n}$. This is a compatible triple.

It follows from Exercise 8.7.13 that the pair $(\omega, \mathbb{J})$ is compatible if and only if ω is invariant under $\mathbb{J}$ and the inequality

$$\omega(v, \mathbb{J} v) > 0$$

holds for every $v \neq 0$.

We need the following lemma:

Lemma 8.2.6. *Let V be a vector space, and ω and $\mathbb{J}$ the symplectic and complex structures on V, respectively. Then, $(\omega, \mathbb{J})$ is compatible if and only if the map,*

$$h : V \times V \longrightarrow \mathbb{C}, \qquad (v, w) \mapsto h(v, w) = \omega(v, \mathbb{J} w) + i\omega(v, w)$$

is a Hermitian inner product on V with respect to the complex scalar multiplication

$$\mathbb{C} \times V \longrightarrow V, \qquad (a + ib, v) \mapsto (a + b\mathbb{J})v.$$

Proof. Let $(\omega, \mathbb{J}, g)$ be a compatible triple on V. Let m be the complex scalar multiplication and define h as above. Let $\dim_{\mathbb{R}}(V) = 2n$. By hypothesis, h is a Hermitian inner product on V. Hence, using Gram-Schmidt orthonormalization process, we can find a basis $e_1, \ldots, e_n$ of V that is unitary with respect to h. Define

$$\Phi : \mathbb{R}^{2n} \longrightarrow V, \quad (q_1, p_1, \ldots, q_n, p_n) \mapsto \sum_{j=1}^{n} (q_j e_j + p_j \mathbb{J} e_j).$$

The identities $h(e_j, e_k) = \delta_{jk}$ and $h(e_j, \mathbb{J}e_k) = ih(e_j, e_k)$ imply that $\Phi^* \omega = \omega_0$ and $\Phi^* \mathbb{J} = \mathbb{J}_0$. Hence the claim follows. $\qquad\qquad\square$

From this Lemma we get the following theorem.

Theorem 8.2.7. *Let V be a vector space and $(\omega, \mathbb{J}, g)$ and $(\omega_1, \mathbb{J}_1, g_1)$ compatible triples on V. Then, there exists an automorphism $\Phi : V \to V$ which intertwines the two triples, i.e. satisfies,*

$$\Phi^* \omega_1 = \omega, \quad \Phi^* \mathbb{J}_1 = \mathbb{J}, \quad and \quad \Phi^* g_1 = g.$$

Compatible complex structures help us to understand the group of automorphisms of a symplectic vector space.

Definition 8.2.8. Now let V be an n-dimensional complex vector space and (ω, J, g) a *compatible* triple on V. We denote by $\mathrm{Aut}(V, A_1, \ldots, A_k) = \mathrm{Aut}(A_1, \ldots, A_k)$, the set of all vector space automorphisms preserving $A_1, \ldots, A_k$. In particular,

- The *linear symplectic automorphisms* $\mathrm{Aut}(\omega) = \mathrm{Sp}(V, \omega)$.

- The *general linear group* $\mathrm{Aut}(\mathbb{J}) = \mathrm{GL}_{\mathbb{C}}(V)$ of the complex space V.

- The *orthogonal group* $\mathrm{Aut}(g) = \mathrm{O}(V, g)$ of (V, g).

- The *unitary group* $\mathrm{Aut}(\omega, \mathbb{J}, g)$ of $(V, \omega, \mathbb{J}, g)$.

Proposition 8.2.9. *Let V be a vector space and $(\omega, \mathbb{J}, g)$ a compatible triple. Then the following identities hold:*

$$\mathrm{Aut}(\omega, \mathbb{J}, g) = \mathrm{Aut}(\omega, \mathbb{J}) = \mathrm{Aut}(\mathbb{J}, g) = \mathrm{Aut}(g, \omega).$$

Proof. It suffices to prove that $\mathrm{Aut}(\omega, \mathbb{J}) \subset \mathrm{Aut}(g)$ and the analogous two other inclusions hold. Observe that for $\Phi \in \mathrm{Aut}(\omega, \mathbb{J})$ we have

$$\Phi^* g = \omega(\Phi_-, \mathbb{J}\Phi_-) = \omega(\Phi_-, \Phi\mathbb{J}_-) = \omega(_-, K_-) = g,$$

i.e. $\Phi \in \mathrm{Aut}(g)$. □

Now, let $\mathrm{U}(n)$ be the unitary group of $\mathbb{C}^n$, $\mathrm{Sp}(n, \mathbb{R})$ the group of symplectic $2n \times 2n$ matrices, $\mathrm{GL}(2n, \mathbb{C}))$ the complex general linear group in dimension $2n$, and $\mathrm{O}(2n, \mathbb{R})$ the group of orthogonal $2n \times 2n$ matrices. We define

$$\phi : \mathrm{End}_{\mathbb{C}}(\mathbb{C}^n) \longrightarrow \mathrm{End}_{\mathbb{R}}(\mathbb{R}^{2n}), \qquad A + iB \mapsto \begin{pmatrix} A & -B \\ B & A \end{pmatrix}.$$

This brings us to the following corollary.

Corollary 8.2.10. *The equalities*

$$\phi(\mathrm{U}(n)) = \mathrm{Sp}(n, \mathbb{R}) \cap \phi(\mathrm{GL}(n, \mathbb{C})) = \phi(\mathrm{GL}(n, \mathbb{C})) \cap O(2n, \mathbb{R})$$
$$= O(2n, \mathbb{R}) \cap \mathrm{Sp}(n, \mathbb{R}),$$

hold.

Since ϕ is injective we can identify a matrix A with its image and rewrite the above equality as

$$\mathrm{U}(n) = \mathrm{Sp}(n, \mathbb{R}) \cap \mathrm{GL}(n, \mathbb{C}) = \mathrm{GL}(n, \mathbb{C}) \cap \mathrm{O}(2n, \mathbb{R})$$
$$= \mathrm{O}(2n, \mathbb{R}) \cap \mathrm{Sp}(n, \mathbb{R}).$$

The group $\mathrm{Sp}(n, \mathbb{R})$ do not depend on which symplectic form we choose because any two symplectic forms are equivalent. (Actually, any two distinct symplectic forms give conjugate symplectic groups.)

8.3   The Topology of the Symplectic Groups

Here we study the topology of the groups $\mathrm{Sp}(n, \mathbb{R})$ and $\mathrm{Sp}(n, \mathbb{C})$. Both are called the *symplectic group*, but as the reader will see, they have somewhat different properties. For one thing, $\mathrm{Sp}(n, \mathbb{C})$ is a *complex* Lie group. Indeed, it is one of the complex classical simple Lie groups, while $\mathrm{Sp}(n, \mathbb{R})$ is a real simple Lie group.

Let $X \in M_{2n}(\mathbb{C})$ or $M_{2n}(\mathbb{R})$. As we saw above, it is useful to use block-matrix notation and write

$$X = \begin{pmatrix} A & B \\ C & D \end{pmatrix}$$

where A, B, C and D are $n \times n$ matrices. One sees easily that X is symplectic if and only if

$$AB^t, \ CD^t \text{ are symmetric, and } AD^t - BC^t = I, \qquad (\ddagger).$$

Exercise 8.3.1. Prove that for $k = \mathbb{R}$ or $\mathbb{C}$, $\mathrm{Sp}(1, k) = \mathrm{SL}(2, k)$.

We now calculate the dimensions of these two groups. To do so we will first find their respective Lie algebras and their dimensions. Then making use of the fact that the exponential map is a local diffeomorphism at 0 in the Lie algebra we see that the dimension of the group (manifold) is the same as that of the Lie algebra.

Proposition 8.3.2. *The Lie algebras $\mathfrak{sp}(n, \mathbb{R})$ and $\mathfrak{sp}(n, \mathbb{C})$ are respectively characterized by $\mathbb{J} \cdot X + X^t \cdot \mathbb{J} = 0$, where X is written in block form. That is, $D = -A^t$, $B = B^t$ and $C = C^t$.*

Proof. Let X be written in block form. When is $\exp(tX) \in \mathrm{Sp}(n, \mathbb{R})$, or $\mathrm{Sp}(n, \mathbb{C})$ for all $t \in \mathbb{R}$? If it were, then, by taking $\frac{d}{dt}|_{t=0}$, we get

$$\omega(Xv, w) + \omega(v, Xw) = 0.$$

That is,

$$\mathbb{J} \cdot X + X^t \cdot \mathbb{J} = 0.$$

Conversely, if this last equation holds, then $\mathbb{J} \cdot X = -X^t \cdot \mathbb{J}$. Hence, $\mathbb{J}X\mathbb{J}^{-1} = -X^t$ and so

$$\exp(\mathbb{J}X\mathbb{J}^{-1}) = \mathbb{J}(\exp X)\mathbb{J}^{-1} = \exp(-X^t).$$

Thus $\exp(X^t)\mathbb{J}\exp(X) = \mathbb{J}$. This means the Lie algebra, $\mathfrak{sp}(n, \mathbb{R})$ and $\mathfrak{sp}(n, \mathbb{C})$ are respectively characterized by $\mathbb{J} \cdot X + X^t \cdot \mathbb{J} = 0$. $\square$

Corollary 8.3.3. $\dim_{\mathbb{R}} \mathrm{Sp}(n, \mathbb{R}) = 2n^2 + n = \dim_{\mathbb{C}} \mathrm{Sp}(n, \mathbb{C})$.

Proof. $\dim_{\mathbb{R}} \mathfrak{sp}(n, \mathbb{R}) = n^2 + 2(n)(n + 1)/2$ and similarly for $\mathbb{C}$. $\square$

We now investigate some questions of topology of $\mathrm{Sp}(n, \mathbb{R})$ and $\mathrm{Sp}(n, \mathbb{C})$. In particular, are they connected and what are the respective fundamental groups. To do this we turn to their *maximal compact subgroups*. These are respectively,

$$\mathrm{Sp}(n, \mathbb{R}) \cap \mathrm{O}(2n, \mathbb{R})$$

and

$$\mathrm{Sp}(n, \mathbb{C}) \cap \mathrm{U}(2n, \mathbb{C}) = \mathrm{Sp}(n).$$

As we saw in Chapter 2 of Volume I, $\mathrm{Sp}(n)$ acts transitively on the sphere S^{4n-1} resulting in the equivariant diffeomorphism,

$$\mathrm{Sp}(n)/\mathrm{Sp}(n - 1) = S^{4n-1}.$$

Then $\mathrm{Sp}(1) = \mathrm{SU}(2, \mathbb{C}) = S^3$, which is both connected and simply connected. Thus, by induction (see [1] p. 61), $\mathrm{Sp}(n)$ is connected and simply connected for all n. It follows from the Iwasawa decomposition theorem that the same is true of $\mathrm{Sp}(n, \mathbb{C})$.

However, for $\mathrm{Sp}(n, \mathbb{R})$ things are somewhat different. Let $g \in \mathrm{Sp}(n, \mathbb{R})$. Then $g^t \cdot \mathbb{J} \cdot g = \mathbb{J}$. If g is also in $\mathrm{O}(2n, \mathbb{R})$, then $g^t = g^{-1}$ and so $\mathbb{J} \cdot g = g \cdot \mathbb{J}$. Therefore,

$$g = \begin{pmatrix} A & B \\ -B & A \end{pmatrix}$$

and $A^t \cdot A + B^t \cdot B = I$ and $A^t \cdot B - B^t \cdot A = 0$.

One sees easily that the map taking

$$g = \begin{pmatrix} A & B \\ -B & A \end{pmatrix} \mapsto A + iB$$

is a multiplicative homomorphism whose kernel is just I_{2n} so that it is injective. By the relations above it maps *onto* $U(n, \mathbb{C})$. Thus, the latter is a maximal compact subgroup of $\mathrm{Sp}(n, \mathbb{R})$ which is known to be connected and have fundamental group $\mathbb{Z}$.

Corollary 8.3.4. $\mathrm{Sp}(n, \mathbb{C})$ *is connected and simply connected.* $\mathrm{Sp}(n, \mathbb{R})$ *is connected and has fundamental group* $\mathbb{Z}$.

Exercise 8.3.5. Let $\mathcal{S}$ denote the *Siegel upper half space*[5]. That is,

$\mathcal{S} = \{A + iB : A,\ B \text{ real } n \times n \text{ symmetric matrices, } B \text{ positive definite}\}$.

Let $\mathrm{Sp}(n, \mathbb{R})$ act on $\mathcal{S}$ by $g \cdot Z = (AZ + B)(CZ + D)^{-1}$, where $Z = A + iB \in \mathcal{S}$.

1. Show this is an action.

2. Show it is transitive.

3. Calculate the isotropy group

$$\mathrm{Stab}_{\mathrm{Sp}(n,\mathbb{R})}(iI) = \mathrm{Sp}(n, \mathbb{R}) \cap O(2n\mathbb{R}) = U_n(\mathbb{C}).$$

4. Since $\mathrm{Sp}(n, \mathbb{R})$ is locally compact and second countable (because it is connected), conclude $\mathrm{Sp}(n, \mathbb{R})/U_n(\mathbb{C}) = \mathcal{S}$ as topological $\mathrm{Sp}(n, \mathbb{R})$-spaces.

5. Show $\dim \mathcal{S} = n^2 + n$.

---

[5]Carl L. Siegel (1896-1981) a German mathematician who made many contributions to celestial mechanics and number theory and its relationship to group theory, particularly to the symplectic group. Its action generalizing hyperbolic space, and the accompanying Siegel domain. As a pacifist during World War I, he was committed to an insane asylum for the duration of that war. He was anti-Nazi during the 12 years Hitler was in power.

6. Notice the similarity of this to the action of $\mathrm{SL}(2,\mathbb{R})$ on the upper half plane given in Chapter 2, Section 2.3.3.4 of Volume I.

Exercise 8.3.6.

1. Show if $Z \in \mathcal{S}$, then $Z + iI = i(i^{-1}Z + I)$ and therefore $\det(Z + iI) \neq 0$ if and only if $\det(i^{-1}Z + I) \neq 0$. But since Z is symmetric, $i^{-1}Z$ is skew-Hermitian.

2. Then use the Cayley transform given in Corollary 4.5.4 of Volume I, to show $Z + iI$ is invertible.

3. Hence, the function $Z \mapsto (Z - iI)(Z + iI)^{-1}$ is well defined for $Z \in \mathcal{S}$.

4. Show this function is a homeomorphism from $\mathcal{S}$ onto an open disk in $\mathbb{R}^{n^2+n}$, providing a way of realizing the unbounded domain, $\mathcal{S}$, as a bounded domain.

8.4 Transvections

8.4.1 Eigenvalues of a Symplectic Matrix

It is easy to see directly from the definition that the determinant of any symplectic matrix (or more generally, the determinant of any automorphism of any non-degenerate bilinear form) is $+1$ or -1. Indeed,

$$\det(A^t \mathbb{J} A) = \det(\mathbb{J}) \Rightarrow (\det A)^2 \det(\mathbb{J}) = \det(\mathbb{J}) \Rightarrow \det(A) = \pm 1.$$

What is not obvious is why this determinant can never be equal to -1 for any symplectic matrix, over any field! This is particularly surprising in view of the situation for other matrix groups such as $\mathrm{O}(n,k)$ or $\mathrm{O}(p,q,k)$, where both $+1$ and -1 determinants are achieved. The following theorem explains why this is so.

Theorem 8.4.1. *Let* $T \in \mathrm{Sp}(n,k)$, *where* $k = \mathbb{R}$ *or* $\mathbb{C}$. *Then* $\det T = +1$.

Proof. We will show that a symplectic map T is orientation preserving (see Chapter 7). Indeed, $\omega^n \neq 0$. This is because if $\{e_i, f_j \mid i, j = 1, \ldots, n\}$ is a basis, i.e. $\omega(e_i, e_j) = \omega(f_i, f_j) = 0$ and $\omega(e_j, f_j) = \delta_{ij}$ for all i, j, then

$$\omega = \sum_{i=1}^{n} e_i^\star \wedge f_i^\star,$$

where $\{e_i^\star, f_i^\star\}$ is the dual basis. So

$$\omega^n = n!\, e_1^\star \wedge f_1^\star \wedge \ldots \wedge e_n^\star \wedge f_n^\star \neq 0.$$

Now, T induces maps

$$T_\star : \bigwedge^n V \longrightarrow \bigwedge^n V.$$

Since $\bigwedge^n V$ is 1-dimensional, $T_\star$ is just the scalar multiplication by $\det T$. Also, the dual map

$$T^\star : \bigwedge^n V^\star \longrightarrow \bigwedge^n V^\star$$

are both multiplications by $\det T$. Hence, since

$$T^\star \omega^n = \omega^n,$$

we get $\det T = 1$.   □

The eigenvalues of a symplectic matrix are of a particular type.

Proposition 8.4.2. *Let $T \in \mathrm{Sp}(n, \mathbb{R})$. Then,*

1. *If λ is an eigenvalue of T, then so are $\bar{\lambda}$ and $1/\lambda$ and hence also $1/\bar{\lambda}$.*

2. *If the eigenvalue λ of T has multiplicity k then so has $1/\lambda$.*

3. *T and T^{-1} have the same eigenvalues.*

Proof. To prove 1 we have to show that the characteristic polynomial $\chi_T(\lambda) = \det(T - \lambda I)$ of the symplectic matrix T satisfies the following equation.

$$\chi_T(\lambda) = \lambda^{2n}\chi_T(\lambda)\left(\frac{1}{\lambda}\right).$$

Then, since real matrices have the eigenvalues in complex conjugate pairs, we get 1. To do this, we will use the facts that, since T is symplectic, $T^t\mathbb{J}T = \mathbb{J}$, so $T = \mathbb{J}^{-1}(T^t)^{-1}\mathbb{J} = -\mathbb{J}(T^t)^{-1}\mathbb{J}$, and $\det T = \det T^t = \det \mathbb{J} = 1$, $\mathbb{J}^2 = -I$. As a result we get:

$$\chi_T(\lambda) = \det(T - \lambda I)$$
$$= \det\left(-\mathbb{J}(T^t)^{-1}\mathbb{J} - \lambda I\right)$$
$$= \det\left(-\mathbb{J}(T^t)^{-1}\mathbb{J} + \lambda\mathbb{J}^2\right)$$
$$= \det\left[\mathbb{J}\left(-(T^t)^{-1} + \lambda I\right)\mathbb{J}\right]$$
$$= \det\left(-(T^t)^{-1} + \lambda I\right)$$
$$= \det(-I + \lambda T) = \lambda^{2n}\det\left(T - \frac{1}{\lambda}I\right)$$
$$= \lambda^{2n}\chi_T(\lambda)\left(\frac{1}{\lambda}\right).$$

$\square$

Exercise 8.4.3. Find a symplectic 2×2 matrix that is not diagonalizable over $\mathbb{C}$. Then for every $\lambda \in i\mathbb{R} - \{0\}$ find a symplectic matrix $A \in \mathbb{R}^{8\times 8}$ for which there exists a matrix $B \in \mathbb{C}^{8\times 8}$, such that

$$BAB^{-1} = \begin{pmatrix} \lambda & 1 & & & & & & \\ & \lambda & & & & & & \\ & & \bar{\lambda} & 1 & & & & \\ & & & \bar{\lambda} & & & & \\ & & & & \lambda^{-1} & 1 & & \\ & & & & & \lambda^{-1} & & \\ & & & & & & \bar{\lambda}^{-1} & 1 \\ & & & & & & & \bar{\lambda}^{-1} \end{pmatrix}.$$

Find a symplectic matrix of smaller size that is not diagonalizable over $\mathbb{C}$ and has λ as an eigenvalue.

8.4.2   Sp(n, ℝ) is generated by Transvections

Definition 8.4.4. Let V be a real vector space. A linear map $\tau : V \to V$ is a *transvection* with a fixed hyperplane W if $\tau|_W = Id_W$ and $\tau(v) - v \in W$ for all $v \in V$.

Geometrically, a transvection maps W onto itself and translates points outside of W parallel to W. For example, a transvection in $\mathbb{R}^2$ is the map $(x, y) \mapsto (x + ay, y)$. Here the fixed hyperplane W is the x-axis and points on opposite sides of this reference line are displaced in opposite directions. Hence, the length of any line segment that is not parallel to the direction of displacement will change under the transvection. In other words, lengths are not preserved. But a transvection preserves area (respectively volume) since its determinant is 1 (For more details see Jacobson [68], p. 391).

Given a vector u and $\lambda \in k$, we can construct a transvection $\tau_{u,\lambda}$ by setting

$$\tau_{u,\lambda}(v) = v + \lambda\omega(v, u)u.$$

This is a transvection because it fixes the hyperplane $W = u^{\perp}$ and $\tau_{u,\lambda}(v) - v = \lambda\omega(v, u)u$ is a scalar multiple of u, which is in $u^{\perp}$. In fact, every transvection can be expressed in this way for some $u \in V$ and some scalar λ. So, we have the equivalent definition:

Let (V, ω) be symplectic vector space over k and x be a non-zero vector in V and let $\lambda \in k$. We define a map

$$\psi : V \to V, \quad \text{such that} \quad \psi(v) := v - \lambda\omega(v, x)x.$$

One sees easily that ψ is a linear map.

Proposition 8.4.5. *The symplectic group* $\mathrm{Sp}(n, \mathbb{R})$ *is generated by transvections[6].*

---

[6] J. Dieudonné, in his paper [32], observing that $\mathrm{Sp}(n, \mathbb{R})$ is generated by its transvections, considered the problem of finding the minimal number of transvections as a function of n that would be required to do this.

Proof. The proof will be done in three steps. Let $\mathcal{T}$ be the group generated by all transvections τ. We will show that $\mathcal{T} = \mathrm{Sp}(n, \mathbb{R})$.

Step 1: $\mathcal{T}$ acts transitively on $V - \{0\}$.

To see this, let $v \neq w \in V - \{0\}$.

Case 1. $\omega(v, w) \neq 0$. Set $\lambda = \frac{1}{\omega(v,w)}$ and $u = v - w$. Then,

$$\tau_{u,\lambda}(v) = v + \lambda\omega(v, u)u = v + \frac{\omega(v, v - w)}{\omega(v, w)}(v - w) = v - (v - w) = w.$$

Case 2. $\omega(v, w) = 0$.

We want to find a vector z such that $\omega(v, z) \neq 0$ and $\omega(w, z) \neq 0$. This is possible because in our symplectic basis $\{u_1, v_1, ..., u_n, v_n\}$ of hyperbolic vectors,

$$v = \sum_1^n (a_i v_i + b_i u_i) \quad \text{and} \quad w = \sum_1^n (c_i v_i + d_i u_i).$$

If for some i, a_i or $b_i \neq 0$ and also c_i or $d_i \neq 0$, then there is some linear combination of v_i and u_i which satisfies the requirement for z. Otherwise, for each i, if a_i or $b_i \neq 0$, then both $c_i = 0 = d_i$ and vice versa. Since neither v nor w is zero, there is a non-zero coefficient a_{i_0} (or b_{i_0}) of the linear combination forming v, and a non-zero c_{j_0} (or d_{j_0}), with $i_0 \neq j_0$. Then let z be an appropriate non-zero linear combination of v_i and u_i as well as an appropriate non-zero linear combination of v_j and u_j. Having constructed the desired vector z, we can now use Case 1 to construct τ_1 and τ_2 such that $\tau_1(v) = z$, $\tau_2(z) = w$ and thus $\tau_2 \circ \tau_1(v) = w$.

Step 2: $\mathcal{T}$ is transitive on hyperbolic pairs.

We want to show that there is a product of transvections sending (u_1, v_1) to (u_2, v_2). By Step 1, there is a transvection τ mapping u_1 to u_2. Thus,

$$\tau : (u_1, v_1) \mapsto (u_2, \tau(v_1)).$$

Set $\tau(v_1) = v_3$. We want to find $\sigma \in \mathcal{T}$ such that $\sigma(u_2) = u_2$ and $\sigma(v_3) = v_2$.

Note. Notice that if $\omega(\alpha, \beta) \neq 0$, and we set $\gamma = \alpha - \beta$ and $a_\gamma = 1/B(\omega\alpha, \beta)$, if $\omega(u_2 n\gamma) = 0$, using Part 1, we get

$$\tau_{\gamma,a_\gamma}(u_2) = u_2 \text{ and } \tau_{\gamma,a_\gamma}(\alpha) = \beta.$$

Case 1. $\omega(v_3, v_2) \neq 0$.

Then let $\gamma = v_3 - v_2$. By the Note, $\tau_{\gamma,a_\gamma}(u_2) = u_2$ and $\tau_{\gamma,a_\gamma}(v_3) = v_2$, as desired.

Case 2. $\omega(v_3, v_2) = 0$. In this case, $\omega(v_3, u_2 + v_3) = -1$. Also, $\omega(u_2 + v_3, v_2) = \omega(u_2, v_2) \neq 0$. Moreover, $\omega(u_2, -u_2) = 0 = \omega(u_2, u_2 + v_3 - v_2)$. So, by Note 1, we can construct σ_1 and σ_2 respectively such that both fix u_2 and $\sigma_1(v_3) = u_2 + v_3$ and $\sigma_2(u_2 + v_3) = v_2$. Thus $\sigma_2\sigma_1\tau \in T$ and $\sigma_2\sigma_1\tau(u_1, v_1) = (u_2, v_2)$, as desired.

Step 3. The symplectic group $\mathrm{Sp}(V)$ is generated by symplectic transvections.

We prove the result by induction on m, where $2m = n = \dim V$. The case $m = 1$ follows from the fact that for $m = 1$, $\mathrm{Sp}(V) = \mathrm{SL}(V)$ and the fact that $\mathrm{SL}(V)$ is generated by transvections.

For the inductive step, chose a hyperbolic pair (u, v) in V and let W be the hyperbolic plane they span. Then $V = W \oplus W^\perp$. Take any $\sigma \in \mathrm{Sp}(V)$. Then $(\sigma(u), \sigma(v))$ is a hyperbolic pair, and by Step 2, there exists $\tau \in \mathcal{T}$ with $\tau \circ \sigma(u) = u$ and $\tau \circ \sigma(v) = v$, so $\tau \circ \sigma|_W = \mathrm{Id}_W$. Moreover, $\tau \circ \sigma|_{W^\perp} \in \mathrm{Sp}(W^\perp)$. By induction, $\tau \circ \sigma|_{W^\perp}$ is a product of symplectic transvections on $W^\perp$. Since any transvection on $W^\perp$ can be extended to a transvection on V which includes W in the fixed hyperplane, we see that in V, $\tau \circ \sigma \in \mathcal{T}$ and hence $\sigma \in \mathcal{T}$. $\square$

We conclude this section with a characterization of the center of $\mathcal{Z}(\mathrm{Sp}(n, \mathbb{R}))$. The same result holds for $\mathcal{Z}(\mathrm{Sp}(n, \mathbb{C}))$.

Proposition 8.4.6. *The center,* $\mathcal{Z}(\mathrm{Sp}(n, \mathbb{R}))$ *is* $\{I, -I\}$.

Proof. Let $\phi \in \mathcal{Z}(\mathrm{Sp}(n, \mathbb{R}))$. It follows from proposition 8.4.5 that ϕ commutes with all transvections. Let ψ be the transvection corresponding to $x \in V$ and $\zeta \in \mathcal{Z}$. Then, for all $y \in V$,

$$\phi(y) + \zeta\langle y, x\rangle\phi(x) = \phi\psi(y) = \psi\phi(y) = \phi(y) + \langle\phi(y), x\rangle x.$$

Let z be another vector in V. In the same way

$$\phi(z) = \zeta_z z \quad \text{and} \quad \phi(x + z) = \zeta_{x+z}(x + z).$$

Hence,

$$\zeta_x x + \zeta_z z = \phi(x + z) = \zeta_{x+z}(x + z).$$

Therefore, $\zeta_x = \zeta_z$ and there is an element $\zeta \in \mathcal{Z}$ such that $\phi(x) = \zeta x$ for all $x \in V$.

Now, choose y such that $\langle y, x \rangle \neq 0$. Then, since $\phi(x) = \zeta x$,

$$\zeta^2 \langle x, y \rangle = \langle \zeta x, \zeta y \rangle = \langle \phi(x), \phi(y) \rangle = \langle x, y \rangle,$$

which implies that $\zeta = \pm I$. $\qquad\qquad\qquad\qquad\qquad\qquad\square$

8.5   J. Williamson's Normal Form

J. Williamson's theorem says that one can diagonalize any positive definite real symmetric matrix A using a symplectic matrix, and that the diagonal matrix D has the very simple form,

$$D = \begin{pmatrix} \Lambda_\omega & 0 \\ 0 & \Lambda_\omega \end{pmatrix},$$

where the diagonal elements of Λ_ω are the moduli of the eigenvalues of $\mathbb{J}A$. One can without exaggeration say that this theorem carries embryonically the recent developments of symplectic topology. It leads immediately to a proof of Gromov's famous non-squeezing theorem in the linear case and has many applications both in mathematics and physics. Williamson proved this result in 1963 and since then it has been rediscovered several times with different proofs.

Let A be a real $n \times n$ symmetric matrix. As we know all the eigenvalues of A are real, and A can be diagonalized using an orthogonal transformation.

Theorem 8.5.1. (J. Williamson's Theorem) *Let A be a positive-definite symmetric real $2n \times 2n$ matrix. Then the following statements hold.*

1. *There exists $S \in \mathrm{Sp}(n)$ such that*

$$S^t A S = \begin{pmatrix} \Lambda & 0 \\ 0 & \Lambda \end{pmatrix}, \quad \text{where } \Lambda \text{ is a diagonal matrix}$$

and the diagonal entries λ_k are defined by the condition

$$\pm i\lambda_k \text{ is an eigenvalue of } \mathbb{J}A.$$

2. *Up to ordering, the sequence $\lambda_1, \ldots, \lambda_n$ does not depend on the choice of S diagonalizing A.*

Proof. A quick examination of the simple case $A = I$ shows the eigenvalues are ± 1 so that it is a good idea to work in the space $\mathbb{C}^{2n}$ and to look for complex eigenvalues and vectors for the matrix $\mathbb{J}A$. We denote by $\langle, \rangle_A$ the scalar product associated with A. That is,

$$\langle z_1, z_2 \rangle_A = \langle A z_1, z_2 \rangle.$$

Since both $\langle, \rangle_A$ and the symplectic form are non-degenerate we can find a unique invertible matrix K of order $2n$ such that

$$\langle z_1, K z_2 \rangle_A = \omega(z_1, z_2)$$

for all z_1, z_2. The matrix K satisfies

$$K^t A = \mathbb{J} = -AK.$$

Since we have a Hermitian inner product, $K = -\widehat{K}$, where $\widehat{K} = -A^{-1}K^t A$. From this it follows that the eigenvalues of $K = A^{-1}\mathbb{J}$ are of the type $\pm i\lambda_j$, $\lambda_j > 0$, and hence so are those of $\mathbb{J}A^{-1}$, the corresponding eigenvectors occurring in conjugate pairs $e'_j \pm if'_j$. We thus obtain a $\langle \cdot, \cdot \rangle_A$-orthonormal basis $\{e'_i, f'_j\}_{1 \le i,\, j \le n}$ of $\mathbb{R}^{2n}_z$ with $Ke'_i = \lambda_i f'_i$ and $Kf'_j = -\lambda_j e'_j$. From these relations it follows that

$$K^2 e'_i = -\lambda_i^2 e'_i, \quad \text{and} \quad K^2 f'_j = -\lambda_j^2 f'_j$$

and that the vectors of the basis $\{e'_i, f'_j\}_{1\leq i,\, j\leq n}$ satisfy the following relations,

$$\omega(e'_i, e'_j) = \langle e'_i, Ke'_j\rangle_A = \lambda_j\langle e'_i, f'_j\rangle_A = 0,$$

$$\omega(f'_i, f'_j) = \langle f'_i, Kf'_j\rangle_A = -\lambda_j\langle f'_i, e'_j\rangle_A = 0,$$

$$\omega(f'_i, e'_j) = \langle f'_i, Ke'_j\rangle_A = \lambda_i\langle e'_i, f'_j\rangle_A = -\lambda_i\delta_{ij}.$$

Setting $e_i = \lambda_i^{-1/2}e'_i$ and $f_j = \lambda_j^{-1/2}f'_j$, the basis $\{e_i, f_j\}_{1\leq i,\, j\leq n}$ is symplectic. Let S be the element of $\mathrm{Sp}(n, \mathbb{R})$ mapping the canonical symplectic basis to $\{e_i, f_j\}_{1\leq i,\, j\leq n}$. The $\langle,\rangle_A$-orthogonality of $\{e_i, f_j\}_{1\leq i,\, j\leq n}$ implies the statement, with $\Lambda = \mathrm{diag}[\lambda_1,\ldots,\lambda_n]$.

To prove the uniqueness statement 2 just above, it suffices to show if there exists $S \in \mathrm{Sp}(n, \mathbb{R})$ such that $S^t L S = L'$ with $L = \mathrm{diag}[\Lambda, \Lambda]$ and $L' = \mathrm{diag}[\Lambda', \Lambda']$, then $\Lambda = \Lambda'$. But because S is symplectic,

$$S^t \mathbb{J} S = \mathbb{J}.$$

Hence, $S^t L S = L'$ is equivalent to $S^{-1}\mathbb{J}LS = \mathbb{J}L'$, from which it follows that $\mathbb{J}L$ and $\mathbb{J}L'$ have the same eigenvalues. These eigenvalues are precisely the complex numbers $\pm i/\lambda_j$. $\qquad\square$

In the above theorem the matrix S is not unique. However, M. de Gosson proved the following result (see [54], p. 245).

Proposition 8.5.2. *Let D be the Williamson's diagonal form of A. If S and S' are in $\mathrm{Sp}(n, \mathbb{R})$ with*

$$A = (S')^t D S' = S^t D S,$$

then

$$S(S')^{-1} \in \mathrm{U}(2n).$$

Proof. Set $U = S(S')^{-1}$. Then, $U^t DU = D$. The proposition will follow once we know $\mathbb{J} = \mathbb{J}U$. Set $R = D^{1/2}UD^{-1/2}$. Then

$$R^t R = D^{-1/2}(U^t DU)D^{-1/2} = D^{-1/2}DD^{-1/2} = I,$$

which shows $R \in O(2n, \mathbb{R})$. Since $\mathbb{J}$ commutes with each power of D, and $\mathbb{J}U = (U^t)^{-1}\mathbb{J}$, it follows that

$$\begin{aligned}
\mathbb{J}R &= D^{1/2}\mathbb{J}UD^{-1/2} = D^{1/2}(U^t)^{-1}\mathbb{J}D^{-1/2} \\
&= D^{1/2}(U^t)^{-1}D^{-1/2}\mathbb{J} = (R^t)^{-1}\mathbb{J}.
\end{aligned}$$

Hence $R \in \mathrm{Sp}(n, \mathbb{R}) \cap O(2n)$ and so, $\mathbb{J}R = R\mathbb{J}$. Now, $U = D^{-1/2}RD^{1/2}$ and therefore

$$\begin{aligned}
\mathbb{J}U &= \mathbb{J}D^{-1/2}RD^{1/2} = D^{-1/2}\mathbb{J}RD^{1/2} \\
&= D^{-1/2}R\mathbb{J}D^{1/2} = D^{-1/2}RD^{1/2}\mathbb{J} \\
&= U\mathbb{J}.
\end{aligned}$$

$\square$

8.6   Wirtinger's Inequality

Here, we pass from $\mathbb{R}^{2n}$ to $\mathbb{C}^n$.

Let V be a complex vector space of complex dimension n, equipped with a Hermitian inner product, which we shall call $H := \langle\, -,\; -\rangle$. Suppose

$$A : V \times V \longrightarrow \mathbb{R} \quad \text{and} \quad \omega : V \times V \longrightarrow \mathbb{R}$$

are such that

$$\langle\, v,\; w\rangle = A(v,w) + i\omega(v,w), \quad \text{for all } v,\, w \in V.$$

In this case, one easily verifies the following:

1. Both A and ω are bilinear over $\mathbb{R}$.

2. A is symmetric and positive definite.

3. ω is antisymmetric.

4. $A(iv,\, iw) = A(v,\, w)$, for all $v,\, w \in V$.

5. $\omega(v\,,\,w) = -\omega(iv\,,\,w)$ whenever $v,\,w \in V$.

This means that the real part $A = \Re(H)$ of H defines an inner product of the *real* vector space V and the imaginary part $\omega = \Im(H)$ defines a 2-form on V, i.e. an alternating real bilinear form. The volume form dV associated with A can be expressed by the n-fold wedge product of ω with itself as

$$dV = \frac{1}{n!}(\omega \wedge \omega \wedge \cdots \wedge \omega) = \frac{1}{n!}\omega^{\wedge n}.$$

Proposition 8.6.1. *(**Wirtinger's Inequality**) Let V be a complex vector space of dimension n equipped with a Hermitian inner product $H := \langle\,-,\,-\,\rangle$. Set $A = \Re(H)$ and $\omega = Im(H)$. Let $W \subseteq V$ be a real $2k$-dimensional subspace and let dV_W be the volume form corresponding to the restriction of A on W. Then, for any $w_1,...,w_{2k}$ in W we have*

$$\frac{1}{k!}\left|\omega^{\wedge k}(w_1,...,w_{2k})\right| \le \left|dV_W(w_1,...,w_{2k})\right|.$$

Proof. It is sufficient to prove the inequality for a basis $w_1,...,w_{2k}$. Choose an orthonormal basis $e_1,\ldots,e_k,f_1,\ldots,f_k$ of W and $\lambda_i \in \mathbb{R}$, $i = 1,...,k$ such that

$$\omega(e_i\,,\,e_j) = 0, \quad \omega(f_i\,,\,f_j) = 0, \quad \omega(e_i\,,\,f_j) = \delta_{ij},$$

for all $1 \le i,j \le k$. Therefore, we can write $\omega = \omega_1 + \cdots + \omega_k$, where each ω_j is the pullback of a 2-form on the subspace

$$V_j := \mathbb{R}e_j \oplus \mathbb{R}f_j$$

via the orthogonal projection. Then $\omega_j \wedge \omega_j = 0$ and since all ω_j are 2-forms, $\omega_m \wedge_l = \omega_l \wedge \omega_m$. Hence,

$$\omega^{\wedge k} = (\omega_1 + \cdots + \omega_k)^{\wedge k} = k!\omega_1 \wedge \cdots \wedge \omega_k.$$

From this it follows that,

$$\frac{1}{k!}\omega^{\wedge k}(e_1, f_1, \ldots, e_k, f_k) = \omega_1(e_1, f_1)\ldots\omega_k(e_k, f_k) = \lambda_1 \cdots \lambda_k.$$

The matrix of the restriction of H to V_j with respect to the basis $\{e_j, f_j\}$ is,

$$H_{V_j} = \begin{pmatrix} 1 & i\lambda_j \\ -i\lambda_j & 1 \end{pmatrix}$$

and since $\det H_{V_j} = 1 - \lambda_j^2 \geq 0$, i.e. $|\lambda_j| \leq 1$, this matrix is positive semi-definite. Hence

$$\frac{1}{k!}\left|\omega^{\wedge k}(e_1, f_1, \ldots, e_k, f_k)\right| \leq \left|\lambda_1 \ldots \lambda_k\right| \leq 1 = \left|dV(e_1, f_1, \ldots, e_k, f_k)\right|.$$

$\square$

8.7   M. Gromov's Non-squeezing Theorem

The non-squeezing theorem, also called Gromov's[7] non-squeezing theorem, is one of the important theorems in symplectic geometry. It was first proved in 1985 in ([56]).

Let

$$\omega = \sum_{i=1}^{n} p_i \wedge q_i$$

be the standard symplectic form on Euclidean space, $\mathbb{R}^n \times \mathbb{R}^n = \mathbb{R}^{2n}$, with coordinates $(p_1, q_1, \ldots, p_n, q_n)$. The non-squeezing theorem states that it is impossible for a symplectic diffeomorphism (i.e. a diffeomorphism that preserves the 2-form ω) to map the $2n$-dimensional ball B_r of radius r into the cylinder Z_R of base radius R when $R < r$. Here

$$Z_R = \{(p_1, q_1, \ldots, p_n, q_n) \in \mathbb{R}^{2n} \mid p_1^2 + q_1^2 < R^2\}.$$

Thus, symplectic diffeomorphisms possess some type of two-dimensional rigidity since the base of the cylinder has dimension two (and not merely the preservation of volume ensured by Liouville's theorem which since it preserves ω, also preserves the standard volume form dV).

---

[7]Mikhael Gromov (1943-) Russian Jewish mathematician. One of the pioneers in Hyperbolic and Symplectic Geometry and their invariants is a mathamatician of great originality and the author of many important contributions.

Gromov's non-squeezing theorem can be restated by saying that the two-dimensional shadow of a symplectic ball of radius r in $\mathbb{R}^{2n}$ has area at least πr^2. More precisely, every symplectic embedding

$$\phi : B_r \longrightarrow \mathbb{R}^{2n}$$

satisfies the inequality

$$\text{area}\big(P(\phi(B_r))\big) \geq \pi r^2,$$

where P denotes the orthogonal projection onto the plane corresponding to the conjugate coordinates (p_1, q_1). In other words, if V is a symplectic plane in $\mathbb{R}^{2n}$ and P is the projection onto V along the symplectic orthogonal complement of P, then

$$\int_{P(\phi(B_r))} \omega \geq \pi r^2$$

for every symplectic embedding $\phi : B_r \longrightarrow \mathbb{R}^{2n}$.

Here we present the proof of the (generalized) *linear* version of Gromov's non-squeezing theorem following A. Abbondandolo, and R. Matveyev, [2].

Let B be the *open* unit ball centered at $0 \in \mathbb{R}^n$. First we will look at the volume of the image of B by a linear surjective transformation. Let n, k be two positive integers, $n \geq k$, and $A : \mathbb{R}^n \to \mathbb{R}^k$ be linear and surjective. We denote by $A^t : \mathbb{R}^k \to \mathbb{R}^n$ the *adjoint* of A with respect to the Euclidean inner product. The linear map $A^t A : \mathbb{R}^n \to \mathbb{R}^n$ is symmetric and positive semi-definite, with a k-codimensional kernel

$$\text{Ker}(A^t A) = \text{Ker}(A) = (\text{Rank}\, A^t)^{\perp}.$$

where, henceforth, by $\text{Rank}\, A^t$ we denote the subspace $A^t(\mathbb{R}^k)$ of $\mathbb{R}^n$. In particular, $A^t A$ restricts to an automorphism of the k-dimensional space

$$(\text{Ker}(A))^{\perp} = \text{Rank}\, A^t,$$

and since this restriction is the composition of the two isomorphisms

$$A\big|_{(\text{Ker}(A))^{\perp}} : (\text{Ker}(A))^{\perp} \longrightarrow \mathbb{R}^k, \quad \text{and} \quad A^t : \mathbb{R}^k \to (\text{Ker}(A))^{\perp},$$

which have the same determinant, one being the adjoint of the other, we conclude

$$\det \left(A^t \, A\Big|_{(\mathrm{Ker}(A))^\perp} \right) = \left| \det \left(A\Big|_{(\mathrm{Ker}(A))^\perp} \right) \right|^2.$$

Here, the absolute value of the determinant of linear maps between different subspaces of the same dimension is induced by the Euclidean inner product. (Since linear subspaces do not have a preferred orientation, the determinant is only defined up to the sign.) Let $\xi_1, \ldots, \xi_k$ be a basis of $(\mathrm{Ker}(A))^\perp$ with

$$\left| \xi_1 \wedge \cdots \wedge \xi_k \right| = 1,$$

where the Euclidean norm of $\mathbb{R}^n$ is extended to multivectors in the standard way (see Theorem 7.3.38.) In particular, $\left| \xi_1 \wedge \cdots \wedge \xi_k \right|$ is the k-volume of the parallelepiped generated by $\xi_1, \ldots, \xi_k$. Since

$$A \circ B = A(B \cap (\mathrm{Ker}(A))^\perp) = A(B \cap \mathrm{Rank}\, A^t),$$

we find

$$\frac{\mathrm{vol}_k(A \circ B)}{\omega_k} = \left| A\xi_1 \wedge \cdots \wedge A\xi_k \right| = \left| \det \left(A\Big|_{(\mathrm{Ker}(A))^\perp} \right) \right|$$

$$= \sqrt{\det \left(A^t \, A\Big|_{(\mathrm{Ker}(A))^\perp} \right)}, \qquad (\star)$$

where ω_k denotes the k-volume of the unit k-ball. Furthermore, the real valued function

$$W \longrightarrow \left| \det A_W \right|, \quad W \in \mathcal{G}(k, n),$$

where $\mathcal{G}(k, n)$ denotes the Grassmannian of k-dimensional subspaces of $\mathbb{R}^n$, has a unique maximum at $(\mathrm{Ker}(A))^\perp = \mathrm{Rank}\, A^t$. Hence

$$\max_{W \in \mathcal{G}(k,n)} \left| \det(A \big|_W) \right| = \left| \det(A \big|_{\mathrm{Rank}\, A^t}) \right|.$$

Let $\mathbb{R}^{2n}$ be $2n$-dimensional Euclidean space, with coordinates

$$(p_1, \, q_1, \, \ldots, \, p_n, \, q_n),$$

and with the complex structure $\mathbb{J}$ corresponding to the identification

$$(p_1, q_1, \ldots, p_n, q_n) \equiv (p_1 + iq_1, \ldots, p_n + iq_n).$$

That is,

$$\mathbb{J}(p_1, q_1, \ldots, p_n, q_n) = (-q_1, p_1, \ldots, -q_n, p_n),$$

and with the symplectic form given by minus the imaginary part of the corresponding Hermitian product, that is

$$dV = \sum_{j=1}^{n} dp_j \wedge dq_j.$$

Now, using Wirtinger's inequality (see 8.6.1), we are ready to prove the linear version of the non-squeezing theorem.

Theorem 8.7.1. (*The Linear Non-Squeezing Theorem*) *Let Φ be a linear symplectic automorphism of $\mathbb{R}^{2n}$, and let $P : \mathbb{R}^{2n} \to \mathbb{R}^{2n}$ be the orthogonal projection onto a complex linear subspace $V \subseteq \mathbb{R}^{2n}$ of real dimension $2k$, $1 \le k \le n$. Then,*

$$\mathrm{vol}_{2k}\left(P \circ \Phi(B_r)\right) \ge \omega_{2k} r^{2k},$$

and equality holds if and only if the linear subspace $\Phi^t V$ is complex.

Proof. By linearity, we may assume $r = 1$. We consider the linear surjection

$$A := P \circ \Phi : \mathbb{R}^{2n} \longrightarrow V.$$

Let $\xi_1, \ldots, \xi_k$ be a basis of $(\mathrm{Ker}(A))^{\perp} = \Phi^t(V)$ such that

$$\left|\xi_1 \wedge \cdots \wedge \xi_k\right| = 1.$$

By Wirtinger's Inequality,

$$\left(\frac{\mathrm{vol}_{2k}(P \circ \Phi(B))}{\omega_{2k}}\right)^2 = \det\left(A^t A \big|_{(\mathrm{Ker}(A))}^{\perp}\right) = \left|A^t A \xi_1 \wedge \cdots \wedge A^t A \xi_{2k}\right|$$

$$\ge \frac{1}{k!}\left|(dV)^k\left(A^t A \xi_1, \ldots, A^t A \xi_{2k}\right)\right|$$

$$= \frac{1}{k!}\left|(dV)^k\left(\Phi^t A \xi_1, \ldots, \Phi^t A \xi_{2k}\right)\right|, \tag{1}$$

and equality holds if and only if the subspace spanned by $A^t A\xi_1, \ldots,$ $A^t A\xi_{2k}$, that is $\Phi^t(V)$, is complex. Since Φ is symplectic, so is Φ^t. Hence

$$(dV)^k \left(\Phi^t A\xi_1, \; \ldots, \; \Phi^t A\xi_{2k} \right) = (dV)^k \left(A\xi_1, \; \ldots, \; A\xi_{2k} \right) \qquad (2)$$

Since the restriction of $(dV)^k$ to the complex subspace V is $k!$ times the standard volume form, we get (using the equation $(\star)$)

$$\frac{1}{k!} \left| dV^k \left(A\xi_1, \; \ldots, \; A\xi_{2k} \right) \right| = \left| dV^k \left(A\xi_1 \wedge \ldots \wedge A\xi_{2k} \right) \right| = \frac{\mathrm{vol}_{2k}(A \circ B)}{\omega_{2k}}$$

$$= \frac{\mathrm{vol}_{2k}(P \circ \Phi(B))}{\omega_{2k}}. \qquad (3)$$

Finally, using (1), (2), and (3) we conclude

$$\mathrm{vol}_{2k}(P \circ \Phi(B)) \geq \omega_{2k},$$

and that the equality holds if and only if the linear subspace $\Phi^t(V)$ is complex. $\qquad\square$

Remark 8.7.2. Unlike orthogonal projections onto complex subspaces, complex sections of the image of the unit ball B by a linear symplectic automorphism Φ has a small $2k$ volume: If $V \subseteq \mathbb{R}^{2n}$ is a complex linear subspace of real dimension $2k$, then,

$$\mathrm{vol}_{2k} \left(V \cap \Phi(B) \right) \leq \omega_{2k},$$

and the equality holds if and only if the subspace $\Phi^{-1}V$ is complex. Indeed, since the restriction of $(dV)^k$ to V is $k!$ times the Euclidean volume form, we have

$$\mathrm{vol}_{2k} \left(V \cap \Phi(B) \right) = \frac{1}{k!} \int_{V \cap \Phi(B)} (dV)^k = \frac{1}{k!} \int_{B \cap \Phi^{-1}(V)} \Phi^\star (dV)^k$$

$$= \frac{1}{k!} \int_{B \cap \Phi^{-1}(V)} (dV)^k$$

$$\leq \mathrm{vol}_{2k} \left(B \cap \Phi^{-1}(V) \right) = \omega_{2k},$$

where the inequality is actually an equality if and only if the restriction of $(dV)^k$ to $\Phi^{-1}(V)$ coincides with the Euclidean volume, that is if and only if $\Phi^{-1}(V)$ is complex.

The Uncertainty Principle and the Non-Squeezing Theorem.

The most famous basic principle in the quantum mechanics is the Heisenberg *uncertainty principle*. It states, roughly, that one can not simultaneously measure the momentum and the position of a particle precisely. This principle can be written as

$$\Delta q \Delta p \geq \hbar,$$

where $\hbar$ is the Planck constant, and Δq, Δp denote the deviation of q and p from the average values $E(q)$ and $E(p)$, respectively. Gromov's non-squeezing theorem is the analogue of the Heisenberg principle in the Hamiltonian formalism of classical mechanics. Indeed, physically speaking, this theorem says that if a collection of particles initially spreads out throughout the unit ball $B_{2n}(r)$, then one cannot squeeze the assemblage into a statistical state in which the momentum and position in the (q_1, p_1)-plane spreads out less than initially [101].

8.7.1   Exercises

Exercise 8.7.3. Let V be a finite-dimensional real vector space and let the 2-form ω_V on the product $V \times V^*$ be given by

$$\omega_V\left(\begin{pmatrix} v \\ \varphi \end{pmatrix}, \begin{pmatrix} v' \\ \varphi' \end{pmatrix}\right) := \varphi'(v) - \varphi(v').$$

Prove that this is a symplectic bilinear form.

Exercise 8.7.4. We define

$$\mathbb{J}_0 = \begin{pmatrix} 0 & -1 & & & \\ 1 & 0 & & & \\ & & \ddots & & \\ & & & 0 & -1 \\ & & & 1 & 0 \end{pmatrix}.$$

Prove that the bilinear form $\Phi : \mathbb{R}^{2n} \longrightarrow \mathbb{R}^{2n}$ is symplectic if and only if

$$\mathbb{J}_0 = \Phi^t \mathbb{J}_0 \Phi.$$

Exercise 8.7.5. Let (V, ω) be a symplectic vector space and $W \subset V$ a hyperplane, i.e., a subspace of dimension $\dim V - 1$. Show that W is co-isotropic.

Exercise 8.7.6. Let (V, ω) be a symplectic vector space. Prove that a map $\varphi : V \to V$ is linear symplectic if and only if its graph

$$\{(v, \varphi(v)) \mid v \in V\}$$

is a Lagrangian subspace of $V \times V$, equipped with the symplectic form $(-\omega) \oplus \omega$.

Exercise 8.7.7. Let (V, ω) be a pre-symplectic vector space and $W \subset V$ a subspace. Prove the following:

1. W is isotropic if and only if W^ω is co-isotropic.

2. If W is co-isotropic then W^ω is isotropic.

3. If ω is non-degenerate then the converse of the above implication is true.

4. Is this converse true in general?

5. If ω is non-degenerate then W is symplectic if and only if W^ω is symplectic.

Exercise 8.7.8. Let (V, ω) be a symplectic vector space, $W \subset V$ a Lagrangian subspace, and $U \subset V$ a subspace that is complementary to U, i.e., satisfies

$$W + U = V, \quad W \cap U = \{0\}.$$

Then the map

$$W \longrightarrow U^* \quad \text{such that} \quad w \mapsto \omega(w, \cdot) \mid U$$

is a vector space isomorphism.

Exercise 8.7.9. Show that a linear subspace of a pre-symplectic vector space is Lagrangian if and only if it is minimal co-isotropic, i.e., it is co-isotropic and does not contain a smaller co-isotropic subspace.

Exercise 8.7.10. Let (V, ω) be a pre-symplectic vector space, $W \subset V$ a co-isotropic subspace, and

$$\overline{U} \subset \overline{W} := W/W^\omega$$

a subspace of the symplectic quotient. We denote by $\pi : W \to \overline{W}$ the canonical projection. Show that $\overline{U}$ is co-isotropic if and only if $U := \pi^{-1}(\overline{U})$ is a co-isotropic subspace of V. What is the codimension of U in V?

Exercise 8.7.11. Classify the isotropic subspaces of a given symplectic vector space.

Exercise 8.7.12. Characterize the complex structures on $\mathbb{R}^2$, i.e., give an equivalent condition for a real 2×2 matrix A to define a complex structure.

Exercise 8.7.13. Let V be a real vector space, ω a symplectic form on V, and $\mathbb{J}$ a complex structure on V. Prove that the following conditions are equivalent:

1. The bilinear form $g_{\mathbb{J}} = \omega(\cdot, \mathbb{J}\cdot)$ on V is symmetric.

2. The form ω is invariant under $\mathbb{J}$, i.e., $\omega(\mathbb{J}\cdot, \mathbb{J}\cdot) = \omega$.

3. The form $g_{\mathbb{J}}$ is invariant under $\mathbb{J}$, i.e., $g_{\mathbb{J}}(\mathbb{J}\cdot, \mathbb{J}\cdot) = g_{\mathbb{J}}$.

Exercise 8.7.14. Prove that on every symplectic vector space (V, ω) there exists a linear complex structure that is compatible with ω.

Exercise 8.7.15. Let V be a vector space and $\mathbb{J}$ a complex structure on V. Prove that the set of $\mathbb{J}$-compatible linear symplectic forms on V is a convex subset of the vector space of bilinear forms on V.

Exercise 8.7.16. Let V be a vector space, and ω and $\mathbb{J}$ symplectic and complex structures on V. Prove $(\omega, \mathbb{J})$ is compatible if and only if the map

$$H : V \times V \longrightarrow \mathbb{C} \text{ given by } H(v, w) = \omega(v, \mathbb{J}w) + i\omega(v, w),$$

is a Hermitian inner product on V with respect to complex scalar multiplication

$$m : \mathbb{C} \times V \longrightarrow V, \text{ given by } m(a + ib, v) = (a + b\mathbb{J})v.$$

Exercise 8.7.17. Let $(\omega, \mathbb{J}, g)$ be a compatible triple and $\phi \in \mathrm{Aut}(V)$ an automorphism. Show that if ϕ preserves two of the structures, then it also preserves the third.

Exercise 8.7.18. Let (V, ω) be a symplectic vector space and $W \subseteq V$ a Lagrangian subspace. Prove that there exists a Lagrangian subspace $W_0 \subseteq V$ that intersects W trivially, by using Exercise 8.7.14.

Exercise 8.7.19. Let (V, g) be an inner product space and $\phi \in \mathrm{Aut}(V)$. We denote by $\phi^\star : V \longrightarrow V$ the g-adjoint map of ϕ. Show that the map $(\phi\phi^\star)^{-1/2}\phi$ is g-orthogonal.

Exercise 8.7.20. Let (ω, g) be a compatible pair on a vector space V. Then the g-adjoint of every symplectic automorphism of V is a symplectic automorphism.

Exercise 8.7.21. Let V be a two-dimensional real vector space and ω, $\mathbb{J}$, and g be a symplectic, complex, and inner product structures on V. Characterize compatibility of $(\mathbb{J}, g)$ and (g, ω).

Exercise 8.7.22. Let V be a vector space. For which $c \in \mathbb{R} - \{0\}$ is there a compatible pair (g, ω) on V consisting of an inner product and a symplectic form such that (cg, ω) is also compatible? What about the same question for a pair consisting of a symplectic form and a complex structure, and a pair consisting of an inner product and a complex structure?

Exercise 8.7.23. How does a complex structure on a vector space V induce an orientation on V? If $(\omega, \mathbb{J})$ is a compatible pair on V show that the orientations induced by ω and $\mathbb{J}$ agree.

Exercise 8.7.24. Is a linear complex structure diagonalizable?

Chapter 9

Commutative Rings with Identity

This chapter is a continuation of the commutative branch of Chapter 5 of Volume I. Here, by ring we shall always mean a *commutative ring with identity*.

In the next sections we will present and discuss the interrelations of the following classes of rings: Integral domains $\supseteq$ UFD's $\supseteq$ PID's $\supseteq$ Euclidean domains $\supseteq$ Fields, as well as ACC and DCC (to be defined shortly). The last two inclusions having been proved in Chapter 5 of Volume I.

9.1 Principal Ideal Domains

This section concerns the place of PID's in the universe. For the reader's convenience, we recall a definition from Chapter 5.

Definition 9.1.1. An ideal I in a ring R is a *principal ideal* if $I = (a)$, i.e. it has a single ideal generator. If all the ideals I of the ring R are principal, we shall call R a *principal ideal domain* (or PID).

Proposition 9.1.2. *A PID has no zero divisors.*

Proof. Suppose $ab = 0$ and $a \neq 0$. If a is a unit then $b = 0$ and we are done. Otherwise, Let I be a maximal ideal containing (a). Since R is a PID $I = (p)$, where p is a prime. But $a = pc$ for some $c \in R$. Hence $0 = ab = pbc$. Therefore $I = (0)$, a contradiction. □

Proposition 9.1.3. *Let R be a PID and p be a prime of R. If $p \mid ab$, then $p \mid a$ or $p \mid b$ (or both).*

Proof. Suppose p does not divide a. Consider the ideal $(p, a) = J$ of R. Evidently, $J \supseteq (p)$. But $J \neq (p)$ because if it were p would divide a. On the other hand since p is a prime and R is a PID, Theorem 5.5.10 of Volume I tells us that (p) is a maximal ideal. Therefore $J = R$. Hence $1 = xp + ya$ for some $x, y \in R$. Hence $b = bxp + yba$. Since $p \mid ab$ and also $p \mid bxp$, we see $p \mid b$. □

We say R is a UFD if every non-zero element and every non-unit of R can be uniquely factored into primes (and a unit).

Theorem 9.1.4. *If R is a PID, then R is a UFD.*

Later we shall see another way to get this result.

Proof. Let $a \in R$, where $a \neq 0$ and a is not a unit. If a is prime we already have a prime factorization. Otherwise $a = bc$, where neither b nor c is a unit. Now by the first isomorphism theorem for rings we get a commutative diagram

where π is surjective, but not injective (and similarly for c). This introduces a partial order $a \succ b$ and $a \succ c$. By transfinite induction we assume b and c are each a product of a finite number of primes of R. Therefore so is a. This proves the existence of a prime factorization of each non-zero, non-unit of R.

Now suppose $p_1^{e_1} \cdots p_n^{e_n} = a = q_1^{f_1} \cdots q_m^{f_m}$. Then $p_1 \mid q_1 \cdot b$, where $b = a/q_1$ up to a unit. By Proposition 9.1.3 $p_1 \mid q_1$ or $p_1 \mid b$. In the latter case by induction and repeated use of Proposition 9.1.3 p_1 divides some q_i. But then after renumbering the q_i, $p_1 \mid q_1$ and therefore $p_1 = q_1$. Since R has no zero divisors by Proposition 9.1.2 we can cancel these and get a product of 1 fewer primes involving the p_i which equals a product of 1 fewer primes involving the q_i. By induction, the number of these primes is the same, and after permutation, each $p_i = q_i$ for $i > 1$. But since $p_1 = q_i$, this proves uniqueness of prime factorization. $\square$

Exercise 9.1.5. Show the relation $a \succ b$ above is a partially ordering.

Corollary 9.1.6. *Let R be a PID and I be a non-zero ideal in R. Then there are only finitely many ideals of J of R with $I \subseteq J \subseteq R$. Hence the ring R/I has only a finite number of non-trivial ideals.*

Proof. Since $I = (a)$ and R is a UFD we know $a = up_1^{e_1} \cdots p_n^{e_n}$, where u is a unit, the p_i are prime in R and the e_i are positive integers. Let $b|a$. Then by uniqueness of prime factorization $b = vp_1^{f_1} \cdots p_n^{f_n}$, where v is a unit and the f_i are non-negative integers $f_i \leq e_i$. Evidently, there are only finitely many possible such b. If J is an ideal between I and R, $J = (b)$, where $b \mid a$. $\square$

Definition 9.1.7. A ring R is said to satisfy the *ascending chain condition* for ideals, written as *ACC*, if for every increasing sequence of ideals,

$$I_1 \subset I_2 \subset \cdots \subset I_i \subset \cdots$$

there is an integer n such that $I_n = I_{n+1} = \dots$, that is, if the chain stabilizes after n steps.

Similarly we have the following definition.

Definition 9.1.8. A ring R is said to satisfy the *descending chain condition* for ideals, written as *DCC*, if for every decreasing sequence of ideals

$$I_1 \supset I_2 \supset \cdots \supset I_i \supset \cdots,$$

there is an integer n such that $I_n = I_{n+1} = ...$, that is, the chain stabilizes after n steps.

Corollary 9.1.9. *If R is a PID, it satisfies both the ACC and DCC.*

Proof. This follows immediately from Corollary 9.1.6.   □

We note that in the case of the ACC, the converse of Corollary 9.1.6 fails. For consider $\mathbb{Z}[x]$, or $k[x,y]$. As we noted in Chapter 5, these are not principal ideal domains. In $k[x,y]$ the ideal generated by (x,y) consists of all the polynomials with 0 constant term. That is

$$(x,y) = \{p(x,y) \ : \ p(0,0) = 0\}.$$

This obviously cannot be generated by a single element only. Hence, (x,y) is not a principal ideal. Similarly if p is a prime in $\mathbb{Z}$, the ideal (p,x) in $\mathbb{Z}[x]$ also cannot have a single generator. However as we shall see below, the Hilbert Basis Theorem tells us both these rings satisfy the ACC.

Exercise 9.1.10. Show (x,y) is a maximal ideal in $k[x,y]$.
Let p be a prime in $\mathbb{Z}$. Show the ideal (p,x) is not principal in $\mathbb{Z}[x]$.

9.1.1   A Principal Ideal Domain which is not Euclidean

As we saw in 5.5.4 of Volume I, a Euclidean ring is always a Principal Ideal Domain. However, the converse of this is false. A counter example was first found by T.S. Motzkin (1949). The following finite list due to Veselin Perić and Mirjana Vuković (see [108]) finds *all* PID of the form $\mathbb{Q}[\sqrt{-d}]$ (imaginary quadratic extensions of $\mathbb{Q}$ with discriminant, D) which are not Euclidean rings. It seems to be an open problem to find all PID which are not Euclidean rings.

Let A be an integral domain. We will denote by $A^\dagger$ the set of units of A together with 0, so that $A - A^\dagger = \varnothing$ if and only if A is a field. That is, in a field every non-zero element is a unit.

Definition 9.1.11. An element a in $A - A^\dagger$ is called a *universal side divisor* if for any $x \in A$ there is some $y \in A^\dagger$ such that a divides $x - y$.

Proposition 9.1.12. *Let A be an integral domain which is not a field, and which has no universal side divisors. Then A cannot be Euclidean.*

Proof. Suppose A were a Euclidean domain. Let g be its grade map and consider the subset

$$S = \{g(u), \mid u \in A - A^\dagger\}$$

of natural numbers. Clearly, $S \neq \varnothing$ since by assumption, A is not a field. Therefore $A - A^\dagger \neq \varnothing$. Now, let $g(u_0)$ be the least element of S. If we divide any element $a \in A$ by this element u_0, then $a = qu_0 + r$, where $r = 0$ or $g(r) < g(u_0)$. If $r = 0$, then $u_0 | a$. If $r \neq 0$, then $g(r) < g(u_0)$. But since $g(u_0)$ is the smallest natural number, r must be a unit. Thus, in either case, $u_0 | a - r$ for some $r \in A^\dagger$ and so u_0 is a universal side divisor, contradicting the hypothesis. $\square$

Corollary 9.1.13. *The following integral domains are PID's $\mathbb{Z}(\sqrt{-19})$, $\mathbb{Z}(\sqrt{-43})$, $\mathbb{Z}(\sqrt{-67})$, $\mathbb{Z}(\sqrt{-163})$, but are not Euclidean rings.*

Proof. One checks that for $d = 19$, 43, 67 and 163, $d \equiv -1 \pmod 4$, i.e. primes of the form $d = 4k - 1$, with $k = 5$, 11, 17 and 41. The ring

$$R = \left\{\zeta = a + b\sqrt{-d}, \ a, \ b \in \mathbb{Z}\right\}$$

is an integral domain with unit element 1 and unit group $U = \{1, -1\}$. Moreover, the map

$$\varphi : R \longrightarrow \mathbb{Z}, \quad \varphi(\zeta) = \zeta\bar{\zeta},$$

has the properties i) and ii) of the Euclidean norm.

If R were a Euclidean domain with grade map g, there would exist a

$$g(\eta) = \min\left\{g(\zeta) \mid \zeta \in R - (U \cup \{0\})\right\}, \quad \eta \in R - \{0\}$$

and for every such $\eta \in R$, we would have,

$$\zeta = \eta\xi + \tau \text{ for some } \xi, \tau \in R, \text{ with } \tau = 0 \text{ or } g(\tau) < g(\eta).$$

In other words $\eta \mid (\zeta - \tau)$, $\tau \in U \cup \{0\} = \{1, -1, 0\}$.

In particular, this would be true for $\zeta = 2$, and so

$$\eta \mid 2 \text{ or } \eta \mid 3 \text{ since } \eta \nmid 2 - 1 = 1.$$

But 2 and 3 are not reducible in R. Namely, if

$$2 = (a + b\sqrt{-d})(c + e\sqrt{-d}), \quad a + b\sqrt{-d} \notin U, \quad c + e\sqrt{-d} \notin U,$$

then $4 = \varphi(2) = \varphi(a + b\sqrt{-d})\varphi(c + e\sqrt{-d})$, i.e.

$$2 = \varphi(a + b\sqrt{-d}) = \varphi(c + e\sqrt{-d}),$$

since $1 \neq \varphi(a + b\sqrt{-d}) \in \mathbb{Z}$ and $1 \neq \varphi(c + e\sqrt{-d}) \in \mathbb{Z}$. Hence,

$$2 = \varphi(a + b\sqrt{-d}) = \left(a + \frac{b}{2} + \sqrt{-d}\frac{b}{2}\right)\left(a + \frac{b}{2} - \sqrt{-d}\frac{b}{2}\right)$$

$$= a^2 + ab + \frac{b^2}{4} + (4k - 1)\frac{b^2}{4} = a^2 + ab + b^2 k.$$

Therefore, for $ab > 0$, in view of $k \geq 5$, $b = 0$, and for $ab \leq 0$,

$$2 = \varphi(a + b\sqrt{-d}) = \varphi(\overline{a + b\sqrt{-d}}) = \varphi\left(a + b - \frac{b}{2} - \sqrt{-d}\frac{b}{2}\right)$$

$$= \left(a + b - \frac{b}{2}\right)^2 + (4k - 1)\frac{b^2}{4}$$

$$= (a + b)^2 - ab + b^2(k - 1).$$

It follows that $b = 0$. Similarly, $e = 0$, and hence $2 = ab$ would be a non-trivial presentation of 2 in $\mathbb{Z}$ as a product of primes, which is impossible.

Similarly one can prove 3 is not reducible in R.

$$\eta \mid 2 \text{ or } \eta \mid 3,$$

and so $\eta = \pm 2$ or $\eta = \pm 3$. But, none of these four numbers can divide $\sqrt{-d}, \sqrt{-d} + 1, \sqrt{-d} - 1$, for otherwise

$$\eta \mid \sqrt{-d} \implies \varphi(\eta) \mid \varphi(\sqrt{-d})$$
$$\eta \mid (\sqrt{-d} + 1) \implies \varphi(\eta) \mid \varphi(\sqrt{-d} + 1)$$
$$\eta \mid (\sqrt{-d} - 1) \implies \varphi\eta \mid \varphi(\sqrt{-d} - 1)$$

This is impossible, since $\varphi(\eta) = 4$ or $\varphi(\eta) = 9$, and

$$\varphi(\sqrt{-d}) = \left(\tfrac{1}{2} + \tfrac{\sqrt{-d}}{2}\right)\left(\tfrac{1}{2} - \tfrac{\sqrt{-d}}{2}\right) = \tfrac{1}{4} + \tfrac{4k-1}{4} = k$$

$$\varphi(\sqrt{-d}+1) = \left(\tfrac{3}{2} + \tfrac{\sqrt{-d}}{2}\right)\left(\tfrac{3}{2} - \tfrac{\sqrt{-d}}{2}\right) = \tfrac{9}{4} + k - \tfrac{1}{4} = k+2$$

$$\varphi(\sqrt{-d}-1) = \left(\tfrac{-1}{2} + \tfrac{\sqrt{-d}}{2}\right)\left(\tfrac{-1}{2} - \tfrac{\sqrt{-d}}{2}\right) = \tfrac{1}{4} + \tfrac{4k-1}{4} = k.$$

$\square$

We remark that for $d = 1, 2, 3, 7, 11$ one actually gets a Euclidean ring.

9.2   Unique Factorization Domains

In order to deal with questions of unique prime factorization we need a more general notion.

Definition 9.2.1. If $a \in R - \{0\}$ is a non-unit element such that it has only trivial factorizations, then a is called *irreducible*.

Definition 9.2.2. Two elements $a, b \in R - \{0\}$ are called *associates*, and we write $a \sim b$, if $a = ub$, where u is a unit of R.

We leave it to the reader to check that $\sim$ is an equivalence relation.

Definition 9.2.3. Let $a, b, b \neq 0$, be two elements of R. We say that b *divides* a (equivalently, b is a *divisor* of a), and write $b \mid a$, if there is a $c \in R$ such that $a = bc$. If b does not divide a, we write $b \nmid a$.

Definition 9.2.4. We say that the element $p \in R - \{0\}$ is a *prime* if p is not a unit and satisfies the following condition:
   If $p \mid ab$, with $a, b \in R$, then $p \mid a$ or $p \mid b$.

The distinction between primes and irreducibles is needed here because, as we shall see, in a ring which is not a UFD these notions are not equivalent: every prime is irreducible, but the converse is not generally true. However, in a UFD, every irreducible is also prime.

Proposition 9.2.5. *Let R be an integral domain. Then every prime element is irreducible.*

Proof. Let x be a prime element of the integral domain R. We want to show x has only trivial divisors. Indeed, let d be a divisor of x, i.e. $x = d \cdot q$, d, $q \in R$. Obviously, x divides $d \cdot q$, and since x is a prime, it must divide d or q. If $x \,|\, d$, then since $d \,|\, x$, it follows that x and d are associates.

If $x \,|\, q$, then there must exist $q' \in R$ such that $q = xq'$, therefore $x = dq = dxq'$, which implies that $dq = 1$, i.e. q is a unit. $\square$

Example 9.2.6. We now present an example showing the converse is false. In other words, an irreducible element is not always a prime. This is given in [6] p. 191). Consider the ring $\mathbb{Z}[\sqrt{-5}]$. In this ring

$$2 \cdot 3 = (1 + \sqrt{-5})(1 - \sqrt{-5}).$$

Therefore, 2 divides $(1 + \sqrt{-5})(1 - \sqrt{-5})$, but 2 does not divide either of these factors. Hence 2 is not a prime in $\mathbb{Z}[\sqrt{-5}]$.

On the other hand, 2 is irreducible, since if $a + b\sqrt{-5}$ divides 2, the norm $N(a + b\sqrt{-5}) = a^2 + 5b^2$ must divide $N(2) = 4$ in $\mathbb{Z}$. But this can happen only if $a = \pm 2$ and $b = 0$, or if $a = \pm 1$ and $b = 0$, showing that 2 has only trivial factorizations in $\mathbb{Z}$ where the only units are ± 1.

Exercise 9.2.7. Let $m \in \mathbb{Z}$ be positive, and not a perfect square. Consider

$$\mathbb{Z}[\sqrt{m}] = \{a + b\sqrt{m} \mid a, b \in \mathbb{Z}\}.$$

One checks easily that $\mathbb{Z}[\sqrt{m}]$ is an integral domain. What are its units?

Let R be an integral domain. We now extending the definition of a UFD.

Definition 9.2.8. We say that R is a *unique factorization domain* (UFD) or simply that R is *factorial*, if and only if every non-zero element r of R which is not a unit, can be written as a (finite) product of irreducible elements. More precisely,

1. If $r \neq 0$ is a non-unit, then $r = p_1 \ldots p_i$ where each p_k is irreducible.

2. If $p_1 \ldots p_i = q_1 \ldots q_j$ then $i = j$ and there is a permutation σ of $\{1, 2, \ldots, i\}$ such that $p_k = u_k q_{\sigma(k)}$ for some units u_k.

Proposition 9.2.9. *If R is a UFD and $p \in R$ is irreducible, then p is prime.*

Proof. Let p be an irreducible element of R. Suppose that $p|ab$. Since R is a UFD we can write a and b as finite products of irreducibles. Now, the product of these two factorizations is the unique factorization of the element ab, and since $p \mid ab$, p must be associate with at least one of the irreducible elements in the factorization of ab, which shows that $p \mid a$ or $p \mid b$. □

Proposition 9.2.10. *If every non-invertible non-zero element of an integral domain R can be written as a finite product of primes, then R is a UFD.*

The proof is almost identical to the corresponding one for PID's.

Proof. As we have seen, any prime is irreducible. Therefore, we only need to prove uniqueness of the factorization. To do this, let $a \in R$ be a non-zero and non-invertible element, which we assume has two different factorizations into irreducible elements: $a = r_1 r_2 ... r_i = s_1 s_2 ... s_j$. Since r_1 is a prime, it must divide some element of the product on the right side, say s_1, after possible rearrangement. This implies that r_1 and s_1 are associates, and since R is an integral domain, we obtain that $r_2 ... r_i$ and $s_2 ... s_j$ are associates. Now, using induction in i we conclude that $i = j$. Hence, $r_2 ... r_i$ is a permutation of associates of $s_2 ... s_j$. Therefore, the factorization is unique. □

9.2.1 Kaplansky's Characterization of a UFD

Here we will give a necessary and sufficient condition for an integral domain to be a UFD (following Kaplansky [71])

Proposition 9.2.11. *Let S be a non-empty subset of a ring R, with $1 \in S$. Also, we assume that S is multiplicatively closed, and does not contain 0. Let I be a maximal ideal of R with respect to the property that $I \cap S = \varnothing$. Then, I is a prime ideal in R.*

Proof. Suppose I is not a prime ideal. Then, there must exist an a and b in R such that $ab \in I$ with $a \notin I$, $b \notin I$. Consider the ideal (I, a). It is clear that $I \subset (I, a)$. Therefore $(I, a) \cap S \neq \varnothing$, so we can find an $x \in (I, a) \cap S$, $x \neq 0$, and $x = m + ra$, $m \in I$, $r \in R$. We do the same with $(I, b) \cap S$, and get a $y \in (I, b) \cap S$, with $y \neq 0$ and $y = n + tb$, with $n \in I$, $t \in R$. Now, $xy = mn + mtb + ran + ratb \in I$. But also, $x, y \in S$ and since S is multiplicatively closed, $xy \in S$. Since this contradicts that $I \cap S = \varnothing$, I is a prime ideal. $\qquad\square$

Theorem 9.2.12. *(**Kaplansky**) An integral domain R is a UFD if and only if every non-zero prime ideal I in R contains a prime element.*

Proof. "$\Longrightarrow$". Suppose that R is a UFD, and I is a non-zero prime ideal in R. Then I must contain an element a which is neither 0 nor a unit. We can therefore factor a as $a = p_1.p_2...p_n$, and since I is prime, at least one of the primes p_i, $i = 1, ..., n$ must be in I.

"$\Longleftarrow$". Assume every non-zero prime ideal I in R contains a prime. Let S be the set of all products of prime elements as well as all units. Then, $0 \notin S$ and S is multiplicatively closed. It follows by induction that if $ab \in S$, then $a \in S$ and $b \in S$.

Now, we want to show that S contains all non-zero elements of R. To do this, let $x \in R$, $x \neq 0$. Suppose that $x \notin S$. Then, if we consider the principal ideal (x), we must have $(x) \cap S = \varnothing$. By Zorn's lemma, x lies in some maximal ideal disjoint from S. By Proposition 9.2.11 this ideal is a prime and contains a prime element, a contradiction. Therefore $x \in S$, and by Proposition 9.2.10 R is a UFD. $\qquad\square$

As an immediate consequence of Kaplansky's theorem we recapture a result above.

Corollary 9.2.13. *Every PID is a UFD.*

Moreover, we can use Kaplansky's theorem to get a criterion for a UFD to be a PID.

Theorem 9.2.14. *Let R be a UFD. Then the following propositions are equivalent:*

1. R is a PID.

2. R has length 1, i.e. every non-zero ideal is maximal.

Later we shall see that this is a statement about Krull dimension 1.

Proof. "(1) $\implies$ (2)". Since any integral domain which is not a field has non-zero prime ideals, the dimension of R must be ≥ 1. We have to prove that in a PID every non-zero prime ideal P is maximal. If not, then there exists a prime ideal Q such that $P \subset Q$. Since R is a PID, there are p and q such that $P = (p)$ and $Q = (q)$. Hence $p \mid q$. But q is prime, and therefore p and q must be associates. Hence $(p) = (q)$, a contradiction.

"(2) $\implies$ (1)". Suppose R is a UFD and R has length 1. That is, each non-zero ideal M is maximal. Then, by Theorem 9.2.12, M contains a prime element p. Hence, $(p) \subset M$. But by assumption $p \neq 0$. Therefore, (p) is maximal, and so $(p) = M$. Thus each maximal ideal is principal. Hence, R is a PID. $\square$

Definition 9.2.15. An *abstract number ring* is an integral domain R which is not a field, and has the property that for any non-zero ideal I of R, the quotient R/I is finite.

Corollary 9.2.16. *An abstract number ring is a UFD if and only if it is a PID.*

Proof. It is sufficient to prove that any abstract number field R has dimension 1. So, let P, $P \neq \{0\}$, be a prime ideal of R. Since R/P is a finite integral domain (by definition), it is a field, 5.8.6 of Volume I. Therefore, P must be maximal and so dim $R = 1$. $\square$

9.3   The Power Series Ring R[[x]]

Let R be an integral domain. The set $R[[x]]$ of *formal power series* in the indeterminate x over R is defined to be the set of all formal infinite sums

$$\sum_{0}^{\infty} a_n x^n = a_0 + a_1 x + a^2 x^2 + a_3 x^3 + \cdots$$

with $a_i \in R$ for all $i = 0, 1, 2, \ldots$. We define the operations between power series as follows:

For the addition and multiplication of power series we extend the usual addition and multiplication rules for polynomials, i.e. term-by-term addition

$$\sum_{0}^{\infty} a_n x^n + \sum_{0}^{\infty} b_n x^n = \sum_{0}^{\infty} (a_n + b_n) x^n,$$

and the *Cauchy-product*

$$\left(\sum_{0}^{\infty} a_n x^n \right) \times \left(\sum_{0}^{\infty} b_n x^n \right) = \sum_{0}^{\infty} \left(\sum_{k=0}^{n} a_k b_{n-k} \right) x^n.$$

Equipped with these two operations, $R[[x]]$ becomes a commutative ring with identity $1 \neq 0$ and no zero divisors, i.e. it is an integral domain. Evidently, if R is $\mathbb{R}$ or $\mathbb{C}$ the ring of formal power series contains as a subring the ring of convergent power series.

Since $R[[x]]$ is a commutative ring with identity, $R[[x, y]]$ can be defined as the set of all formal power series in y over $R[[x]]$, i.e.

$$R[[x, y]] := R[[x]] [[y]]$$

and by induction we can define the *ring of multivariate formal power series*

$$R[[x_1, \ldots, x_n]] := R[[x_1, \ldots, x_{n-1}]] [[x_n]].$$

For any ring R let R^2 be the ideal generated by the set of all products. In the next example (see [81]) we have $R \neq R^2 \neq 0$. This example does not contradict the fact that in a ring with identity there always exist maximal ideals. Indeed it shows why the identity is crucial to having maximal ideals.

Example 9.3.1. Let $k[[x]]$ be the formal power series ring over a field k of characteristic zero, and $R = xk[[x]]$. Then R has no maximal ideals.

Proof. We note that $R^2 = x^2 k[[x]]$. Let I be an alleged maximal ideal of R. Recall that in $k[[x]]$ any element with non-zero constant term has an inverse in $k[[x]]$. We first show that $R^2 \subseteq I$. Indeed, either $I \subseteq R^2$ (so that equality holds since I is maximal), or I contains some element xf of R that is not in R^2. But in that case f is invertible and any element $x^2 g \in R^2$ is $x^2 g = (xf)(xf^{-1}g) \in I$, and so $R^2 \subseteq I$.

Now R/R^2 has trivial multiplication and is a rational vector space. Hence R/R^2 is a divisible group that has no maximal subgroups by 1.14.29 of Volume I. But since $R^2 \subseteq I$, I/R^2 is a maximal ideal of R/R^2. This contradiction shows that R has no maximal ideals. $\square$

A ring R is said to be *factorial* if any element can be factored into primes and unit. For example, if R is factorial so is $R[x]$.

Proposition 9.3.2. *A formal power series $a(x) \in R[[x]]$ is invertible if and only if a_0 is invertible in R.*

Proof. If $a(x) \in R[[x]]$ is a unit, there exists $b(x) \in R[[x]]$ such that $a(x)b(x) = 1$. In particular, $a_0 b_0 = 1$, and so a_0 is a unit in R.

Conversely, assume that a is a unit in R. In order for $a(x)$ to be invertible, we must find $b(x) = \sum_0^\infty b_j x^j \in R[[x]]$ satisfying the relation $a(x)b(x) = 1$. Thus the following infinite system of equations must hold:

$$
\begin{aligned}
1 &= a_0 b_0 \\
0 &= a_0 b_1 + a_1 b_0 \\
0 &= a_0 b_2 + a_1 b_1 + a_2 b_0 \\
&\vdots \\
0 &= a_0 b_n + \sum_{j=1}^n a_j b_{n-j} \\
&\vdots
\end{aligned}
$$

Hence, if we define

$$b_0 = a_0^{-1}$$
$$\vdots \quad \vdots \quad \vdots \quad \vdots$$
$$b_n = -a_0 \sum_{j=1}^{n} a_j b_{n-j} \quad \text{with } n = 1, 2, \ldots$$

we get $a(x)b(x) = 1$, so $a(x)$ is a unit. $\square$

Since the only units in $\mathbb{Z}$ are ± 1, we get the following corollary.

Corollary 9.3.3. *The formal power series $\sum a_n x^n \in \mathbb{Z}[[x]]$ is a unit if and only if $a_0 = \pm 1$.*

The reciprocal of a formal power series $p(x)$ is usually denoted by $\frac{1}{p(x)}$. From Calculus we know the geometric series converges for $|x| < 1$, and

$$\sum_{n=1}^{\infty} x^n = \frac{1}{1-x}.$$

Now, we can prove this identity in a more general setting, that of the ring $R[[x]]$. Indeed, using the rule of multiplication in $R[[x]]$ we see

$$(1-x)(1 + x + x^2 + \cdots) = 1 + 0x + 0x^2 + \cdots = 1,$$

which shows that $1 + x^2 + x^3 + \cdots$ is the reciprocal of $1 - x$.

In general a prime integer is no longer a prime if we pass to some integral extension of the ring $\mathbb{Z}$. However, an element in R is prime if and only if it is a prime of $R[x]$ (Gauss' Lemma). The same is true for the formal power series ring $R[[x]]$.

Proposition 9.3.4. *If p is prime in the integral domain R, then p is prime in $R[[x]]$.*

Proof. Let $c(x) = a(x)b(x)$ with $a(x)$, $b(x) \in R[[x]]$. Assume that $p \mid c(x)$, but that p does not divide $b(x)$. Let m be the smallest power of x such that p does not divide the coefficient b_m. Since

$c_m = a_0 b_m + \sum_{j=1}^{m} a_j b_{m-j}$, and $p \mid c_m$ and $p \mid \sum_{j=1}^{m} a_j b_{m-j}$, we conclude that $p \mid a_0$. By induction, suppose p divides a_j for all $j < k$. Now, the coefficient $c_{m+k} = \sum_{j=0}^{m+k} a_j b_{m+k-j}$ can be rearranged as

$$c_{m+k} = a_k b_m + \sum_{\substack{i+j=m+k \\ i<k}} a_i b_j + \sum_{\substack{i+j=m+k \\ j<m}} a_i b_j.$$

By hypothesis, $p \mid c_{m+k}$, $p \mid a_i$ for all $i < k$, and $p \mid b_j$ for all $j < m$. Therefore, $p \mid a_k b_m$ and thus $p \mid a_k$. Hence, $p \mid a(x)$.   □

We now specialize to $\mathbb{Z}[[x]]$.

Proposition 9.3.5. *Let $\sum a_n x^n \in \mathbb{Z}[[x]]$ be a formal power series with a_0 a prime integer. Then $\sum a_n x^n$ is irreducible.*

Proof. Suppose the contrary, i.e. $\sum a_n x^n = \left(\sum b_n x^n \right) \left(\sum c_n x^n \right)$ in $\mathbb{Z}[[x]]$. Since a_0 is a prime, either $b_0 = \pm 1$, or $c_0 = \pm 1$, and one of the two factors must be a unit. Hence, by Corollary 9.3.3, $\sum a_n x^n$ is irreducible.   □

Lemma 9.3.6. *Let $\sum a_n x^n \in \mathbb{Z}[[x]]$. Then, if $|a_0| \neq 0, 1$, and a_0 is not a prime power, then $\sum a_n x^n$ is reducible.*

Proof. Since $|a_0| \neq 0, 1$, and a_0 is not a prime power, it can be factorized into two relatively prime, non-unit factors, say b_0 and c_0. Since $(b_0, c_0) = 1$, there exist two integers β and δ such that $b_0 \beta + c_0 \delta = 1$. Multiplying by a_1 we get

$$a_1 = b_0 \beta a_1 + c_0 \delta a_1.$$

Setting $b_1 = \delta a_1$ and $c_1 = \beta a_1$, we have $a_1 = b_0 c_1 + c_0 b_1$.

Now, assume that we have found all the b_i's and c_i's up to $i = n-1$. Then, we require either

$$a_n = b_0 c_n + \sum_{i=1}^{n-1} b_i c_{n-i} + b_n c_0,$$

or

$$a_n - \sum_{i=1}^{n-1} b_i c_{n-i} = b_0 c_n + b_n c_0,$$

which we obtain if we set

$$b_n = \delta\left(a_n - \sum_{i=1}^{n-1} b_i c_{n-i}\right) \text{ and } c_n = \beta\left(a_n - \sum_{i=1}^{n-1} b_i c_{n-i}\right).$$

Hence by induction,

$$\sum a_n x^n = \left(\sum b_n x^n\right)\left(\sum c_n x^n\right).$$

$\square$

Corollary 9.3.7. *Each irreducible formal power series in $\mathbb{Z}[[x]]$ (up to associates of x) has constant term equal to $\pm p^k$ for some prime p and some positive integer k.*

A sufficient, but not necessary condition for the irreducibility of a formal power series with integer coefficients is the following.

Proposition 9.3.8. *Let $\sum a_n x^n \in \mathbb{Z}[[x]]$ such that $a_0 = \pm p^k$ for some prime p and $k \geq 2$. If $p \nmid a_1$, then $\sum a_n x^n$ is irreducible in $\mathbb{Z}[[x]]$.*

Proof. Let $p \nmid a_1$. Suppose the contrary, i.e.

$$\sum a_n x^n = \left(\sum b_n x^n\right)\left(\sum c_n x^n\right),$$

where neither factor is a unit. Since $a_0 = \pm p^k$ and $a_0 = b_0 c_0$ it follows that $b_0 = \pm p^s$ and $c_0 = \pm p^t$, with $s + t = k$ and $s, t \geq 1$. Now,

$$a_1 = b_0 c_1 + b_1 c_0 = p^s c_1 + p^t c_0$$

which implies that $p \mid a_1$, a contradiction. $\square$

To our knowledge, there is not yet criteria to handle all cases of irreducibility. For more information see Birmajer and Gil in [17] and [18]

Theorem 9.3.9. *The ring* $\mathbb{Z}[[x]]$ *is a UFD.*

Proof. We will show that every non-zero prime ideal in $\mathbb{Z}[[x]]$ is generated either by x, or by p and x, with p prime, or by $a(x)$ with $a(x)$ prime. Hence every non-zero prime ideal in $\mathbb{Z}[[x]]$ contains a prime element, and therefore the statement of the theorem is a direct consequence of the theorem that an integral domain is a UFD if and only if every non-zero prime ideal in R contains a prime element with $R = \mathbb{Z}[[x]]$. Let P be a prime ideal in $\mathbb{Z}[[x]]$ different from (0). We will study two cases: $x \in P$ or $x \notin P$.

Assume first that $x \in P$. If $P = (x)$, we are done. Otherwise, there exists $a(x) \in P$ with $a_0 \neq 0$. Then

$$a_0 = a(x) - x(a_1 + a_2 x + a_3 x^2 + \cdots) \in P.$$

Since P is a proper ideal, a_0 is not a unit in $\mathbb{Z}$ and can be factorized as a product of primes, with at least one of them in P. Call this prime p. Then $(p, x) \subset P$.

On the other hand, if $b(x) \in P$ and $b_0 \neq 0$ then, as above, $b_0 \in P$. If p does not divide b_0, then $\gcd(b_0, p) = 1 \in P$, a contradiction. It follows that $p \mid b_0$ and $P = (p, x)$.

Suppose now that $x \notin P$. If $a(x)$ is a non-zero element in P, we write $a(x) = x^m(b_0 + b_1 x + b_2 x^2 + \cdots)$ with $b_0 \neq 0$. Since P is prime and $x \notin P$, it follows that $b_0 + b_1 x + b_2 x^2 + \cdots \in P$. Then the image P^* of P under the natural projection

$$\mathbb{Z}[[x]] \longrightarrow \mathbb{Z}, \quad \text{such that} \quad a(x) \mapsto a_0,$$

is a non-zero ideal in $\mathbb{Z}$, say $P^* = (q)$.

Choose $q(x) \in P$ with $q_0 = q$. We first claim that q is a prime power. Arguing by contradiction, suppose not. Then $q = st$ with s and t non invertible and $\gcd(s, t) = 1$. Following the reasoning in Lemma 9.3.6, we see $q(x) = s(x)t(x)$ with $s_0 = s$ and $t_0 = t$. One of these factors, say $s(x)$, must lie in P. But then, $q \mid s_0$, which is impossible by our assumption on q. Thus q must be a prime power.

We now show that $P = (q(x))$. Let $b_1(x) \in P$. Then

$$b_1(x) - k_1 q(x) = x b_2(x) \in P$$

for some $k_1 \in \mathbb{Z}$. Since $x \in P$, we have that $b_2(x) \in P$. Similarly, there exists $k_2 \in \mathbb{Z}$ such that $b_2(x) - k_2 q(x) = x b_3(x) \in P$. Continuing in this way we obtain,

$$b_1(x) = k_1 q(x) + x b_2(x) = k_1 q(x) + x\big(k_2 q(x) + x b_3(x)\big)$$

$$= q(x)(k_1 + k_2 x + k_3 x^2 + \cdots).$$

$\square$

Exercise 9.3.10. Here we extend the notion of formal derivative to the ring of formal power series as follows: If $f(x) = \sum_{n=0}^{\infty} a_n x^n$, we define the *formal derivative* of $f(x)$ to be the formal series,

$$\frac{df(x)}{dx} = f'(x) = \sum_{n=0}^{\infty} n a_n x^{n-1}.$$

Prove:

1. $(fg)' = f'g + fg'$.

2. $(f^n)' = nf^{n-1}f'$.

3. $f' = 0$ if and only if f is a constant.

4. $(f^{-1})' = -f' f^{-2}$.

5. $(f^{-n})' = -nf' f^{-n-1}$.

6. $(1 - x)^{-n} = \sum_{m=0}^{\infty} \binom{n+m-1}{m} x^m$.

Exercise 9.3.11. Prove that if R satisfies the ACC condition, then $f = \sum_{n=0}^{\infty} a_n x^n \in R[[x]]$ is nilpotent if and only if there is an $m \geq 1$ such that $a_n^m = 0$ for all $n \geq 0$.

9.3.1 Exercises

Exercise 9.3.12. Prove that if k is a field and x, y, z, w are indeterminates over k, then the ring $k[x, y, z, w]/(xy - zw)$ is not a unique factorization domain.

Exercise 9.3.13. Prove that $1 - x$ is a unit in $R[[x]]$ with inverse $1 + x + x^2 + \dots$.

Exercise 9.3.14. Prove that $\sum_{n=0}^{\infty} a_n x^n$ is a unit in $R[[x]]$ if and only if a_0 is a unit in R.

Exercise 9.3.15. Let $k(x)$ be the meromorphic functions in the variable x with coefficients in the field k. Show that $k(x)$ is actually a field. Indeed, it is isomorphic the quotient field of $k[[x]]$.

9.4 Prime Ideals

Definition 9.4.1. A subset S of a ring R is said to be *multiplicatively closed* if for any x, y in S, $xy \in S$.

Proposition 9.4.2. *The complement S of a prime ideal I of R is multiplicatively closed.*

Proof. Let x and y be two elements of $S = R - I$. Then, if xy is not in S, it must be in I. Since I is a prime ideal, either $x \in I$, or $y \in I$, a contradiction. $\qquad \square$

Here is a partial converse of the above proposition:

Proposition 9.4.3. *Let $S \subset R$ be a multiplicatively closed subset, and I be an ideal of R, maximal among those not meeting S. Then I is a prime ideal.*

Proof. Suppose $xy \in I$. Then we will show either $x \in I$, or $y \in I$. Otherwise both $x, y \notin I$. Consider the ideals $(x) + I$ and $(y) + I$. Using the maximality of I we conclude that both ideals must meet S. So, there exist $z = ax + i_1$ and $w = by + i_2$, where i_1, $i_2 \in I$ and a, $b \in R$, with z, $w \in S$. Then $zw = abxy + axi_2 + byi_1 + i_1 i_2$. So, since $xy \in I$,

$zw \in I$, and, because S is multiplicatively closed, $zw \in S$, i.e. $I \cap S \neq \emptyset$, a contradiction. $\qquad\square$

Definition 9.4.4. A *chain* of prime ideals of R is a finite strictly increasing sequence of prime ideals

$$\mathfrak{p}_0 \subsetneq \mathfrak{p}_1 \subsetneq \cdots \subsetneq \mathfrak{p}_n$$

of R. The integer n is called the *length* of the chain.

Definition 9.4.5. The maximum of the lengths of *all* chains of prime ideals of R is called the *Krull dimension* of R, and it is denoted by $\dim(R)$.

Definition 9.4.6. Let $\mathfrak{p}$ be a prime ideal of the ring R. The *height* of $\mathfrak{p}$ is denoted by $\mathrm{ht}(\mathfrak{p})$, and is defined as the maximum of the lengths of chains of prime ideals of R

$$\mathfrak{p}_0 \subsetneq \mathfrak{p}_1 \subsetneq \cdots \subsetneq \mathfrak{p}_n = \mathfrak{p}$$

ending at $\mathfrak{p}$. In addition, for any ideal I of R we define

$$\mathrm{ht}(I) := \{\mathrm{ht}(\mathfrak{p}) \mid I \subset \mathfrak{p}\}.$$

9.4.1 The Prime Avoidance Lemma

In the following important lemma we will show that an ideal contained in the finite union of prime ideals is contained in at least one of them. Its name, the "prime avoidance lemma" comes from the fact that if an ideal I is not contained in any of a finite number of prime ideals $\mathfrak{p}_1,...,\mathfrak{p}_n$, then there is an element $x \in I$ that *avoids* being in any of the $\mathfrak{p}_j$'s.

Lemma 9.4.7. *(Prime avoidance lemma) Let R be a ring and I be an ideal of R. Let $\mathfrak{p}_1,...,\mathfrak{p}_n$ be prime ideals of R such that*

$$I \subset \bigcup_{i=1}^{n} \mathfrak{p}_i.$$

Then, $I \subset \mathfrak{p}_k$ for some k, $1 \leq k \leq n$.

Proof. We will show, using induction on n, that if I is not contained in any of the ideals $\mathfrak{p}_1,...,\mathfrak{p}_n$, then I has an element which is not in any of the $\mathfrak{p}_n$.

If $n = 1$, the statement is trivial. If $n = 2$, then I contains an element $x \notin \mathfrak{p}_1$ and similarly another element $y \notin \mathfrak{p}_2$. Therefore $x \in \mathfrak{p}_2$ and $y \in \mathfrak{p}_1$. But $x + y \in I$ and so is in either $\mathfrak{p}_1$ or $\mathfrak{p}_2$. Hence $x + y \in \mathfrak{p}_1 \cup \mathfrak{p}_2$.

Now suppose the statement is true for $n - 1$, $n \geq 3$. Then for each i we can assume that

$$I \not\subseteq \bigcup_{j \neq i} \mathfrak{p}_j$$

(otherwise $I \subseteq \mathfrak{p}_i$). Therefore, I contains an element, x_i, which is not in $\mathfrak{p}_j$ for any $j \neq i$ and we may assume that $x_i \in \mathfrak{p}_i$ (if not, we would be done). Since $\mathfrak{p}_n$ is prime, the product $x_1 \cdots x_{n-1}$ is not an element of $\mathfrak{p}_n$. However, it is an element of $\mathfrak{p}_1 \mathfrak{p}_2 \cdots \mathfrak{p}_{n-1}$. So, $x_1 \cdots x_{n-1} + x_n$ is an element of I which is not contained in any $\mathfrak{p}_i$. $\square$

We now give a slightly stronger version of the avoidance lemma.

Lemma 9.4.8. *(Second Version) If $\mathfrak{p}$ is a prime ideal in R and $\mathfrak{a}_1, ..., \mathfrak{a}_n$ are just ideals of R such that*

$$\bigcap_{i-1}^{n} \mathfrak{a}_i \subset \mathfrak{p},$$

then $\mathfrak{p} \supset \mathfrak{a}_j$ for some j.

If $\mathfrak{p} = \bigcap_i \mathfrak{a}_i$, then $\mathfrak{p} = \mathfrak{a}_j$ for some j.

Proof. Assume the first statement is false. Then, for each i there is an $x_i \mathfrak{a}_i$ but $x_i \notin \mathfrak{p}$. Hence,

$$x_1 x_2 \cdots x_n \in \mathfrak{a}_1 \mathfrak{a}_2 \cdots \mathfrak{a}_n \subset \mathfrak{a}_1 \cap \mathfrak{a}_2 \cap \cdots \cap \mathfrak{a}_n \subset \mathfrak{p}.$$

This is a contradiction since $\mathfrak{p}$ is a prime ideal.

Now, if $\mathfrak{p} = \bigcap_{i=n}^{n} \mathfrak{a}_i$, then $\mathfrak{p} \subset \mathfrak{a}_i$ for all i. So, if $\mathfrak{p} \supset \mathfrak{a}_j$ for some j, then, $\mathfrak{p} = \mathfrak{a}_j$. $\square$

Proposition 9.4.9. *Let R be a ring and $I = (a_1, ..., a_r)$ be a finitely generated ideal of R. Let $\mathfrak{p}_1, ..., \mathfrak{p}_n$ be prime ideals of R such that $I \not\subseteq \mathfrak{p}_i$ for each i, $1 \leq i \leq n$. Then, there are $b_2, ..., b_r \in R$ such that*

$$c = a_1 + b_2 a_2 + \cdots + b_r a_r \notin \bigcup_{i=1}^{n} \mathfrak{p}_i.$$

Proof. Without loss of generality, assume that $\mathfrak{p}_i \not\subseteq \mathfrak{p}_j$ for $i \neq j$. We prove the statement by induction on n. The case $n = 1$ is obvious. Assume the result for $n-1$ prime ideals. Then, there exist $c_2, ..., c_r \in R$ such that

$$d = a_1 + c_2 a_2 + \cdots + c_r a_r \notin \bigcup_{i=1}^{n-1} \mathfrak{p}_i.$$

If $d \notin \mathfrak{p}_n$, we are done. So, let $d \in \mathfrak{p}_n$. If $a_2, ..., a_r \in \mathfrak{p}_n$, then from the expression for d we see that $a_1 \in \mathfrak{p}_n$, which contradicts the fact that $I \not\subseteq \mathfrak{p}_n$. So, $a_i \notin \mathfrak{p}_n$ for some i, say for $i = 2$. Since $\mathfrak{p}_i \not\subseteq \mathfrak{p}_j$ for $i \neq j$, we can find

$$x \in \bigcap_{i=1}^{n-1} \mathfrak{p}_i$$

such that $x \notin \mathfrak{p}_n$. Then,

$$c = a_1 + (c_2 + x)a_2 + \cdots + c_r a_r \notin \bigcup_{i=1}^{n} \mathfrak{p}_i$$

completing the proof. □

Exercise 9.4.10. For an ideal $I \subseteq R$, where R is a commutative ring with identity, let $I[[x]]$ denote the set of power series with coefficients in I, and let $IR[[x]]$ denote the ideal of $R[[x]]$ generated by I.

1. Prove that both $I[[x]]$ and $IR[[x]]$ are ideals and they are equal if I is finitely generated.

2. Give an example showing that $I[[x]] \neq IR[[x]]$.

3. Prove that if $P \subseteq R$ is a prime ideal, then $P[[x]]$ is a prime ideal in $R[[x]]$.

9.5   Local Rings

Definition 9.5.1. A *local ring*[1] is a ring R with a unique maximal ideal I. In this case we call R/I the *residue field* of R.

Example 9.5.2.

- Any field k is a local ring with $\{0\}$ the maximal ideal.

- The ring of p-adic numbers (to be discussed in Chapter 10).

- If C_2 is the cyclic group of order 2, $C_2 = \{x, \ x^2 = 1\}$, then the group ring
$$\mathbb{Z}_2[C_2] = \{1, \ x, \ 0, \ 1 + x\}$$
is a local ring with maximal ideal $\{0, \ 1 + x\}$ and residue field isomorphic to $\mathbb{Z}_2$.

- The ring
$$R = \left\{ \frac{a}{b} \ \middle| \ a, \ b \in \mathbb{Z}, \ 2 \nmid b \right\}$$
is a local ring with residue field $\mathbb{Z}_2$.

Exercise 9.5.3. Of course $\mathbb{Z}$ is not a local ring. Let k be a field of characteristic 0. Show $k[x]$ is also not a local ring.

Here is an element wise characterization of a local ring.

Proposition 9.5.4. *Let R be a non-zero ring. Then the following propositions are equivalent:*

1. *R has a unique maximal ideal.*

2. *If $x \in R$, then either x or $1 - x$ is invertible.*

Proof. (2) $\Rightarrow$ (1). Assume that for each x in R either x or $1 - x$ is invertible. We will find the maximal ideal. To do this, let $\mathfrak{N}$ be the set of all non-invertible elements of R. It is clear that $1 \notin \mathfrak{N}$, and that $\mathfrak{N}$

---

[1]Zariski in [136] attributes the theory of local rings to Krull (in [73]) and Chevalley (in [25]).

is closed under multiplication. Also, any proper ideal of R must be a subset of $\mathfrak{N}$, otherwise it must contain an invertible element. It remains to prove that $\mathfrak{N}$ is closed under addition. Suppose the contrary, i.e. there exist x and y in $\mathfrak{N}$ such that $(x+y) \notin \mathfrak{N}$. Hence $x+y$ invertible. So, we can consider the element $z = \frac{x}{x+y}$. We have:

$$1 = z + (1-z) = \frac{x}{x+y} + \frac{y}{x+y}.$$

Then, by hypothesis, one of the two elements z and $1-z$ must be invertible, which implies that one of the two

$$\frac{x}{x+y} \quad \text{or} \quad \frac{y}{x+y}$$

is invertible, so either x or y is invertible. This is a contradiction.

$(1) \Rightarrow (2)$. Suppose R has a unique maximal ideal $\mathfrak{N}$. Then, $\mathfrak{N}$ must contain all non-invertible elements. Indeed, since $\mathfrak{N}$ is a proper ideal, it can not contain any invertible element. Also, if x is a non-invertible element, then (x) is a proper ideal of R and so, by Proposition 5.4.2 of Volume I, it must be contained in the unique maximal ideal $\mathfrak{N}$, and so $x \in \mathfrak{N}$. To finish the proof, let x be any element of R. Then, we can write

$$1 = x + (1-x),$$

and since $1 \notin \mathfrak{N}$, either x or $1-x$ is not in $\mathfrak{N}$. $\square$

Exercise 9.5.5. Prove a ring R is a local ring if and only if for any finite sum which happens to be invertible at least one summand is invertible.

Proposition 9.5.6. *A local ring R does not contain idempotent elements other than* $0, 1$.

Proof. First solution. If a is an idempotent element $\neq 0, 1$, then

$$a(a-1) = a^2 - a = 0.$$

This shows that both a and $a-1$ are zero divisors hence non-invertible. Hence they must be contained in the maximal ideal. But then $1 = a - (a-1)$ is also in the maximal ideal, a contradiction.

Second solution. Assume $a^2 = a$. Then $a(1-a) = 0$. Since $a+(1-a)$ is invertible then using the above characterization a or $1-a$ is invertible. In the first case, this implies $1 - a = 0$. In the second, $a = 0$. □

Proposition 9.5.7. *A ring is local if and only if all the non-units form an ideal.*

Proof. Suppose first that R is local with maximal ideal I. Let x be any element not in I. Then x must be a unit. Otherwise, x generates an ideal (x) which is contained in a maximal ideal other than I.

Conversely, suppose that R is a ring in which the non-units form an ideal I. Then every ideal in R must be contained in I since ideals cannot contain units. □

Proposition 9.5.8. *If R is a ring and I is a maximal ideal such that every element of $1 + I$ is a unit, then R is a local ring.*

Proof. Since I is a maximal ideal, for any $x \in R - I$ we have

$$R = (I, \{x\}) = \{ax + b \mid a \in R, \ b \in I\}.$$

In particular, $ax + b = 1$ for some $a \in R$, $b \in I$, and therefore $ax = 1 - b \in 1 + I$. Now, if ax is a unit, x must be a unit, and hence all elements of $R - I$ are units, which implies that R is a local ring. □

We now define the *Jacobson radical* in the case of a commutative ring with identity. In Chapter 13 we will consider this concept when the ring is not necessarily commutative. We shall do this by considering 2-sided maximal ideals.

Definition 9.5.9. When R is a commutative ring with 1, we call the *Jacobson radical*, $J(R)$, the intersection of all maximal ideals of R.

Exercise 9.5.10. Show that in $\mathbb{Z}$, $J(\mathbb{Z}) = (0)$. Such a ring is called *semi-simple*.

Corollary 9.5.11. *If R is a local ring, evidently $J(R)$ is its unique maximal ideal. $J(R)$ is the complement of the set of all invertible elements.*

Proof. We show that $x \in R$ is invertible if and only if $x \notin J(R)$. Suppose first that x is invertible. Then, $x \notin J(R)$ since $xy = 1$ and $x \in J(R)$ imply $1 \in J(R)$ because $J(R)$ is an ideal. The converse follows from Proposition 9.5.7.   □

Corollary 9.5.12. *If R is a local ring and $J(R)$ is its unique maximal ideal, then $R/J(R)$ is a field.*

Proof. $R/J(R)$ has no ideals except $\{0\}$. Hence every non-zero element is invertible.   □

9.6   Localization

Localization is a method of adding multiplicative inverses to a ring, i.e. is a formal way to introduce denominators to a given ring R. It consists of fractions a/b where the denominators b come from a given subset S of R. The basic construction follows that of constructing the quotient field, $\mathbb{Q}$, of $\mathbb{Z}$ so that every non-zero element of $\mathbb{Z}$ becomes invertible (see Chapter 5). However, in many cases we might not want to go so far. For example, if we are interested in a single prime number p, we might want to simply invert all numbers not divisible by p. Thus, we would consider the subring of $\mathbb{Q}$ consisting of all rational numbers a/b with $\gcd(a, b) = 1$ such that b is relatively prime to p. In this ring, the only kind of divisibility that counts is divisibility by p and its powers. This process of inverting a specialized class of elements is called *localization*.

Definition 9.6.1. A subset $S \subset R - \{0\}$ is called a *multiplicative set* if S is closed under multiplication, and $1 \in S$.

The reason we want $0 \notin S$ will be clear in a moment.

Example 9.6.2. Here are some multiplicative subsets of $\mathbb{Z}$:

1. $S = \mathbb{Z} - \{0\}$.

2. $S = \{n \in \mathbb{Z} \mid p \nmid n\}$, where p is a prime number. Then $S = \mathbb{Z} - \langle p \rangle$.

Let S be a multiplicative set in R. We consider the following set of formal symbols:

$$\left\{ \frac{r}{s} \;\middle|\; r \in R, \; s \in S \right\}.$$

We shall say two symbols $\frac{r_1}{s_1}$ and $\frac{r_2}{s_2}$ are equivalent, and denote this by $\frac{r_1}{s_1} \sim \frac{r_2}{s_2}$, if there is an $s \in S$ such that

$$s(r_1 s_2 - r_2 s_1) = 0.$$

This is an equivalence relation. Since R is commutative one checks easily that it is reflexive and symmetric. To prove transitivity, suppose

$$\frac{a}{r} \sim \frac{b}{t}, \quad \text{and} \quad \frac{b}{t} \sim \frac{c}{u} \qquad a, b, c \in R \text{ and } r, t, u \in S.$$

Then, for some u, w in S,

$$v(at - br) = 0 = w(bu - ct).$$

Thus

$$vat - vbr = 0$$

$$wbu - wct = 0.$$

Multiplying by uw in the first equation, and by rv in the second, we get

$$uwvat - uwubr = 0$$

and

$$-wctru + wburv = 0.$$

Adding these yields,

$$uwuat - wctrv = 0 \quad \text{or} \quad (twu)(au - cr) = 0,$$

which since t, w and v are in S shows that $\frac{a}{r} \sim \frac{c}{u}$.

We denote by $\left[\frac{r}{s}\right]$ the class of $\frac{r}{s}$ and by $S^{-1}R$ the set of all distinct equivalence classes.

Next, we equip the set $S^{-1}R$ with the operations of addition and multiplication, by defining:

$$\left[\frac{r_1}{s_1}\right] + \left[\frac{r_2}{s_2}\right] = \left[\frac{r_1 s_2 + r_2 s_1}{s_1 s_2}\right]$$

and

$$\left[\frac{r_1}{s_1}\right] \cdot \left[\frac{r_2}{s_2}\right] = \left[\frac{r_1 r_2}{s_1 s_2}\right].$$

Proposition 9.6.3. *Under these two operations, $S^{-1}R$ is a commutative ring. Furthermore there is a natural homomorphism*

$$\varphi : R \longrightarrow S^{-1}R \quad \text{given by} \quad \varphi(r) = \left[\frac{r}{1}\right].$$

We leave to the reader to prove first that the two operations are well defined (i.e. they do not depend on the choice of the representative) and, second, that the axioms of a commutative ring are satisfied.

Definition 9.6.4. The ring $S^{-1}R$ is called the *ring of quotients*[2] of R with respect to S.

This results in the following proposition.

Proposition 9.6.5.

1. *If $0 \in S$, then $S^{-1}R = \{0\}$.*

2. *The homomorphism φ is injective if and only if S has no zero divisors.*

Proof. 1. One sees easily that $[0/1]$ is the zero element in $S^{-1}R$. So, if $0 \in S^{-1}R$ then, for any $[r/s] \in S^{-1}R$, $0(r1 - 0s) = 0$, i.e. $[r/s] = [0/1]$, which means $S^{-1}R = \{0\}$.

2. To prove the second statement

$$r \in \operatorname{Ker} \varphi \Leftrightarrow \varphi(r) = \left[\frac{r}{1}\right] = 0 \in S^{-1}R \Leftrightarrow \exists\, s \in S \; : \; sr = 0.$$

Thus, $\operatorname{Ker} \varphi = \{0\}$ if and only if S has no zero divisors. $\qquad\square$

---

[2]$S^{-1}R$ means fractions with denominators from S.

Of course, when $R = \mathbb{Z}$ and S is all of $\mathbb{Z} - \{0\}$, then

$$S^{-1}\mathbb{Z} = \mathbb{Q}.$$

Proposition 9.6.6. *An element $\frac{x}{s}$ in $S^{-1}R$ is equal to zero if and only if there is an element $t \in S$ so that $xt = 0$.*

Proof. If such a t exists then $\frac{xs}{st} = \frac{0}{st} \sim \frac{0}{1} = \varphi(0)$ since $0 \cdot 1 = 0 = st \cdot 0$. Conversely, if $\frac{x}{s} \sim \frac{0}{1} = \varphi(0)$, then there is a $t \in S$ so that $xt = x1t = st0 = 0$. □

It is then easy to verify that:

Proposition 9.6.7. *An element $\frac{x}{s} \in S^{-1}R$ is equal to $1 = \frac{1}{1} = \varphi(1)$ if and only if there exists $t \in S$ so that $tx = ts$.*

Proposition 9.6.8. *(The Universal Property of $S^{-1}R$) Suppose R is a commutative ring with identity and $\phi : R \to T$ is a ring homomorphism, where T is a commutative ring with identity, with the property that $\phi(s)$ is a unit in T for all $s \in S$. Then, there is a unique ring homomorphism $\overline{\phi} : S^{-1}R \to T$ such that the following diagram*

commutes.

Proof. First notice that if $\psi : S^{-1}R \to T$ is a ring homomorphism and $\psi \circ \pi = \phi$, then for $r \in R$, $\psi\left(\frac{r}{1}\right) = \phi(r)$. Hence, since ψ is a ring homomorphism,

$$\psi\left(\frac{r}{s}\right) = \psi\left(\frac{r}{1}\left(\frac{s}{1}\right)^{-1}\right) = \psi\left(\frac{r}{1}\right)\psi\left(\frac{s}{1}\right)^{-1} = \phi(r)\phi(s)^{-1}.$$

Hence $\psi\left(\frac{r}{s}\right) = \phi(r)\phi(s)^{-1}$.

Suppose (r, s), (r', s') are in $R \times S$ and $(r, s) \sim (r', s')$, say $t(rs' - r's) = 0$. Then $\phi(t)(\phi(r)\phi(s') - \phi(r')\phi(s)) = 0$ and so $\phi(r)\phi(s') =$

$\phi(r')\phi(s)$ since $\phi(t)$ is a unit in T. Moreover, $\phi(s)$ and $\phi(s')$ are units in R, so that $\phi(r)\phi(s)^{-1} = \phi(r')\phi(s')^{-1}$.

Now suppose C is an equivalence class in $R \times S$. The preceding calculation shows the element $\phi(r)\phi(s)^{-1}$ in T depends only on C and not on the choice of (r, s) in C. Therefore we can define

$$\overline{\phi} : S^{-1}R \to T \quad : \quad \overline{\phi}\left(\frac{r}{s}\right) = \phi(r)\phi(s^{-1}).$$

It is straightforward to check that

$$\overline{\phi}\left(\frac{r}{s} + \frac{r'}{s'}\right) = \overline{\phi}\left(\frac{r}{s}\right) + \overline{\phi}\left(\frac{r'}{s'}\right)$$

and that

$$\overline{\phi}\left(\frac{r}{s} \cdot \frac{r'}{s'}\right) = \overline{\phi}\left(\frac{r}{s}\right) \cdot \overline{\phi}\left(\frac{r'}{s'}\right)$$

for $\frac{r}{s}$ and $\frac{r'}{s'}$ in $S^{-1}R$. Hence $\overline{\phi}$ is a ring homomorphism. If $r \in R$, then

$$\overline{\phi} \circ \pi(r) = \overline{\phi}\left(\frac{r}{1}\right) = \phi(r)\phi(1)^{-1} = \phi(r),$$

so $\overline{\phi} \circ \pi = \phi$.

Finally, if $\psi : S^{-1}R \to T$ is a ring homomorphism and $\psi \circ \pi = \phi$, then the calculation above shows that $\psi\left(\frac{r}{s}\right) = \phi(r)\phi(s)^{-1}$, so $\psi = \overline{\phi}$.   □

Example 9.6.9. (Laurent Series) Let k be a field and consider the polynomial ring $k[x]$. Then $x^n \neq 0$ for any $n \geq 0$, so $\{x^n \mid n \in \mathbb{N}\}$ is a multiplicative subset of $k[x]$, and hence we may consider the corresponding localization of $k[x]$. This localization, usually denoted by $k[x, x^{-1}]$, is called the *k-algebra of the Laurent series in x*. If we write x^{-n} instead of $\frac{s}{1}x^n$ for $n \in \mathbb{N}$, then a typical element in $k[x, x^{-1}]$ has the form

$$\frac{1}{x^m}(a_0 + a_1 x + \cdots + a_k x^k) = a_0 x^{-m} + a_1 x^{-m+1} + \cdots + a_k x^{-m+k}$$

$$= \sum_{i=-m}^{n} b_i x^i,$$

where $n = -m + k$ and $b_i = a_{i+m}$ for $m \leq i \leq n$. It is easy to see that $\{ x^i \mid i \in \mathbb{Z} \}$ is a basis of $k[x, x^{-1}]$ considered as a k-vector space, since if

$$\sum_{i=-m}^{n} a_i x^i = \sum_{i=-m}^{n} b_i x^i,$$

then multiplying by x^m gives

$$\sum_{i=0}^{m+n} a_{i-m} x^i = \sum_{i=0}^{m+n} b_{i-m} x^i$$

and both sides are in $k[x]$. Therefore, $a_i = b_i$ for $-m \leq i \leq n$.

Example 9.6.10. (Rings of germs of functions) Suppose X is a topological space and $\mathcal{C}(X)$ is the $\mathbb{R}$-algebra of all continuous functions $X \to \mathbb{R}$. Fix $x_0 \in X$ and define $\mathfrak{m}_{x_0} = \{ f \in \mathcal{C}(X) \mid f(x_0) = 0 \}$. Then, prove that $\mathfrak{m}_{x_0}$ is a maximal ideal and hence a prime ideal in $\mathcal{C}(X)$. Let

$$S = \mathcal{C}(X) - \mathfrak{m}_{x_0} = \{ f \in \mathcal{C}(X) \mid f(x_0) \neq 0 \}.$$

Then S is a multiplicative subset of $\mathcal{C}(X)$. Elements in the localization, $S^{-1}\mathcal{C}(X)$, have the form $\frac{f}{g}$ where f and g are continuous real-valued functions on X and $g(x_0) \neq 0$.

Now let X be a compact, Hausdorff space and $\mathcal{C}(X)$ be the ring of continuous, real or complex valued functions on X. For $A \subseteq X$, let $I(A)$ be the ideal

$$I(A) = \{ f \in \mathcal{C}(X) : f(x) = 0, \text{ for all } x \in A \}.$$

Proposition 9.6.11. *Let $\mathfrak{p}$ be a prime ideal in $\mathcal{C}(X)$, then*

$$V(\mathfrak{p}) = \{ x \in X \mid f(x) = 0, \text{ for all } x \in \mathfrak{p} \}$$

consists of a single point.

Proof. Since $\mathfrak{p}$ is proper, $V(\mathfrak{p})$ consists of at least one point. So we can assume that x, $y \in V(\mathfrak{p})$ with $x \neq y$. Since X is Hausdorff, we can find neighborhoods U_x and U_y of x and y respectively, such that $U_x \cap U_y = \varnothing$.

Now, X is compact and Hausdorff, therefore it is completely regular, which implies that there are bump functions f and g supported in U_x and U_y respectively, with $f(x) = g(y) = 1$. From this we get that f and $g \notin \mathfrak{p}$ but $fg = 0 \in \mathfrak{p}$, which is a contradiction. $\qquad\square$

If $x \in X$, let

$$J(x) = \{f \in C(X) : f \,|_{U_x} = 0 \text{ for some open } U_x \text{ of } x\}$$

be the ideal of functions that vanish in a neighborhood of x, so $C(X)/J(x)$ is the *ring of germs* of continuous functions at x.

Proposition 9.6.12. *The prime ideals contained in $I(x)$ are in bijection with the prime ideals of $C(X) = J(x)$.*

Proof. It suffices to show that $J(x) \subseteq \mathfrak{p}$ for all primes $\mathfrak{p} \subseteq I(x)$. If $f \in J(x)$ then f vanishes in a neighborhood U_x of x. Let g be a function supported in U_x with $g(x) = 1$. Then $g \notin I(x)$, so $g \notin \mathfrak{p}$, but $fg \in \mathfrak{p}$, so $f \in \mathfrak{p}$. $\qquad\square$

We give a second proof, based on the following fact:

Proposition 9.6.13. *The ring of germs $C(X)/J(x)$ is isomorphic to the localization $C(X)_{I(x)}$.*

Proof. First we show that any $f \notin I(x)$ becomes a unit in $C(X)/J(x)$. It suffices to show that f^2 becomes a unit. Let $y = f(x)^2$ and $g = \max\{f^2, \frac{y}{2}\}$. Then g never vanishes, hence is a unit, and $g - f^2 \in J(x)$. Second, we show that the natural map $C(X) \longrightarrow C(X)_{I(x)}$ sends any $f \in J(x)$ to 0. Since $f \in J(x)$, f vanishes on a neighborhood U_x of x. Let g be a function supported in U_x with $g(x) = 1$. Then $g \notin I(x)$ and $fg = 0$, so $\frac{f}{1} = \frac{0}{g}$. $\qquad\square$

Proposition 9.6.14. *Every maximal ideal of $C(X)$ is of the form I_x for some $x \in X$.*

Proof. Let $\mathfrak{m}$ be a proper, maximal ideal of $C(X)$. If $\mathfrak{m} \subseteq I_x$ then $\mathfrak{m} = I_x$ and we are done. So, we assume that $\mathfrak{m} \not\subseteq I_x$, for all $x \in X$. This implies that for each $x \in X$ there is a function $f_x \in \mathfrak{m}$ such that

$f_x \notin I_x$. Now, pick such a function f_x and let $|f_x(x)| = k > 0$. Then, since f_x is continuous, there is an open neighborhood $U_x \subseteq X$ such that $|f_x(y)| > \frac{k}{2}$ for each $y \in X$. Now, we remark that

$$X = \bigcup_{x \in X} U_x,$$

and since X is compact, this covering must have a finite sub-covering, i.e. there exists $n \in \mathbb{N}$ such that

$$X = \bigcup_{x_i = 1}^{n} U_{x_i}.$$

Now, consider the function

$$g(x) = \sum_{i=1}^{n} |f_{x_i}|^2(x). \tag{1}$$

It is clear that g is continuous on X and $g(x) > 0$, since each of the terms in (1) is positive on a subspace of X. In addition, $g \in \mathfrak{m}$. Since $g(x) \neq 0$ on X, the function $\frac{1}{g(x)}$ is well defined on X and it is also continuous, which shows that $\frac{1}{g(x)} \in \mathcal{C}(X)$. But then, $g(x)\frac{1}{g(x)} = 1 \in \mathfrak{m}$, which is a contradiction since $\mathfrak{m}$ is a proper ideal. $\square$

We leave to the reader to check the following alternative construction of the ring of germs. Define

$$\mathcal{F} = \{\, (U, f) \mid U \text{ is a neighborhood of } x_0 \text{ and } f : U \to \mathbb{R} \text{ is continuous} \,\}.$$

For (U, f) and (V, g) in $\mathcal{F}$, and r in $\mathbb{R}$, define

$$(U, f) + (V, g) = (U \cap V, f + g)$$
$$(U, f) \cdot (V, g) = (U \cap V, fg)$$
$$r(U, f) = (U, rf),$$

where on the right hand side we consider $f + g$ and fg as functions from $U \cap V$ to $\mathbb{R}$, that is "$f + g$" is really $f|_{U \cap V} + g|_{U \cap V}$ and "$f\,g$" is really $f|_{U \cap V}\, g|_{U \cap V}$.

Exercise 9.6.15. Show that $\mathcal{F}$ with these operations is a commutative $\mathbb{R}$-algebra with identity.

Define

$$I = \{ (U, f) \in \mathcal{F} \mid f|_V \equiv 0 \text{ on some neighborhood of } x_0 \text{ in } U \}$$
$$= \{(U, f) \in \mathcal{F} \mid \exists \text{ a neighborhood } V, \text{ of } x_0 : f(x) = 0 \, \forall x \in U \cap V \}.$$

Exercise 9.6.16. Show that I is an ideal in $\mathcal{F}$ and that $(U, f) + I = (V, g) + I$ if and only if f and g agree on some neighborhood of x_0.

Define $\mathcal{F}_{x_0} = \mathcal{F}/I$. The ring, $\mathcal{F}_{x_0}$, is called the *ring (ℝ-algebra) of germs of continuous functions defined near x_0*. Denote the coset $(U, f) + I$ by $[U, f]$. The ring $\mathcal{F}_{x_0}$ is the rigorous formalization of the idea of "the ring of all continuous functions that are defined on some neighborhood of x_0".

9.6.1   Localization at a Prime Ideal

The technique of localization at a prime ideal gives us the possibility of verifying certain properties of rings or modules locally. Then that property holds if and only if it holds for the localization at *every* prime/maximal ideal.

Definition 9.6.17. Let R be a commutative ring with identity and $\mathfrak{p}$ be a prime ideal of R. Then, $S = R - \mathfrak{p}$ is a multiplicative set. The ring $S^{-1}R$ is called the *localization* of R at $\mathfrak{p}$ and is denoted by $R_{\mathfrak{p}}$.

Let S be a multiplicatively closed subset of a ring R and consider the ring homomorphism $\phi : R \longrightarrow S^{-1}R$. If I is an ideal of R let I^e be the ideal in $S^{-1}R$ generated by $\phi(I)$. As $\frac{1}{s} \in S^{-1}R$ for $s \in S$, the ideal I^e must contain the elements

$$\frac{1}{s}\phi(a) = \frac{a}{s}$$

for any $a \in I$. One can easily check that $\{\frac{a}{s} : a \in I, s \in S\}$ is already an ideal in $S^{-1}R$, and so it has to be equal to I^e.

We will now prove $R_{\mathfrak{p}}$ is a local ring.

Theorem 9.6.18. *Let R be a ring and $\mathfrak{p}$ be a prime ideal of R. Then, $\frac{r}{s} \in R_{\mathfrak{p}}$ is a unit if and only if $r \notin \mathfrak{p}$. Therefore, $R_{\mathfrak{p}}$ is a local ring with unique maximal ideal, $\mathfrak{m} = \mathfrak{p}R_{\mathfrak{p}}$, namely,*

$$\mathfrak{m} = \{\frac{a}{s} \ : \ a \in \mathfrak{p} \ s \notin \mathfrak{p}\}.$$

In particular, any prime ideal of $R_{\mathfrak{p}}$ is maximal.

Proof. "⇒". Suppose that $\frac{r}{s} \in R_{\mathfrak{p}}$ is a unit. Then, there exists an element $\frac{t}{v} \in R_{\mathfrak{p}}$ such that

$$\frac{r}{s} \cdot \frac{t}{v} = \frac{1}{1},$$

from which we get $u(rt - sv) = 0$ for some $u \in R - \mathfrak{p}$. Hence $r \notin \mathfrak{p}$.

"⇐". If $r \notin \mathfrak{p}$, then $r \in R - \mathfrak{p}$, and therefore $\frac{s}{r} \in R_{\mathfrak{p}}$, for each $s \in R - \mathfrak{p}$. This implies

$$\frac{r}{s} \cdot \frac{s}{r} = \frac{1}{1},$$

and thus $\frac{r}{s}$ is a unit.

It follows that each element which is not in the ideal $\mathfrak{p}R_{\mathfrak{p}}$ is a unit. Therefore, $\mathfrak{p}R_{\mathfrak{p}}$ is the unique maximal ideal of $R_{\mathfrak{p}}$. Thus $R_{\mathfrak{p}}$ is a local ring.

Since the set of all prime ideals of R contained in $\mathfrak{p}$ has a maximal element, $\mathfrak{p}^e$, it follows that every prime ideal in $R_{\mathfrak{p}}$ is contained in $\mathfrak{p}^e$. In particular, this is true for maximal ideals in $R_{\mathfrak{p}}$. Hence, $\mathfrak{p}^e$ is the only maximal ideal. □

For example, the localization of $\mathbb{Z}$ at the prime ideal (p) is

$$\mathbb{Z}_{(p)} = \{\frac{a}{b} \in \mathbb{Q} \mid a, b \in \mathbb{Z}, p \nmid b\}.$$

Also, the localization of $k[x]$ at the prime ideal $(x - a)$, $a \in k$, is

$$k[x]_{(x-a)} = \{\frac{f(x)}{g(x)} \in k(x) \mid a \text{ is not a root of } g(x)\}.$$

We now apply the Localization Principle to R-modules.

9.6.2   Localization of Modules

Definition 9.6.19. Let M be an R-module and S be a multiplicatively closed subset of R. Then

$$(m, s) \sim (m_1, s_1) \text{ if and only if there is a } u \in S \, : \, u(s_1 m - s m_1) = 0$$

is an equivalence relation on $M \times S$. We denote the equivalence class of $(m, s) \in M \times S$ by $\frac{m}{s}$.

Definition 9.6.20. The set of all equivalence classes

$$S^{-1}M = \left\{ \frac{m}{s}, \, m \in M, \, s \in S \right\}$$

is called the *localization* of M at S.

One checks easily that $S^{-1}M$ equipped with the addition and scalar multiplication

$$\frac{m}{s} + \frac{m_1}{s_1} = \frac{s_1 m + s m_1}{s s_1} \text{ and } \frac{r}{t} \cdot \frac{m}{s} = \frac{rm}{ts},$$

where m, $m_1 \in M$ and s, s_1, $t \in S$, becomes an $S^{-1}R$-module. If $S = R - \mathfrak{p}$, where $\mathfrak{p}$ is a prime ideal of R, we will write $M_\mathfrak{p}$ instead of $S^{-1}M$.

If $f : M \longrightarrow N$ is a homomorphism of the R-modules M and N, we get an induced homomorphism of the $S^{-1}R$-modules

$$S^{-1}f : S^{-1}M \longrightarrow S^{-1}N$$

given by

$$S^{-1}f\left(\frac{m}{s}\right) = \frac{f(m)}{s}.$$

As before, if $S = R - \mathfrak{p}$, $\mathfrak{p}$ a prime ideal, we will write $f_\mathfrak{p}$ instead of $S^{-1}f$.

We shall now show how one can regard the localization of an R-module as a tensor product.

Proposition 9.6.21. *Let R be a ring, M an R-module, and S a multiplicatively closed subset of R. Then,*

$$S^{-1}M \cong M \otimes_R S^{-1}R.$$

Proof. Consider the following R-module homomorphism:

$$\varphi : S^{-1}M \longrightarrow M \otimes_R S^{-1}R \text{ such that } \frac{m}{s} \mapsto m \otimes \frac{1}{s}$$

This homomorphism is well defined. Indeed, if $m_1 \in M$ and $s_1 \in S$ with $\frac{m}{s} = \frac{m_1}{s_1}$, which means that there is some $u \in S$ such that $u(s_1 m - s m_1) = 0$, we have

$$m \otimes \frac{1}{s} - m_1 \otimes \frac{1}{s_1} = u(s_1 m - s m_1) \otimes \frac{1}{u s s_1} = 0.$$

Now, using the universal property of the tensor product we get a homomorphism

$$\phi : M \otimes_R S^{-1}R \longrightarrow S^{-1}M \text{ such that } \phi\left(m\frac{t}{s}\right) = \frac{tm}{s}$$

which is also well defined. Indeed, if $t_1 \in R$ and $s_1 \in S$ with $u(s_1 t - s t_1) = 0$ for some $u \in S$ we also get $u(s_1 t m - s t_1 m) = 0$, i.e.

$$\frac{tm}{s} = \frac{t_1 m}{s_1}.$$

By construction, φ and ϕ are inverse to each other, and thus φ is an isomorphism. $\qquad\square$

One of the most important properties of the localization is that it preserves exact sequences, as the following proposition shows.

Proposition 9.6.22. *For any short exact sequence*

$$0 \longrightarrow M_1 \xrightarrow{\varphi} M_2 \xrightarrow{\chi} M_3 \longrightarrow 0$$

of R-modules, and any multiplicatively closed subset $S \subset R$, the localized sequence

$$0 \longrightarrow S^{-1}M_1 \xrightarrow{S^{-1}\varphi} S^{-1}M_2 \xrightarrow{S^{-1}\chi} S^{-1}M_3 \longrightarrow 0$$

is also exact.

Proof. By Proposition 9.6.21, localization at S is the same as tensoring with $S^{-1}R$. But, as we learned in Chapter 6, tensor products are right exact. Hence it only remains to prove the injectivity of the map $S^{-1}\varphi$.

To do this, let $\frac{m}{s} \in S^{-1}M_1$ with $S^{-1}\varphi\left(\frac{m}{s}\right) = 0$. This implies that there is an element $u \in S$ with $u\varphi(m) = \varphi(um) = 0$. But φ is injective, thus $um = 0$. Therefore,

$$\frac{m}{s} = \frac{1}{us}um = 0.$$

So, $S^{-1}\varphi$ is injective. $\square$

The next proposition illustrates the use of localization at *maximal* ideals of R to obtain a property of R-modules. It is always used in the form, to prove $M = (0)$ it is sufficient to show $M_\mathfrak{p} = (0)$ for every maximal ideal $\mathfrak{p}$.

Theorem 9.6.23. *Let R be a ring and M be an R-module. Then $M = (0)$ if and only if $M_\mathfrak{p} = (0)$ for every maximal ideal $\mathfrak{p}$ of R.*

Proof. Suppose $M \neq (0)$. Choose any non-zero $x \in M$. Let

$$I = \mathrm{Ann}_R(x) := \{r \in R \mid rx = 0\}.$$

Since $1 \notin I$, I is proper. Hence, I is contained in a maximal ideal $\mathfrak{p}$ of R. Now, suppose $M_\mathfrak{p} = (0)$. Then

$$\frac{x}{1} = \frac{0}{1}$$

in $M_\mathfrak{p}$. This means there is a $y \in R - \mathfrak{p}$ with $yx = 0$, and therefore $y \in I$, contradicting the fact that $I \subset \mathfrak{p}$. Therefore, $M_\mathfrak{p} \neq (0)$. The converse is trivial. $\square$

Corollary 9.6.24. *Let $f : M \longrightarrow N$ be an R-module homomorphism, and $\mathfrak{p}$ be a prime ideal of R. Then, the following statements are equivalent:*

1. f is injective (resp. surjective).

2. $f_{\mathfrak{p}}$ *is injective (resp. surjective) for all prime ideals* $\mathfrak{p}$.

3. $f_{\mathfrak{m}}$ *is injective (resp. surjective) for all maximal ideals* $\mathfrak{m}$.

Proof. We will prove the injective case, leaving the surjective case as an exercise.

$(1 \Rightarrow 2)$. If f is injective, then the sequence

$$0 \longrightarrow M \xrightarrow{\ f\ } N$$

is exact. Therefore, for each prime ideal $\mathfrak{p}$

$$0 \longrightarrow M_{\mathfrak{p}} \xrightarrow{\ f_{\mathfrak{p}}\ } N_{\mathfrak{p}}$$

is exact (by Proposition 9.6.22), i.e. $f_{\mathfrak{p}}$ is injective.

$(2 \Rightarrow 3.)$ Obvious.

$(3 \Rightarrow 1)$. Put $M' = \mathrm{Ker}(f)$. Then, the sequence

$$0 \longrightarrow M' \xrightarrow{\ i\ } M \xrightarrow{\ f\ } N,$$

where i is the inclusion map, is exact. For each maximal ideal $\mathfrak{m}$ of R the sequence

$$0 \longrightarrow M'_{\mathfrak{m}} \xrightarrow{\ i\ } M_{\mathfrak{m}} \xrightarrow{\ f_{\mathfrak{m}}\ } N_{\mathfrak{m}}$$

is exact, and so (by 3)

$$M' \cong \mathrm{Ker}\, f_{\mathfrak{m}} = 0,$$

since $f_{\mathfrak{m}}$ is injective. Therefore, by the theorem above $M' = 0$, i.e. f is injective. $\qquad\square$

9.6.3   Exercises

Exercise 9.6.25. Suppose that every $x \in R$ satisfy the equation $x^n = x$ for some $n \geq 2$ (depending on x). Prove that every prime ideal of R is maximal.

Exercise 9.6.26. A multiplicatively closed set $S \subseteq R$ is said to be *saturated* if whenever $uv \in S$, then $u \in S$ and $v \in S$. Prove that S is a saturated multiplicatively closed set if and only if S is the complement of a union of prime ideals.

Exercise 9.6.27. Let $x \in R$ be a non-zero divisor and set $S := \{1, x, x^2, \ldots\}$. Prove that R is an integral domain if and only if $S^{-1}R$ is an integral domain.

Exercise 9.6.28. Let $I \subseteq R$ be an ideal and $S \subseteq R$ a multiplicatively closed set. Prove that $S^{-1}R = S^{-1}I$ is isomorphic to $\overline{S}^{-1}(R/I)$, where $\overline{S}$ denotes the image of S in R/I. In other words, localization commutes with the formation of factor rings.

Exercise 9.6.29. If G is a finite Abelian group and $\mathbb{Z}_{(p)}$ is the localization of $\mathbb{Z}$ at the prime ideal (p), then

$$\mathbb{Z}_{(p)} \otimes_{\mathbb{Z}} G$$

is a p-group.

9.7 Noetherian Rings

Noetherian[3] rings and modules occur in many different areas of mathematics.

---

[3] Amalie Emmy Noether, one of the outstanding mathematicians of the twentieth century, was born in Erlangen, Germany in 1882. She entered the University of Erlangen at the age of 18 as one of only two women among the university's 986 students. Noether went to Göttingen to study in 1916 where David Hilbert and Felix Klein tried unsuccessfully to secure her an appointment. Some of the faculty objected to women lecturers, saying, "How can we allow a woman to attend the Faculty Senate", to which Hilbert responded: "I do not see that the sex of a candidate is an argument against her admission as a Privatdozent. After all, the Senate is not a bathhouse". However, she had three strikes against her: She was a woman, a Jew, and a pacifist. In 1922, Noether became a Privatdozent at Göttingen. Over the next 11 years she used axiomatic methods to develop an abstract theory of rings and ideals. One of her many students was B. L. van der Waerden, author of the first text on abstract algebra from a modern point of view called "Modern Algebra" which we would now call Algebra. Some of the other mathematicians Noether influenced or closely worked with were Alexandroff, E. Artin, Brauer, Courant, Hasse, Hopf,

Over a hundred years ago, Hilbert[4] used properties of what now are called Noetherian rings to settle a long-standing problem of invariant theory. Later, it was realized that commutative Noetherian rings are one of the building blocks of modern algebraic geometry, leading to their study both abstractly and in examples. The other was the Hilbert Nullstellensatz. It was not until the late 1950s, with the appearance of Goldie's theorem, that it became clear that non-commutative Noetherian rings constitute an important class of rings in their own right.

Classic examples of Noetherian rings include the coordinate rings of affine varieties, rings of differential operators on smooth algebraic varieties, universal enveloping algebras of finite dimensional Lie algebras and group algebras of polycyclic-by-finite groups. More recently, quantum groups have provided a new class of Noetherian rings and there has also been much interest in non-commutative geometry and the non-commutative Noetherian rings that arise, such as Sklyanin algebras.

Definition 9.7.1. A ring R is called *Noetherian* if it satisfies the ascending chain condition (see 9.1.7).

Pontryagin, von Neumann, and Weyl. She was never promoted to full professor by the Prussian academic bureaucracy. In 1933, as a Jew, Noether ,was banned from participation in all academic activities in Germany. She then emigrated to the United States and took a position at Bryn Mawr College. She became a member of the Institute for Advanced Study at Princeton. Noether died suddenly on April 14, 1935.

[4]David Hilbert (1862-1943) German, was one of the two most important figures in mathematics of the late 19th-early 20th century (the other being Poincaré). He is truly the last mathematician who could be said to understand all the mathematics of his day. For this reason we will not attempt to describe all his achievements, but rather concentrate on the issue at hand. His first work on invariant functions and one of his first papers was his proof of Theorem 9.7.10 now called the Hilbert basis theorem which he proved for $R = k[x_1, \ldots, x_n]$ in 1888. This was submitted to Matematishe Annalen and was refereed by Gordon who had proved a special case 20 years earlier by computational methods. Hilbert did this by using a completely different approach, proving the existence of a finite set of ideal generators without actually constructing them. As a result, the paper was rejected with the comment: "This is not Mathematics. This is Theology". However, Klein (an editor of MA) recognized the significance of the work, and guaranteed that it would be published as is, writing to Hilbert saying, "Without doubt this is the most important work on general algebra that the Annalen has ever published".

The first section of this chapter shows any PID is Noetherian. The idea here is that Noetherian rings are not too big. The following are some examples of rings that are too big:

Example 9.7.2. The big ring R of continuous functions $f : \mathbb{R} \longrightarrow \mathbb{R}$, is not Noetherian. Indeed, if we consider the ideals

$$I_n = \{f \mid f(x) = 0, \text{ for all } x \geq n\},$$

then the sequence

$$I_1 \subseteq I_2 \subseteq I_3 \subseteq ...$$

never terminates.

Example 9.7.3. Let X be an infinite set, and let $\mathcal{F}(X)$ be the set of all its finite subsets. Now, for any A, B in $\mathcal{F}(X)$ define two operations " $+$ " and " $\cdot$ " as follows:

$$A + B = (A \cup B) - (A \cap B)$$

$$A \cdot B = A \cap B$$

We can check that $(\mathcal{F}(X), +, \cdot)$ is a commutative ring, but it is not Noetherian. Indeed, if S is an infinite subset of X, then

$$I = \{A \mid A \subset S, \ A \text{ finite}\}$$

is an ideal of $\mathcal{F}(X)$, but it is not finitely generated.

Definition 9.7.4. We say that the ring R has the *maximal condition* if every not empty set $\mathcal{S}$ of its ideals, partially ordered by set inclusion, has a maximum element. That is, there is an $I \in \mathcal{S}$ such that if $J \in \mathcal{S}$ and $J \supset I$, then $I = J$.

Theorem 9.7.5. *For any commutative ring R with identity the following statements are equivalent:*

1. Each ideal I of R is finitely generated.

2. R is Noetherian.

3. The maximal condition for ideals holds in R.

Proof. "1 $\Rightarrow$ 2". Take an increasing sequence

$$I_1 \subset I_2 \subset I_3 \subset \cdots$$

of ideals in R, and let

$$I = \bigcup_{n=1}^{\infty} I_n.$$

Obviously, I is an ideal in R, and so it is generated by a finite number $a_1, ..., a_k$ of elements of R. Since $a_i \in I$ for all $i = 1, ..., k$, $a_i \in I_{n_i}$ for some n_i. Now, let

$$m = \max\{n_1, ..., n_k\}.$$

Then, $a_i \in I_m$ for all $i = 1, ..., k$. Therefore $I \subset I_m$, and thus

$$I \subset I_m \subset \cdots \subset I.$$

As a result $I_i = I_m$ for all $i \geq m$.

"2 $\Rightarrow$ 3". Let $\mathcal{S}$ be a set of ideals of R, and let I_1 be in $\mathcal{S}$. If I_1 is not maximal, then there is an $I_2 \in \mathcal{S}$ such that $I_1 \subsetneq I_2$. If by repeating the process we do not arrive at a maximal ideal after a finite number of steps, then we will have constructed an infinite ascending sequence of distinct ideals which does not stabilize, a contradiction.

"3 $\Rightarrow$ 1". Let I be an ideal in R, and $\mathcal{S}$ be the set of all ideals properly contained in I and finitely generated. This set is not empty since $(0) \in \mathcal{S}$. By assumption the set $\mathcal{S}$ has a maximal ideal that is finitely generated, say the ideal $(a_1, ..., a_n)$. Now, if $I \neq (a_1, ..., a_n)$, there must exist an $a \in I$ such that $a \notin (a_1, ..., a_n)$. In that case,

$$(a_1, ..., a_n) \subsetneq (a_1, ..., a_n, a) \subseteq I,$$

and so $(a_1, ..., a_n, a) \in \mathcal{S}$. This contradicts the fact that $(a_1, ..., a_n)$ is maximal in $\mathcal{S}$. Therefore $I = (a_1, ..., a_n)$. $\square$

Exercise 9.7.6. Prove that an ideal I in the Noetherian ring R is Noetherian.

Indeed, starting with a Noetherian ring, many elementary constructions again lead to Noetherian rings.

Proposition 9.7.7. *If R is a Noetherian ring and I is any ideal, then R/I is Noetherian.*

Proof. Since any ideal of R/I is of the form J/I for some ideal J of R, an ascending chain of ideals in R/I corresponds to an ascending chain of ideals in R, and therefore it must be stationary.   □

For a ring, R, being Noetherian is preserved by localization.

Proposition 9.7.8. *Let R be a commutative ring with 1 and $S \subset R$ a multiplicatively closed subset. If R is Noetherian (resp. Artinian), so is $S^{-1}R$.*

Proof. We will prove the Noetherian case, the Artinian being quite similar. Let $\phi : R \to S^{-1}R$ be the canonical map, and let $I \subset S^{-1}R$ be an ideal. Then since $\phi^{-1}(I) \subset R$ is an ideal, it is finitely generated. It follows that

$$\phi^{-1}(I)(S^{-1}R) \subset S^{-1}R$$

is finitely generated as an ideal in $S^{-1}R$.
    We claim that

$$\phi^{-1}(I)(S^{-1}R) = I.$$

If $\frac{x}{s} \in I$, then $x \in \phi^{-1}(I)$, and so

$$x = (1/s)x \in (S^{-1}R)\phi^{-1}(I).$$

This proves the claim and implies that I is finitely generated (see Theorem 9.7.5).   □

As we know, ideals in a Noetherian ring are Noetherian, but the same is not true for subrings.

Example 9.7.9. As we shall see shortly, the polynomial ring $k[x, y]$ is Noetherian (by Hilbert's Basis Theorem 9.7.10). However, the subring generated by

$$\{xy^n : n \geq 0\}$$

is evidently not finitely generated over k.

9.7.1 Hilbert's Basis Theorem

The following proof is an adaptation from [112] and [55].

Theorem 9.7.10. (Hilbert's Basis Theorem) *If the commutative ring R with identity is Noetherian, then $R[x]$ is also Noetherian.*

Proof. We will show any ideal I in $R[x]$ is finitely generated as an ideal. Since we only have information about the ideals in R, we must somehow relate I to an ideal in R. Given a polynomial $p \in I$, a natural way to obtain elements of R is to look at its coefficients, and the most significant one to look at is the leading coefficient since it is related to the $\deg p$. We denote by $\lambda(p)$ the leading coefficient of $p \in R[x]$.

We could define

$$J = \{a \in R : \text{ there is a } p \in I \text{ such that } a = \lambda(p)\}.$$

However, this is not in general an ideal of R, since if a and $b \in J$, with $a = \lambda(p)$, $b = \lambda(q)$, $p, q \in I$, then $\lambda(p + q) = a + b \in J$ *only if* $\deg(p) = \deg(q)$. The remedy for this is to define, for each $m \in \mathbb{N}$

$$J_m = \{a \in R : \text{ there is } p \in I \text{ with } \deg(p) = m \text{ such that } a = \lambda(p)\} \cup \{0\}.$$

One sees easily that J_m is indeed an ideal of R and since R is Noetherian by Theorem 9.7.5, J_m is finitely generated, i.e.

$$J_m = (a_{m1}, a_{m2}, ..., a_{mr_m})$$

for some $a_{mj} \in R$, $j = 1, ..., r_m$. In particular, given $m \in \mathbb{N}$, for each $j = 1, ..., r_m$, we get some $p_{mj} \in I$ so that:

1. $\deg p_{mj} = m$, and

2. $\lambda(p_{mj}) = a_{mj}$.

For each m, set

$$S_m = \{p_{mj} \ : \ 1 \leq j \leq r_m\}.$$

We will prove the set

$$S = \bigcup_{m \in \mathbb{N}} S_m$$

generates I and is finite.

Indeed, let $p \in I$, with $\deg(p) = n$ and $\lambda(p) = a$. Then, $a \in J_n$, so there are $s_1, s_2, ..., s_{r_n}$ in I satisfying

$$a = s_1 a_{n1} + s_2 a_{n2} + \cdots + s_{r_n} a_{n r_n}.$$

Thus, letting

$$q = \sum_{k=1}^{r_n} s_k p_{nk} \ \in \ (S_n),$$

it follows that $\lambda(q) = a$. In particular, $p - q \in I$ with $\deg(p - q) < n$, and by induction on n we obtain $t \in (S_0, S_1, ..., S_n)$ such that $p = t$, which proves S is a generating set for I. We shall now show

$$J_m \subset J_{m+1} \quad \text{for any } m \in \mathbb{N}.$$

Let $a \in J_m$. Then there is some $p \in I$ with $\deg(p) = m$ and such that $\lambda(p) = a$. Hence,

1. $xp \in I$, since I is an ideal.

2. $\deg xp = m + 1$.

3. $\lambda(xp) = a$.

Thus, $a \in J_{m+1}$. But since R is Noetherian its ideals satisfy the ACC. This means there is an N such that

$$J_N = J_{N+k} \quad \text{for any } k$$

and so the set, S, of generators of I is finite. $\qquad \square$

It follows by induction that $R[x_1, \ldots x_n]$ is Noetherian, if R is. In particular, $\mathbb{Z}[x_1, \ldots x_n]$ and $k[x_1, \ldots x_n]$, where k is a field, are both Noetherian.

Example 9.7.11. Since $\mathbb{C}[x, y]$ is Noetherian it would be interesting to find ideal generators for an important ideal. Let $V \subset \mathbb{C}^2$ be any subset. We will use the fact that $\mathbb{C}[x, y] = \mathbb{C}[x][y]$ to find the generators of the ideal,

$$I(V) := \left\{ f(x, y) \in \mathbb{C}[x, y] \mid f(p, q) = 0 \text{ for all } (p, q) \in V \right\}.$$

For example, if we take

$$V = \{(0, 0), (0, 1), (1, 0)\},$$

then

$$I_0 = (x^2 - x) \text{ since } p(x) \in I_0 \text{ iff } p(0) = p(1) = 0.$$

$I_1 = (x)$ because $a_0(x) + a(x)y \in I_1$ if and only if $a(0) = a_0(0) = 0$ and $a_0(1) = 0$ and we can choose $xy \in I(V)$.
$I_2 = (1)$ and we can choose $y^2 - y \in I(V)$ and we stop here because I_2 is as large as it can get. So,

$$I(V) = (x^2 - x, xy, y^2 - y).$$

The Hilbert Basis Theorem can be extended somewhat further. The following is due to Matsumura [82].

Theorem 9.7.12. *If R is Noetherian, then so is $R[[x]]$.*

Proof. Set $B = R[[x]]$, and let I be an ideal of B. We will prove that I is finitely generated. Write $I(r)$ for the ideal of R formed from the leading coefficients a_r of

$$f = a_r x^r + a_{r+1} x^{r+1} + \cdots$$

as f runs through $I \cap x^r B$. Then

$$I(0) \subset I(l) \subset I(2) \subset \cdots$$

Since R is Noetherian, there is an s such that $I(s) = I(s+1) = \cdots$. Moreover, each $I(k)$ is finitely generated. For each k with $0 \leq k \leq s$ we take finitely many elements $a_{k\nu} \in R$ generating $I(k)$, and choose $g_{k\nu} \in I \cap x^k B$ with $a_{k\nu}$ as coefficient of x^k. These $g_{k\nu}$ generate I. Indeed, for $f \in I$ we can take a linear combination g_0 of the $g_{0\nu}$ with coefficients in R such that $f - g_0 \in I \cap xB$, then take a linear combination g_1 of the $g_{1\nu}$ with coefficients in R so that $f - g_0 - g_1 \in x^2 B$, and proceeding in the same way we get

$$f - g_0 - g_1 - \cdots - g_s \in I \cap x^{s+1} B.$$

Now, at some point $I(s+1) = I(s)$, so we can take a linear combination g_{s+1} of the $xg_{s\nu}$ with coefficients in R so that

$$f - g_0 - g_1 - \cdots - g_{s+1} \in I, cap x^{s+2} B.$$

We now proceed as before to get $g_{s+2}, \ldots$ For $k \leq s$, each g_k is a linear combination of the $g_{k\nu}$ with coefficients in R, and, for $k > s$, g_k is a combination of the elements $x^{k-s} g_{s\nu}$. For each $k \geq s$ write

$$g_k = \sum_{\nu} a_{k\nu} x^{k-s} g_{s\nu},$$

and then for each ν we set

$$h_\nu = \sum_{k=s}^{\infty} a_{k\nu} x^{k-s}.$$

Now h_ν is an element of B, and

$$f = g_0 + \cdots + g_{s-1} + \sum_{\nu} h_\nu g_{s\nu}.$$

$\square$

9.8 Artinian Rings

Definition 9.8.1. A ring R is called *Artinian*[5] if it satisfies the descending chain condition (see 9.1.8).

---

[5]Emil Artin (1898-1962) was an Austro-German mathematician who made fundamental contributions to analytical and arithmetic theory of quadratic number fields

Definition 9.8.2. We say that the ring R satisfies the *minimal condition* if every not empty set $\mathcal{S}$ of its ideals, partially ordered by set inclusion, has a minimal element. That is, there is an $I \in \mathcal{S}$ such that if $J \in \mathcal{S}$ and $J \subset I$, then $I = J$.

As mentioned earlier it is important to understand that $R[x]$ is not Artinian since $(x) \supset (x^2) \supset \cdots$ is an unending chain.

Proposition 9.8.3. *If an Artinian ring R is an integral domain, then it is a field.*

Proof. Let $0 \neq x \in R$ and consider the decreasing chain of ideals

$$(x) \supset (x^2) \supset \cdots .$$

Since R is Artinian, this chain stabilizes. So, there is an $n > 0$ such that $(x^n) = (x^{n+1})$. This implies $x^n \in (x^{n+1})$, so there is a $y \in R$ such that $x^n = x^{n+1}y$, and since R is an integral domain, after cancellation $xy = 1$. Thus x is invertible and R is a field. □

Proposition 9.8.4. *An Artinian ring R has only a finite number of maximal ideals.*

Proof. If there were infinitely many maximal ideals we would get a descending chain of ideals,

$$\mathfrak{m}_1 \supsetneq \mathfrak{m}_1 \cap \mathfrak{m}_2 \supsetneq \mathfrak{m}_1 \cap \mathfrak{m}_2 \cap \mathfrak{m}_3 \supsetneq \cdots$$

These are proper inclusions since

$$\mathfrak{m}_1 \cap \cdots \cap \mathfrak{m}_n = \mathfrak{m}_1 \cap \cdots \cap \mathfrak{m}_{n+1}$$

---

and abstract algebra in 1926 and the following year used the theory of formal-real fields to solve the Hilbert problem of positive polynomials. In 1927 he also made notable contributions in theory of associative algebras. In 1944 he discovered rings with minimum conditions for right ideals, now known as Artinian rings. He presented a new foundation for and extended the arithmetic of semi-simple algebras over the rational number field, class field theory and a general law of reciprocity. After one year at the University of Göttingen, Artin joined the staff of the University of Hamburg in 1923. Because his wife was Jewish, he emigrated to the US in 1937 where he taught at Notre Dame University, Indiana University and Princeton University (1946-1958). In 1958 he returned to the University of Hamburg.

$$\Rightarrow \quad \mathfrak{m}_1 \cap \cdots \cap \mathfrak{m}_n \subseteq \mathfrak{m}_{n+1}$$

$$\Rightarrow \quad \mathfrak{m}_i \subseteq \mathfrak{m}_{n+1}.$$

Since these maximal ideals are prime, this contradicts the Prime Avoidance Lemma 9.4.7).   □

Just as with the Noetherian rings, the property of a ring being Artinian is preserved under localization.

Proposition 9.8.5. *Let R be a commutative ring, $S \subset R$ be a multiplicatively closed subset. If R is Artinian, then $S^{-1}R$ is Artinian.*

The proof is similar to the Noetherian case (see Proposition 9.7.8) and is left to the reader.

9.8.1   The Hopkins-Levitzki Theorem

We prove below that Artinian rings are always Noetherian. To do so we will need Nakayama's Lemma which is dealt with in the next section. However, as we shall see in *An Artinian Module but not Noetherian* this is not true for modules.

Theorem 9.8.6. *(The Hopkins-Levitzki Theorem)*[6]. *An Artinian ring R is Noetherian.*

Proof. As per the title of this chapter (as well as because of the significantly increased difficulty in the non-commutative case) our proof is done only for the commutative case. If R is an Artinian ring it is enough to prove it has finite length as an R-module as R acts on itself by left multiplication. First we will show that R has only finitely many

---

[6]This theorem was proved for commutative rings by Akizuki ([5]) in 1935. Four years later it was generalized independently by Hopkins ([62]) and Levitzki ([76]) to non-commutative rings in 1939.

maximal ideals. Indeed, if $\mathfrak{p}_1$, $\mathfrak{p}_2$, ... is an infinite set of distinct maximal ideals, then it is easy to see that

$$\mathfrak{p}_1 \supset \mathfrak{p}_1\mathfrak{p}_2 \supset \mathfrak{p}_1\mathfrak{p}_2\mathfrak{p}_3 \supset \cdots$$

is an infinite descending chain of ideals, which contradicts our assumption. Thus, let $\mathfrak{p}_1$, $\mathfrak{p}_2$, ..., $\mathfrak{p}_r$ be all the maximal ideals of R and set

$$I = \mathfrak{p}_1\mathfrak{p}_2 \ldots \mathfrak{p}_r = \text{rad}(R).$$

By the descending chain condition $I \supset I^2 \supset \cdots$ stops after finitely many steps, and so there is an s with $I^s = I^{s+1}$. If we set $(0 : I^s) = J$, then

$$(J : I) = ((0 : I^s) : I) = (O : I^{s+1}) = J.$$

We will show that $J = R$. Suppose that $J \neq R$. Then, there exists an ideal J_1 which is minimal among all ideals strictly larger than J. For any $x \in J_1 - J$, $J_1 = Rx + J$. Now $I = \text{rad}(R)$ and $J \neq J_1$, so by Nakayama's Lemma see 9.11.1 $J_1 \neq Ix + J$, and hence by the minimality of J_1, $Ix + J = J$, and hence $Ix \subset J$. Thus $x \in (J : I) = J$, which is a contradiction. Therefore $J = R$, and as a result $I^s = 0$. Now consider the chain of ideals

$$R \supset \mathfrak{p}_1 \supset \mathfrak{p}_1\mathfrak{p}_2 \supset \cdots \supset \mathfrak{p}_1 \ldots \mathfrak{p}_{r-1} \supset I \supset I\mathfrak{p}_1 \supset I\mathfrak{p}_1\mathfrak{p}_2 \ldots$$
$$\ldots \supset I^2 \supset I^2\mathfrak{p}_1 \supset \cdots \supset I^s = 0.$$

Let M and $M\mathfrak{p}_i$ be any two consecutive terms in this chain. Then, $M/M\mathfrak{p}_i$ is a vector space over the field $R/\mathfrak{p}_i$, and since it is Artinian, it must be finite dimensional. Hence, the length of $M/M\mathfrak{p}_i$ is $< \infty$, and therefore the sum $l(R)$ of these terms is also finite. $\quad\square$

9.9 Noetherian and Artinian Modules

Here R is any commutative ring with identity.

Definition 9.9.1. Let M be an R-module with R always operating from the left.

1. M is called *Noetherian* if every ascending chain

$$M_0 \subseteq M_1 \subseteq M_2 \subseteq \cdots$$

   of submodules of M stabilizes, i.e. if for every such chain there is an index $n \in \mathbb{N}$ such that $M_k = M_n$ for all $k \geq n$.

2. Similarly, M is called *Artinian* if every *descending* chain

$$M_0 \supseteq M_1 \supseteq M_2 \supseteq \cdots$$

   of submodules stabilizes.

Example 9.9.2. Let V be a finite dimensional vector space over a field k, viewed as a k-module, V is both Noetherian and Artinian. Similarly, a finitely generated $\mathbb{Z}$-module is a Noetherian module, as is a finitely generated $k[x]$-module.

Example 9.9.3. If $p \in \mathbb{Z}$ is a prime, the $\mathbb{Z}$-module $\mathbb{Z}(p^\infty)$ is not Noetherian since it has an infinite ascending chain of subgroups. For the same reason it is Artinian.

Exercise 9.9.4. Given a prime number p, the $\mathbb{Z}$-module

$$\mathbb{Z}(p^\infty) := \left\{ \frac{\overline{a}}{b} \in \mathbb{Q}/\mathbb{Z} \ \middle| \ a,\ b \in \mathbb{Z} \ \text{ and } b = p^i \text{ for some } i \geq 0 \right\}$$

is Artinian but not Noetherian.

Proposition 9.9.5. *If*

$$0 \longrightarrow K \overset{i}{\longrightarrow} M \overset{\pi}{\longrightarrow} Q \longrightarrow 0$$

is a short exact sequence of modules, then M is Noetherian (resp. Artinian) if and only if both K and Q are Noetherian (resp. Artinian).

Proof. We will consider the Noetherian case, the Artinian being similar. ("$\Longrightarrow$"). If M is Noetherian, then K, as a submodule of M, is Noetherian. Similarly, Q is Noetherian as the homomorphic image of the submodule $\pi^{-1}(Q)$ of M.

(" $\Longleftarrow$ "). To prove the converse, suppose both K and Q are Noetherian. To show M is Noetherian let

$$M_1 \subset M_2 \subset \ldots$$

be an ascending chain of submodules of M. Set

$$K_i = i^{-1}\big(i(K) \cap M_i\big), \text{ and } Q_i = \pi(M_i).$$

Then,

$$K_1 \subseteq K_2 \subseteq \cdots$$

is an ascending chain of submodules of K and

$$Q_1 \subseteq Q_2 \subseteq \cdots$$

an ascending chain of submodules of Q. Since both are Noetherian, there must exist some positive integer n such that

$$K_n = K_{n+1} = \ldots \text{ and } Q_n = Q_{n+1} = \ldots$$

Now, for any $m \geq n$ we obtain the following commutative diagram,

$$
\begin{array}{ccccccccc}
0 & \longrightarrow & K_n & \xrightarrow{\ i_n\ } & M_n & \xrightarrow{\ \pi_n\ } & Q_n & \longrightarrow & 0 \\
 & & \alpha \downarrow & & \beta \downarrow & & \gamma \downarrow & & \\
0 & \longrightarrow & K_m & \xrightarrow[\ i_m\ ]{} & M_m & \xrightarrow[\ \pi_m\ ]{} & Q_m & \longrightarrow & 0
\end{array}
$$

where α, β, and γ are the inclusion maps, and α, γ are isomorphisms. Then, the Five lemma (see Volume I, Lemma 6.25. of Volume I) implies that γ is surjective, and so $M_m = M_n$ for all $m \geq n$. $\square$

Corollary 9.9.6. *Suppose $N \subseteq M$ is a submodule,*

1. *M is Noetherian if and only if both N and M/N are Noetherian.*

2. *M is Artinian if and only if both N and M/N are Artinian.*

Corollary 9.9.7. *Let R_1, and R_2 be two Artinian rings. Then $R_1 \times R_2$ is also Artinian.*

Proof. We leave to the reader the proof that both R_1, and R_2 can be seen as $R_1 \times R_2$-modules under the actions $(a, b) \cdot c = ac$ and $(a, b) \cdot c = bc$, respectively. It is clear from the definition of the action that each ring R_i, $i = 1, 2$ has the same submodules as an $R_1 \times R_2$-module that it does as an R_i-module. Therefore, both R_1 and R_2 are Artinian $R_1 \times R_2$-modules. Now, the obvious short exact sequence

$$0 \longrightarrow R_1 \overset{i}{\longrightarrow} R_1 \times R_2 \overset{\pi}{\longrightarrow} R_2 \longrightarrow 0$$

we find that $R_1 \times R_2$ is an Artinian $R_1 \times R_2$-module and hence an Artinian ring. $\qquad\square$

Corollary 9.9.8. *Let M be an R-module. If M is both Artinian and Noetherian, then it has a composition series.*

Proof. Given any non-zero R-module Q, consider the set of all its proper submodules. By Zorn's Lemma there is a maximal such submodule Q_1 hence Q/Q_1 is a simple module. Iterating we get $M = M_0 \supsetneq M_1 \supsetneq \cdots$ which terminates since M is Artinian. $\qquad\square$

Proposition 9.9.9. *Let M be an R-module. If R is Artinian and M is finitely generated, then M has finite length.*

Proof. First we note that if M has n generators it can be placed in a short exact sequence

$$0 \longrightarrow \mathrm{Ker}(\phi) \longrightarrow R^n \overset{\phi}{\longrightarrow} M \longrightarrow 0,$$

where ϕ sends each basis element of R^n to the appropriate generator of M. In addition, a finite direct sum of Artinian rings is Artinian and for the short exact sequence of R-modules

$$0 \longrightarrow N \longrightarrow M \overset{\pi}{\longrightarrow} M/N \longrightarrow 0$$

M is Artinian if N and M/N are. $\qquad\square$

Our next result is an extension of the fact that a commutative ring with identity is Noetherian if and only if each ideal in it is finitely generated (see Theorem 9.7.5).

Proposition 9.9.10. *Let M be an R-module, where R is a commutative ring with identity. Then M is Noetherian if and only if every submodule of M is finitely generated. In particular, M is Noetherian if and only if it is finitely generated. The analogous statement fails for Artinian modules.*

Proof. "(1) $\implies$ (2)". Take a chain $M_0 \subseteq M_1 \subseteq \cdots$ of submodules of M. $\bigcup M_i$ is finitely generated so there is some j such that all the generators (finitely many) are in M_j. Hence $M_j \subseteq M_{j+1} \subseteq \cdots \subseteq \bigcup M_i \subseteq M_j$ thus $M_j = M_{j+1} = \cdots$
"(2) $\implies$ (1)". Suppose N is a submodule which is not finitely generated. Consider the set of finitely generated submodules of N. This is nonempty since it contains 0. By Zorn's Lemma it contains a maximal element call it N', $N' \subsetneqq N$ since N is assumed not finitely generated. Let $a \in N - N'$. Then $N' + Ra$ is a submodule of N and is finitely generated and properly contains N', contradicting the maximality of N'. $\square$

The following result shows modules over a Noetherian ring, although more general, still somewhat resemble $\mathbb{Z}$-modules (Abelian groups).

Theorem 9.9.11. *Let M be a finitely generated module over a Noetherian ring R. Then any submodule N of M is finitely generated.*

Proof. Let $\{x_1, \ldots x_n\}$ generate M. We will prove this by induction on n. If $n = 1$, then M being cyclic is R/I where I is an ideal in R, and N being a submodule of M is an ideal in R/I. Since R/I is Noetherian by 9.7.7 and N is an ideal in it, by 9.7.5 N is finitely generated. If $n > 1$, look at M_1, the submodule of M generated by $\{x_1, \ldots x_{n-1}\}$ and let $N_1 = N \cap M_1$. Then N_1 is a submodule of M_1 so by inductive hypothesis N_1 is finitely generated by say $\{y_1, \ldots y_k\}$. However, N_1 is also a submodule of N and so N/N_1 is a submodule of the cyclic module M/M_1. The case $n = 1$ then tells us N/N_1 is finitely generated say by $\{z_1 + N_1, \ldots z_l + N_1\}$. Hence $\{y_1, \ldots y_k, z_1 \ldots z_l\}$ generates N. $\square$

9.9.1   An Artinian but not Noetherian Module

As we have seen by the Hopkins-Levitzki theorem, an Artinian ring is Noetherian. Here we will show this is not true for modules.

Let p be a prime and consider the set

$$G = \left\{ \frac{a}{p^n} \in \mathbb{Q} \;\middle|\; a \in \mathbb{Z}, \; n > 0 \right\}.$$

Then G is a $\mathbb{Z}$-module under the standard operations of addition and multiplication. For a fixed n consider

$$G_n = \left\{ \frac{a}{p^n} \in \mathbb{Q} \;\middle|\; a \in \mathbb{Z} \right\}.$$

Then $G_n \subset G$ and one sees easily that G_n is a submodule and

$$\mathbb{Z} = G_0 \subsetneq G_1 \subsetneq G_2 \subsetneq G_3 \subsetneq \cdots$$

We will prove $G/\mathbb{Z}$ is Artinian but not Noetherian.

Consider the canonical projection $\pi : G \longrightarrow G/\mathbb{Z}$. Then $\overline{G}_n = \pi(G_n)$ is a submodule of $G/\mathbb{Z}$ and

$$0 = \overline{G}_0 \subsetneq \overline{G}_1 \subsetneq \overline{G}_2 \subsetneq \overline{G}_3 \subsetneq \cdots$$

By the third isomorphism theorem for modules, these inclusions are proper, since for any n,

$$\overline{G}_{n+1}/\overline{G}_n \cong (G_{n+1}/\mathbb{Z})/(G_n/\mathbb{Z}) \cong G_{n+1}/G_n \neq 0.$$

Hence $G/\mathbb{Z}$ is not Noetherian. To see that $G/\mathbb{Z}$ is Artinian we will first show that any of its proper submodules is of the form $\overline{G}_n$. Let N be a proper submodule of $G/\mathbb{Z}$, and suppose that for some $a \in \mathbb{Z}$, with $\gcd(a, p^n) = 1$, $\frac{a}{p^n} + \mathbb{Z} \in N$. Then, there exists an α, $\beta \in \mathbb{Z}$ such that $1 = \alpha a + \beta p^n$, and since N is a $\mathbb{Z}$-module,

$$\frac{\alpha a}{p^n} = \mathbb{Z} \in N.$$

Therefore,

$$\frac{1}{p^n} + \mathbb{Z} = \frac{\alpha a + \beta p^n}{p^n} + \mathbb{Z} \in N.$$

Let m be the smallest integer such that $\frac{1}{p^m} + \mathbb{Z} \notin N$. We previously showed

$$N = \overline{G}_{m-1} = \pi(G_{m-1})$$

because for each n with $0 \leq n \leq m - 1$ we get $\frac{1}{p^n} + \mathbb{Z}$. It follows that N is the image of a submodule of G generated by $\frac{1}{p^n}$, which is precisely G_{m-1}.

Now, consider the chain

$$N_1 \supset N_2 \supset N_3 \supset \cdots$$

of these submodules of $G/\mathbb{Z}$ and let $n_1, n_2, \ldots$ be chosen so that $N_i = \overline{G}_{n_i}$. Then $\overline{G}_k \supset \overline{G}_l$ if and only if $k \geq l$. Since the sequence $n_1 \geq n_2 \geq \cdots$ eventually terminates, the decreasing chain also terminates, showing that G is Artinian.

9.10   Krull Dimension

Definition 9.10.1. Let R be a commutative ring with identity. The *spectrum* of R, $\mathrm{Spec}(R)$, is the set of prime ideals of R.

Definition 9.10.2. A *chain* in $\mathrm{Spec}(R)$ is an ascending chain

$$P_0 \subset \cdots \subset P_\lambda$$

of length λ. A prime ideal P has *height* λ if there is a chain of length λ in $\mathrm{Spec}(R)$ with P as the largest element, but no such chain of length $\lambda + 1$.

Since by the 2nd Isomorphism Theorem, there is a bijection of sets

$$\{\text{Ideals of } R/I\} \leftrightarrow \{\text{Ideals of } R \text{ containing } I\}$$

with $(R/I)/(J/I) \cong R/J$, where J is an ideal of R containing I. Hence J/I is maximal if and only if J is maximal. Also, J/I is prime if and only if J is prime. Thus we have a bijection of sets at the level, both of maximal and of prime ideals. In particular, $\mathrm{Spec}(R/I)$ is naturally contained in $\mathrm{Spec}(R)$.

Definition 9.10.3. The *Krull dimension* of a commutative ring R, $K\dim(R)$, if it exists, is the maximal height of any prime ideal of R.

By convention, the Krull dimension of the trivial ring is set to be -1. $K\dim(R)$ need not be finite. For example the ring $R = k[x_1, \ldots, x_n]$ has chains of arbitrary length. Thus even when R is a Noetherian ring, an example due to Nagata shows that $K\dim(R)$ need not be finite. However, it is *finite* if R is a *local ring*.

Example 9.10.4.

1. $K\dim(\mathbb{Z}) = 1$, since $(0) \subset (p)$, and more generally the Krull dimension of a PID is 1.

2. $K\dim(\mathbb{Z}[x]) = 2$, since the maximal chains of prime ideals are of the form
$$(0) \subset (p) \subset (p) + (f)$$
or
$$(0) \subset (f) \subset (p) + (f)$$
for some prime number p and some non-constant polynomial f.

3. $\dim(k[x_1, \ldots, x_n]) = n$ since
$$(0) \subset (x_1) \subset (x_1, x_2) \subset \cdots \subset (x_1, \ldots, x_n).$$

Although the following theorem concerns Artinian rings, its main corollaries tell us important things about Krull dimension.

Proposition 9.10.5. *In an Artinian ring A, every prime ideal is maximal. (And of course in general, every maximal ideal is prime.)*

Proof. Let $\mathfrak{p}$ be a prime ideal in the Artinian ring A, so $A/\mathfrak{p}$ is also an Artinian ring (because any infinite descending chain of ideals in $A/\mathfrak{p}$ can be lifted to an infinite descending chain of ideals in A). Pick $0 \neq x \notin A/\mathfrak{p}$. Since the sequence of ideals $(x^n)_{n \in \mathbb{N}}$ is a decreasing sequence, we conclude $(x^m) = (x^{m+1})$ for some m. Then, $x^m = rx^{m+1}$, i.e. $x^m(1 - rx) = 0$, and since $x^m \neq 0$, $1 - rx = 0$, and so x is invertible. Therefore, any non-zero element of $A/\mathfrak{p}$ is invertible, i.e. $A/\mathfrak{p}$ is a field so $\mathfrak{p}$ is maximal. $\square$

Corollary 9.10.6. *An Artinian ring has Krull dimension zero.*

We can now characterize Artinian rings in terms of Krull dimension. In order to prove Theorem 9.10.9 below we first require the following proposition.

Proposition 9.10.7. *If V is a vector space over a field k, then the following statements are equivalent:*

1. *V is a finite dimensional vector space over k.*

2. *V is a Noetherian k-module.*

3. *V is an Artinian k-module.*

Proof. "1 $\Rightarrow$ 2". If V is a finite dimensional vector space over k, then every k-submodule of V is a subspace of V which is finite dimensional and, hence finitely generated. Therefore, V is a Noetherian k-module.

"2 $\Rightarrow$ 3". If V is a Noetherian k-module, then V is finitely generated, so it has a finite basis. Given any nonempty collection, $\mathcal{M}$, of k-submodules (i.e. vector subspaces) of V, we can choose a subspace of least dimension, which serves as a minimal element of $\mathcal{M}$. Thus, V is Artinian.

"3 $\Rightarrow$ 1". Suppose V is Artinian, but is infinite dimensional as a vector space over k. Then, we can find a countable subset $\{e_n \mid n \in \mathbb{N}\}$ of linearly independent vectors in V such that

$$\sum_{i \geq 1} ke_i \supseteq \sum_{i \geq 2} ke_i \supseteq \sum_{i \geq 3} ke_i \supseteq \cdots$$

is an infinite strictly decreasing chain of k-submodules of V, a contradiction. □

Corollary 9.10.8. *Let R be a ring and $\mathfrak{m}_1, ..., \mathfrak{m}_n$ be maximal ideals of R (not necessarily distinct). Suppose M is an R-module such that*

$$\mathfrak{m}_1 \cdot \mathfrak{m}_2 \cdots \mathfrak{m}_n M = (0).$$

Then M is Noetherian if and only if M is Artinian.

Proof. "1 $\Rightarrow$ 2". Suppose M is Noetherian. Consider the chain

$$0 \subseteq \mathfrak{m}_1 \cdots \mathfrak{m}_n M \subseteq \mathfrak{m}_1 \cdots \mathfrak{m}_{n-1} M \subseteq \cdots \subseteq \mathfrak{m}_1 M \subseteq M,$$

and set $M_i = \mathfrak{m}_1 \cdots \mathfrak{m}_n M$. Since M is Noetherian, M_i/M_{i+1} is also Noetherian for all i. But M_i/M_{i+1} is an $R/\mathfrak{m}_{i+1}$ module, that is, a vector space over the field $R/\mathfrak{m}_{i+1}$. Thus, M is Artinian by Lemma 9.10.7. The proof of the converse is similar and is left to the reader. $\square$

A Noetherian ring may not be Artinian as for example $\mathbb{Z}$ which is Noetherian, but not Artinian. Theorem 9.10.9 tells us that an Artinian ring is always Noetherian, so the descending chain condition (DCC) is stronger than the ascending chain condition (ACC).

As we shall see, a ring of Krull dimension zero is actually very close to being a field. A field has Krull dimension 0, because its zero ideal is the only prime ideal. More generally, the dimension of a ring R is 0 if and only if there are no strict inclusions among prime ideals of R, i.e. if and only if all prime ideals are already maximal. This gives us the possibility of sharpening Hopkins Theorem from an inclusion to an exact statement.

We are now ready to prove Theorem 9.10.9 concerning the main characterization of Artinian rings.

Theorem 9.10.9. *A ring R is Artinian if and only if it is Noetherian and of Krull dimension zero.*

Proof. Since a Noetherian ring of Krull dimension zero is automatically Artinian we only need to prove the converse. Let R be an Artinian ring. Then, by Proposition 9.10.5, every prime ideal of R is maximal. Hence the Krull dimension $K\dim(R) = 0$. By Proposition 9.8.4, R has only finitely many maximal ideals, $\mathfrak{m}_1, ..., \mathfrak{m}_n$. Proposition 9.13.12 below then tells us that there is a positive integer l such that

$$\mathcal{N}(A)^l = \mathrm{Jac}(A)^l = \left(\bigcap_{i=1}^{n} \mathfrak{m}_i\right)^l = (\mathfrak{m}_1 \cdots \mathfrak{m}_n)^l = 0.$$

By Corollary 9.10.8 and the above equation it follows that R is Noetherian. $\square$

Here is an example of a non Noetherian ring R of Krull dimension $K \dim(R) = 0$, which of course is not Artinian. Let

$$R = k[x_1, x_2, \ldots,]/(x_1, x_2, \ldots,)^2 \cong k[\nu_1, \nu_2, \ldots],$$

where all ν_i's are nilpotent elements. Here there is only one prime ideal, the zero ideal, but, as one can see, this ring is not Noetherian since

$$(0) \subset (\nu_1) \subset (\nu_1, \nu_2) \subset \ldots$$

is an infinite ascending chain.

Exercise 9.10.10. Let A_1, A_2 be Artinian rings. Then $A_1 \times A_2$ is also Artinian.

Exercise 9.10.11. Any Artinian ring decomposes uniquely (up to isomorphism) as a direct product of finitely many local Artinian rings. Suggestion: use the Chinese Remainder Theorem.

We remark when R is an Artinian ring, $\operatorname{Spec}(R)$ is just a finite set of points, and as we shall see in the next to last section of this chapter. In this case, the Zariski topology is discrete.

9.11 Nakayama's Lemma

Lemma 9.11.1. *(Nakayama's[7] Lemma)[8] Let R be a local Noetherian ring, $\mathfrak{r}$ be a proper ideal in R, M be a finitely generated R-module, and define*

$$\mathfrak{r}M = \Big\{ \sum r_i m_i \mid r_i \in \mathfrak{r}, \ m_i \in M \Big\}.$$

---

[7]Tadashi Nakayama (1912-1964) Japanese mathematician who made major contributions to non-commutative ring theory and algebras and their representations, was a professor at Nagoya University and editor of the Nagoya Mathematics Journal. Nakayama contracted tuberculosis as a young man and was in ill health for most of the second half of his rather short life. During the difficult times of World War II he continued his pioneering works in mathematics. Indeed, his intense activity continued until his untimely death in 1964.

[8]This theorem is usually referred to as Nakayama's Lemma, but the late Professor Nakayama maintained that it should be referred to as the theorem of Krull and Azumaya; it is in fact difficult to determine which of the three did this first. Many authors refer to it as NAK.

Then,

1. *If* $\mathfrak{r}M = M$, *then* $M = 0$.

2. *If* N *is a submodule of* M *such that* $N + \mathfrak{r}M = M$, *then* $N = M$.

Proof. (1). Suppose $M \neq 0$. Among the finite sets of generators for M, choose one, say $\{m_1, ..., m_k\}$ having the fewest elements. By hypothesis, we can write

$$m_k = r_1 m_1 + r_2 m_2 + \cdots + r_k m_k, \quad \text{with some } r_i \in \mathfrak{r}.$$

Then,

$$(1 - r_k)m_k = r_1 m_1 + r_2 m_2 + \cdots + r_{k-1} m_{k-1}.$$

As $1 - r_k$ is not in $\mathfrak{r}$, it is a unit, and so $\{m-1, ..., m_{k-1}\}$ generates M. This contradicts our choice of $\{m_1, ..., m_k\}$, and so $M = 0$.

(2). We shall now show $\mathfrak{r}(M/N) = M/N$, and then apply the first part of the lemma to deduce that $M/N = 0$. Consider $m + N$, $m \in M$. By assumption,

$$m = n + \sum r_i m_i,$$

with $r_i \in \mathfrak{r}$, and $m_i \in M$. Therefore,

$$m + N = \sum r_i m_i + N = \sum r_i (m_i + N),$$

where the last equality holds by the definition of the action of R on M/N, and so $m + N \in \mathfrak{r}(M/N)$. $\qquad\square$

The hypothesis in the lemma that M is finitely generated is crucial. For example, when R is a local integral domain with maximal ideal $\mathfrak{m} \neq 0$, then $\mathfrak{m}M = M$ for any field M containing R, but $M \neq 0$.

9.11.1   Applications of Nakayama's Lemma

The following surprising result is due to Vasconcelos ([123]). Its proof is crucially based on Nakayama's Lemma.

Proposition 9.11.2. *Let M be a finitely generated R-module and let $f : M \longrightarrow M$ be a surjective module homomorphism. Then f is an isomorphism.*

Proof. M can be viewed as an $R[x]$-module if for $p(x) \in R[x]$ and $m \in M$ we define the product by setting

$$p(x)m = p(f)(m).$$

In particular $xm = f(m)$ and since f is surjective $xM = M$, that is $(x)M = M$ for the principal ideal $I = (x) = xR[x]$. Clearly, the $R[x]$-module M is finitely generated. By Nakayama's Lemma there is a $p \in R[x]$ such that $pM = 0$ and $p \equiv 1 \pmod{x}$. Hence $p = 1 + xq$ for some polynomial $q \in R[x]$ and $(1 + xq) \cdot M = 0$. If $a \in \operatorname{Ker} f$, then

$$0 = (1 + xq)a = a + q \cdot f(a) = a.$$

Hence, $\operatorname{Ker} f = (0)$ and f is injective.   □

Using Proposition 9.11.2 in the particular case of $\mathbb{Z}$-modules, that is, Abelian groups, we get several corollaries:

Corollary 9.11.3. *Let G be a finitely generated Abelian group and $f : G \longrightarrow G$ be an endomorphism. If f is surjective, then f is an isomorphism.*

Similarly, a striking result in Noetherian rings is that the surjectivity of a ring homomorphism also implies injectivity.

Proposition 9.11.4. *Let R be a Noetherian ring and $f : R \to R$ be a ring homomorphism. If f is surjective, then it is an isomorphism.*

Proof. Set $f^1 = f$, $f^2 = f \circ f$,..., $f^n = f^{n-1} \circ f$. Since f is surjective all these iterates of f are surjective maps. In addition, their kernels $\operatorname{Ker}(f^k)$ form an ascending chain of ideals in R:

$$\operatorname{Ker}(f) \subseteq \operatorname{Ker}(f^2) \subseteq \cdots \subseteq \operatorname{Ker}(f^n) \subseteq \cdots .$$

Since R is a Noetherian ring this chain must stabilize, i.e. there is an $n \in \mathbb{N}$ such that

$$\mathrm{Ker}(f^n) = |Ker(f^{n+1}).$$

Now, let $x \in \mathrm{Ker}(f)$. Since f^n is surjective, there is a $y \in R$ such that $f^n(y) = x$. So, we get

$$0 = f(x) = f\left(f^n(y)\right) = f^{n+1}(y),$$

which shows that $y \in \mathrm{Ker}(f^{n+1}) = \mathrm{Ker}(f^n)$, and thus $x = f^n(y) = 0$. In other words, f is an injective map.   □

Remark 9.11.5. Even if the ring R is viewed as a free R-module of rank 1, the proof of the above theorem does not follow from Proposition 9.11.2. This is because a ring endomorphism of R is not in general an endomorphism of the R-module R (and vice versa). Indeed, for the ring endomorphism f we have

$$f(rs) = f(r)f(s), \quad r, \, s \in R$$

and for the R-module endomorphism we have

$$f(rs) = rf(s), \quad r, \, s \in R.$$

Thus, in general, the two maps do not act in the same way on the products of elements in R.

The following corollary generalizes the situation for vector spaces.

Corollary 9.11.6. (Isomorphism Theorem for Finitely Generated Free Modules.) *Suppose R is a commutative ring with 1. $R^n \cong R^m$ as R-modules if and only if $n = m$.*

Proof. Suppose $n \geq m$ and $R^n \cong R^m$. Fix a basis $e_1, ..., e_n$ of R^n, and identify R^m with the submodule of R^n generated by $e_1, ..., e_m$. Therefore, $R^m \subset R^n$. Let

$$\pi : R^n \longrightarrow R^m$$

the projection map. Since, by assumption $R^n \cong R^m$, there must be some isomorphism

$$\sigma : R^m \longrightarrow R^n.$$

But then

$$\pi \circ \sigma : R^m \longrightarrow R^m,$$

is surjective, and therefore by 9.11.2 is also an isomorphism. But σ is surjective, therefore the only way $\pi \circ \sigma$ can be an isomorphism is if π is injective. This means $n = m$. $\qquad\qquad\qquad\qquad\qquad\qquad\square$

Here is another application of Nakayama's Lemma.

Proposition 9.11.7. *Let R be a Noetherian local ring with maximal ideal $\mathfrak{m} \subset R$ and $\mathfrak{m}^{n+1} = \mathfrak{m}^n$. Then $\mathfrak{m}^n = (0)$. If R is a Noetherian integral domain and $P \subset R$ is a prime ideal, then the powers $(P^n)_{n \geq 1}$ are distinct.*

Proof. The first statement is an immediate consequence of Nakayama Variant 1 with $M = \mathfrak{m}^n$. For the second, since R is a domain all of whose localization morphisms are injective. If $P^{n+1} = P^n$, localize at P and conclude from the first statement that $P^n = (0)$. But this cannot be since R is a domain. $\qquad\qquad\qquad\qquad\qquad\qquad\square$

9.11.2   Three Variants of Nakayama's Lemma

The next proposition characterizes the Jacobson radical in a commutative ring.

Proposition 9.11.8. *If R is a commutative ring, with invertible elements $R^\star$, then*

$$J(R) = \{x \in R \mid 1 + rx \in R^\star \text{ for all } r \in R\}.$$

Proof. If x is in every maximal ideal of R, then $1 + rx$ is in no maximal ideal. Hence $1 + rx$ is invertible. Conversely, if x is not in some maximal ideal I, then $1 = y - rx$ for some $y \in I$, $r \in R$ and then $1 + rx = y \in I$ is not invertible. $\qquad\qquad\qquad\qquad\qquad\qquad\square$

Variant 1. If $I \subset R$ is an ideal contained in all maximal ideals of R and if M is a finitely generated R-module with $M = IM$, then $M = (0)$.

Variant 2. Let R be a local ring with maximal ideal $\mathfrak{m} \subset R$ and suppose $N \subset M$ are R-modules with M finitely generated and $N + \mathfrak{m}M = M$. Then $N = M$.

We remark that since only finitely many elements of N appear in formulas for the generators of M, we can assume that N is already finitely generated by replacing it by what these finitely many elements generate.

Variant 3. Let R be a local ring with maximal ideal $\mathfrak{m} \subset R$ and $\{x_1, \ldots, x_n\}$ be finitely many elements of a finitely generated R-module M such that the residue classes $\{\bar{x}_1, \ldots, \bar{x}_n\}$ span the vector space $M/\mathfrak{m}M$ over $R/\mathfrak{m}$. Then $\{x_1, \ldots, x_n\}$ generate M as an R-module.

Proposition 9.11.9. *All three variants above are equivalent.*

Proof. By taking N to be the submodule generated by $\{x_1, \ldots, x_n\}$ and using the remark just above one sees Variant 2 implies Variant 3. Next, Variant 1 implies Variant 2 because

$$\frac{N + \mathfrak{m}M}{N} = m\left(\frac{M}{N}\right).$$

Then apply Variant 1 to the module M/N.

To show Variant 2 implies Variant 1, and it suffices to show $M_{\mathfrak{p}} = (0)$, for all maximal ideals $\mathfrak{p} \subset R$. But $I \subset \mathfrak{p}$, so if $M = IM$, then localizing gives

$$M_{\mathfrak{p}} = \mathfrak{p}M_{\mathfrak{p}} = (0) + \mathfrak{p}M_{\mathfrak{p}}.$$

Now apply Variant 2 with $N = (0)$.   □

Here we will give a module style form of the Cayley-Hamilton Theorem (see 6.5.11 of Volume I) which we will use to obtain Version 1 of Nakayama's lemma and also get an alternative proof of the Cayley-Hamilton Theorem.

Theorem 9.11.10. *(Cayley-Hamilton) Suppose $I \subset R$ is an ideal and $\varphi : M \to IM$ is an R-module homomorphism, where M is an R-module generated by $x_1,..., x_n$. Then φ satisfies*

$$\varphi^n + a_1\varphi^{n-1} + \cdots + a_n = 0 \in \mathrm{End}_R(M),$$

with $a_j \in I^j$. Furthermore, version 1 of Nakayama's Lemma follows by taking

$$\varphi = I = multiplication\ by\ 1 : M \longrightarrow M = IM.$$

Proof. Set

$$\varphi(x_i) = \sum_j a_{ij}x_j,$$

and consider the $n \times n$ matrix $\left((a_{ij}) - \varphi I_{n \times n}\right)$ over the ring $R[\varphi]$, where $(a_{ij}) = A$. This matrix acts on $M(n, R)$, by

$$((a_{ij}) - \varphi I_{n \times n})(x_1, \ldots, x_n)^t = (0)^t.$$

There is the transposed matrix $\left((a_{ij}) - \varphi I_{n \times n}\right)^t$ with

$$\left((a_{ij}) - \varphi I_{n \times n}\right)^t \left((a_{ij}) - \varphi I_{n \times n}\right) = \det\left((a_{ij}) - \varphi I_{n \times n}\right)I_{n \times n}.$$

It follows that $\det\left((a_{ij}) - \varphi I_{n \times n}\right)$ annihilates all x_j and hence is the 0 endomorphism of M. Expanding the determinant shows the coefficient of φ^{n-j} belongs to I^j. Of course, we recognize $\chi_A(\lambda) = \det\left((a_{ij}) - \varphi I_{n \times n}\right)$ is exactly the characteristic polynomial of the endomorphism φ, with respect to the generating set $x_1,..., x_n$ of M.

Therefore $(1 + a)M = (0)$, for some $a \in I$. But $(1 + a) \in R^t$ is invertible if I is contained in every maximal ideal of R, so $M = (0)$. $\square$

9.12   The Radical of an Ideal

Definition 9.12.1. An ideal I in a commutative ring R with identity is called a *primary ideal* if whenever $ab \in I$ and $a \notin I$, $b^n \in I$ for some $n \in \mathbb{Z}^+$.

Definition 9.12.2. Let $\mathfrak{a}$ be an ideal of R. We call the *radical* of $\mathfrak{a}$, and we denoted by $\sqrt{\mathfrak{a}}$, the set of all $a \in R$ such that, $a^n \in \mathfrak{a}$ for some positive integer n. An ideal $\mathfrak{a}$ is called a *radical ideal* if $\sqrt{\mathfrak{a}} = \mathfrak{a}$.

Example 9.12.3. Let $R = \mathbb{Q}[x, y]$, and consider the ideal $\langle x^2 + 3xy, 3xy + y^2 \rangle$. This is not a radical ideal since

$$(x + y)^3 = x(x^2 + 3xy) + y(3xy + y^2) \in \langle x^2 + 3xy, \; 3xy + y^2 \rangle,$$

$x + y \in \sqrt{\langle x^2 + 3xy, \; 3xy + y^2 \rangle}$. However,

$$x + y \notin \langle x^2 + 3xy, \; 3xy + y^2 \rangle$$

since each of the two generators of this ideal is homogeneous of degree 2.

Proposition 9.12.4. *Let R be a commutative ring with identity. The radical of an ideal $\mathfrak{a}$ is itself an ideal.*

Proof. Suppose a and b are in $\sqrt{\mathfrak{a}}$, with a^n and b^m in $\mathfrak{a}$. Then,

$$(a + b)^{n+m} = \sum_{i=0}^{m+n} c_i a^i b^{m+n-i},$$

where c_i are coefficients in R. In any term of the above sum, either $i \geq n$ or $m + n - i \geq m$. In the first case $a^i \in \mathfrak{a}$ and in the second $b^{m+n-i} \in \mathfrak{a}$. Since R is a commutative ring and $\mathfrak{a}$ is an ideal, the sum of all these terms is in $\mathfrak{a}$. Hence $a + b \in \mathfrak{a}$.

To see that $\mathfrak{a}$ is closed under multiplication, let $k = \max(m, n)$. Then $(ab)^k = a^k b^k \in \mathfrak{a}$. $\qquad\square$

Proposition 9.12.5. *The radical of $\sqrt{\mathfrak{a}}$ is $\sqrt{\mathfrak{a}}$.*

Proof. Obviously, $\sqrt{\mathfrak{a}}$ is contained in the radical of $\sqrt{\mathfrak{a}}$. Conversely let x be an element of the radical of $\sqrt{\mathfrak{a}}$. Then, there is an integer n such that $x^n \in \sqrt{\mathfrak{a}}$. Therefore, by definition of the radical, there is an integer m such that $(x^n)^m \in \mathfrak{a}$, and, since $x^{nm} \in \mathfrak{a}$, it follows that $x \in \sqrt{\mathfrak{a}}$.   □

Proposition 9.12.6. *If I is an ideal of R, then,*

$$\sqrt{I} = \bigcap_{\substack{I \subset \mathfrak{p} \\ \mathfrak{p} \in \mathrm{Spec}(R)}} \mathfrak{p}.$$

Proof. Let $r \in \sqrt{I}$. Then $r^n \in I$ for some $n \in \mathbb{N}$. So, $r^n \in \mathfrak{p}$ for each prime ideal $\mathfrak{p}$, such that $I \subset \mathfrak{p}$. This implies that $r \in \mathfrak{p}$ for any prime ideal $\mathfrak{p} \supset I$, i.e. that

$$\sqrt{I} \subseteq \bigcap_{\substack{I \subset \mathfrak{p} \\ \mathfrak{p} \in \mathrm{Spec}(R)}} \mathfrak{p}.$$

On the other hand, let

$$r \in \bigcap_{\substack{I \subset \mathfrak{p} \\ \mathfrak{p} \in \mathrm{Spec}(R)}} \mathfrak{p}.$$

If $r \notin \sqrt{I}$, then the set

$$S = \{1, r, r^2, \ldots\}$$

is a multiplicative set with the property $S \cap I = \varnothing$. By Proposition 9.2.11, there is a prime ideal $\mathfrak{p} \supseteq I$ such that $\mathfrak{p} \cap S = \varnothing$. But then $r \in \mathfrak{p} \cap S$, a contradiction. So $r \in \sqrt{I}$, i.e.

$$\bigcap_{\substack{I \subset \mathfrak{p} \\ \mathfrak{p} \in \mathrm{Spec}(R)}} \mathfrak{p} \subseteq \sqrt{I}.$$

□

9.13   The Nilradical and the Jacobson Radical of a Ring

The idea behind a *radical* of a ring R is to obtain an ideal I which contains the *nasty* subsets of R in order to make the quotient R/I well behaved. Actually, here we will consider two different radicals: the *nil* radical, $\mathcal{N}(R)$, and the *Jacobson* radical, $J(R)$. These are respectively related to the intersections of the prime and maximal ideals of R. In the commutative case the quotients $R/\mathcal{N}(R)$ and $R/J(R)$ are respectively integral domains or fields.

9.13.1   The Jacobson Radical

Here R will denote any ring with 1.

Definition 9.13.1. The *Jacobson*[9] *radical* of R, denoted by $J(R)$, is

---

[9]Nathan Jacobson (1910-1999), born in the Jewish ghetto of Warsaw, Poland, where he and his brother Solomon began their elementary schooling. There, Jewish children were required to sit on special benches separated from Polish children. Shortly thereafter his father emigrated to the United States leaving his family behind until he had earned sufficient money for them to make the trip. He bought a small grocery store in Nashville, Tennessee, where he lived in a room at the rear. Two years later he had saved enough to pay for his family's passage. The family moved several times in the South and Jacobson got his BA at the University of Alabama in 1930. He went to graduate school at Princeton and wrote a dissertation under Wedderburn. Then, due to the depression and anti-Semitism, he had a series of jobs at several universities. As discrimination diminished after World War II he moved to Yale in 1947 and became the first Jew to hold a tenured position in the mathematics department there. From 1972 to 1974 Jacobson and Pontryagin were both vice-presidents of the International Mathematical Union. They argued over the Soviet Union's refusal to allow invited Jewish speakers to attend and lecture at conferences in the West. In 1978 Pontryagin made the following statement: "There was an attempt by Zionists to take the International Mathematical Union into their hands. They attempted to raise N. Jacobson, a mediocre scientist but an aggressive Zionist, to the presidency. I managed to repel this attack." Over his career Jacobson developed a deep structure theory for non-commutative rings with identity. He also made substantial contributions to Lie and Jordan algebras and algebras satisfying a polynomial identity. He has written sixteen algebra books, presenting important results. These have aged very well. He was a member of the National Academy of Sciences and the American Academy of Arts and Sciences. He served as president of

the intersection of all maximal left-ideals of R. If $J(R) = (0)$ we shall call R *semi-simple*.

Examples: $J(\mathbb{Z}) = \{0\}$.
Clearly, the maximal ideals of $R/J(R)$ have the form $\mathfrak{m}/J(R)$ where $\mathfrak{m}$ is a maximal ideal of R, and so $\mathfrak{m}/J(R))$ is semi-simple.

Proposition 9.13.2. *For any $x \in R$ the following statements are equivalent:*

1. x lies in the intersection of all maximal left ideals of R.

2. $x \in J(R)$, that is, x acts trivially in every simple left R-module.

3. $1 - rx$ is left-invertible for all $r \in R$.

Proof. ("1 $\Longleftrightarrow$ 2".) Let M be a simple left R-module. For any non-zero element $m \in M$, the left module homomorphism $R \ni r \mapsto rm \in M$ is surjective by simplicity, and so its kernel is a maximal left ideal. All maximal left ideals occur in this way (if I is such an ideal, take $M = R/I$, $m = 1$). Hence $x \in J(R)$ if and only if x lies in the intersection of all maximal left ideals.

("2 $\Longleftrightarrow$ 3".) Suppose x misses some maximal left ideal I. Then, the left ideal generated by x has the property that its sum with I must be the entire ring. Hence there is an $r \in R$ such that $rx + I$ contains 1. Hence $1 - rx \in I$ is contained in a maximal left ideal and can not be left-invertible. Conversely, suppose there exists r such that $1 - rx$ is not left-invertible. Then, it is properly contained in some maximal left ideal I which can not contain rx, and so cannot contain x. $\square$

At this point we still distinguish between the left and right Jacobson radicals, where the right Jacobson radical consists of elements which act trivially in all simple right modules (or equivalently, which are in the

the American Mathematical Society from 1971 to 1973. The University of Alabama designated him Sesquicentennial Honorary Professor in 1981. He received the Leroy P. Steele Prize for Lifetime Achievement from the American Mathematical Society in 1998.

intersection of all maximal right ideals or are such that $1 - xr$ is right-invertible for all $r \in R$). However, the next proposition shows that this is not necessary.

Proposition 9.13.3. *$x \in J(R)$ if and only if $1 - rxs$ is a unit (has a two-sided inverse) for all r, $s \in R$.*

Proof. (" $\implies$ ".) Just set $s = 1$.

(" $\impliedby$ ".) If $x \in J(R)$, then $xs \in J(R)$ for all $s \in R$, and hence $1 - rxs$ is left-invertible, say with left inverse u. This gives $u(1 - rxs) = 1$ or $u = 1 + urxs$. Now, $rxs \in J(R)$, and therefore $1 + urxs = u$ is also left-invertible. Since u has both a left and a right inverse, they must agree, and so u (and its inverse) are invertible.

The condition that $1 - rxs$ is a unit is left-right symmetric, and so we conclude that $J(R) =$ the right $J(R)$. One way to restate the above equivalent definition is that $J(R)$ is the largest ideal of R such that $1 + J(R) \in U(R)$ (the unit group of R). $\square$

Proposition 9.13.4. *Let I be an ideal contained in $J(R)$. Then $J(R/I) \cong J(R)/I$.*

Proof. By assumption, I is contained in every maximal left ideal M of R, so the pullback map from maximal left ideals of R/I to maximal left ideals of R is a bijection (hence R and R/I have the same simple modules). The conclusion follows. $\square$

Now, we have the following structure theorem of (left) Artinian rings.

Theorem 9.13.5. *If R is a left Artinian ring with identity, then $R/J(R)$ is a semi-simple left Artinian ring.*

Proof. Since left ideals in $R/J(R)$ correspond bijectively with left ideals in R that contain $J(R)$, it follows that $R/J(R)$ is left Artinian.

Let

$$\mathcal{S} = \{I \subseteq R \mid \text{there are maximal left ideals } J\colon I = J_1 \cap \cdots \cap J_n\}.$$

Consider $\mathcal{S}$ as ordered by inclusion. Suppose $\mathcal{S}$ contains no minimal element. Then for every I in $\mathcal{S}$ there is an I' in $\mathcal{S}$ with $I' \subseteq I$ and

$I' \neq I$. Let J_1 be any maximal left ideal in R, and so J_1 is in $\mathcal{S}$. Choose J_2 in $\mathcal{S}$ with $J_2 \subseteq J_1$, $J_2 \neq J_1$. Choose J_3 in $\mathcal{S}$ with $J_3 \subseteq J_2$, $J_3 \neq J_2, \ldots$. Continuing this process we get a descending chain of left ideals, $J_1 \supseteq J_2 \supseteq \ldots$, that never stabilizes. Since R is left Artinian, this is a contradiction. Therefore, we may conclude that $\mathcal{S}$ contains a minimal element, say $I = J_1 \cap \cdots \cap J_n$, where each J_i is a maximal left ideal.

Notice that if J is any maximal left ideal, then $J \cap I \subseteq I$ and $J \cap I$ is in $\mathcal{S}$, so by the minimality of I, $J \cap I = I$. Therefore, $I \subseteq J$. It follows that $I \subseteq J(R)$. Conversely, $J(R) \subseteq J_1 \cap \cdots \cap J_n = I$, so $I = J(R)$.

Define

$$\phi : R \to R/J_1 \oplus \cdots \oplus R/J_n, \text{ such that } \phi(r) = (r + J_1, \ldots, r + J_n).$$

It is easy to check that ϕ is an R-module homomorphism and $\mathrm{Ker}\,\phi = J_1 \cap \cdots \cap J_n = J(R)$. Since each J_i is a maximal left ideal, $R/J_1 \oplus \cdots \oplus R/J_n$ is a completely reducible R-module. Also, $\mathrm{Im}(\phi)$ is a submodule of $R/J_1 \oplus \cdots \oplus R/J_n$, and therefore $\mathrm{Im}(\phi)$ is also a completely reducible R-module. Since $\mathrm{Im}(\phi) \cong R/J(R)$ it follows that $R/J(R)$ is a completely reducible R-module, and so $R/J(R)$ is a completely reducible $R/J(R)$-module, hence $R/J(R)$ is a semi-simple ring. $\square$

9.13.2   The Nilradical

We begin with the nilradical. For its relationship to finite dimensional associative algebras see Chapter 7.

Definition 9.13.6. In a ring R with identity, an element x is called *nilpotent* if $x^n = 0$ for some $n \in \mathbb{Z}^+$.

Definition 9.13.7. We denote by $\mathcal{N}(R)$ the set of all nilpotent elements of R, i.e.

$$\mathcal{N}(R) = \{x \in R \mid \exists\, n \in \mathbb{Z}^+ : x^n = 0\}.$$

Exercise 9.13.8. Prove that $\mathcal{N}(R) = \sqrt{(0)}$. In particular, $\mathcal{N}(R)$ is an ideal in R.

Proposition 9.13.9. *$R/\mathcal{N}(R)$ has no non-zero nilpotent elements.*

Proof. Let $r + \mathcal{N}(R)$ be a nilpotent element of $R/\mathcal{N}(R)$. Then, for some k

$$(r + \mathcal{N}(R))^k = r^k + \mathcal{N}(R) = \mathcal{N}(R),$$

and therefore, $r^k \in \mathcal{N}(R)$, which implies that for some positive integer l we have

$$r^{kl} = (r^k)^l = 0,$$

i.e. $r \in \mathcal{N}(R)$ and thus $r + \mathcal{N}(R) = \mathcal{N}(R)$.   □

Corollary 9.13.10.

$$\mathcal{N}(R/\mathcal{N}(R)) = (0).$$

An alternative description of the nilradical is the following.

Theorem 9.13.11. *The nilradical of R is the intersection of all the prime ideals of R. In particular, if R is commutative $R/\mathcal{N}(R)$ is an integral domain.*

Proof. Obviously, any nilpotent element of R is contained in every prime ideal. Conversely, suppose $r \in R$ is not a nilpotent element. Then

$$S = \{1,\ r,\ r^2,\ ...\}$$

is a multiplicative set and (0) is an ideal disjoint from S. Consider an ideal $\mathfrak{a}$ such that $\mathfrak{a} \cap S = \varnothing$ and maximal with this property. Then, by Proposition 9.4.3, $\mathfrak{a}$ is prime. Hence,

$$r \notin \bigcap \mathfrak{p}.$$

□

Proposition 9.13.12. *When R is a Noetherian ring $\mathcal{N}(R)$ is actually a nilpotent ideal.*

Proof. Since R is Noetherian we can assume that $x_1,...,x_n$ are the generators of $\mathcal{N}(R)$. Each of these elements is nilpotent so we may choose $k \in \mathbb{N}$ large enough so that $x_i^k = 0$ for each $i = 1,...,n$. Since the x_i's are generators, any element of $\mathcal{N}(R)$ can be written as

$$y = a_1 x_1 + \cdots + a_n x_n.$$

Now, if we multiply nk such elements together the result will be a linear combination of monomials of the form

$$ax_1^{i_1}x_2^{i_2}\cdots x_n^{i_n} \quad \text{with } i_1 + i_2 + \cdots + i_n = nk.$$

By the Pigeon Hole Principle of Chapter 0 of Volume I, we must have $i_j \geq k$ for some j, which means that the result of the multiplication is zero, i.e.

$$\mathcal{N}(R)^{nk} = 0.$$

$\square$

Since all maximal ideals are prime, we see immediately that

$$\mathcal{N}(R) \subseteq J(R).$$

Proposition 9.13.13. *Let R be an Artinian ring. Then, $J(R)$ is a nilpotent ideal of R (i.e. $J(R)^n = 0$ for some $n > 0$).*

Proof. Set $J = J(R)$ and consider the chain

$$J^0 = R \supseteq J^1 \subseteq J^2 \subseteq J^3 \subseteq \cdots$$

This is a descending chain of two-sided ideals of R and, since R is Artinian, the chain stabilizes. Hence, for some integer $k > 0$

$$J^k = J^{k+1} = J^{k+1} = \ldots.$$

Suppose $J^k \neq 0$. Let $\mathcal{K}$ be the set of all non-zero left ideals I such that $JI = I$. Obviously, $\mathcal{K} \neq \varnothing$ since $J \in \mathcal{K}$, and since R is Artinian, $\mathcal{K}$ has a minimal element M. Since $m \in \mathcal{K}$,

$$J^n M = M$$

and so there is some $x \in M$ such that $J^k x \neq 0$. We have

$$J(J^k x) = J(J^k Rx) = J^{k+1} Rx = J^k Rx = J^k x.$$

Therefore, since M is minimal, $J^k x = M$, i.e. $M = Rx$. This shows that M is finitely generated. Hence by Nakayama's lemma, $\mathcal{K} = 0$, a contradiction. $\square$

Corollary 9.13.14. *In particular, a commutative Artinian ring R is semi-simple if and only if it has no non-zero nilpotent elements.*

Proof. Since R is a commutative ring, if $x \in R$ is a nilpotent element, the ideal (x) it generates is a nilpotent ideal. $\qquad \square$

Example 9.13.15. Let R be a commutative ring with identity. In $R[x]$ the Jacobson radical is equal to the nilradical. Indeed, since every maximal ideal is a prime ideal, we get

$$\mathcal{N}(R) \subseteq J(R). \qquad (1)$$

Now, if $f = a_0 + a_1 x + \cdots + a_n x^n$ is in $J(R)$, then $1 - fg$ is a unit for all $g \in R[x]$. In particular,

$$1 - fx = -a^n x^{n+1} - a_{n-1} x^n - \cdots - a_0 x + 1, \qquad (2)$$

is a unit, and so $a_0,...,a_n$ are nilpotent. This implies that f is nilpotent. Therefore, $J(R) \subseteq \mathcal{N}(R)$. From (1) and (2) we get the result.

Theorem 9.13.16. *Every Artinian ring R is isomorphic to a finite direct product of Artinian local rings R_i.*

Proof. By Proposition 9.8.4, R has only finitely many maximal ideals $\mathfrak{m}_1, \ldots, \mathfrak{m}_n$. The intersection of the $\mathfrak{m}_i$ is the Jacobson radical $J(R)$, which is nilpotent by Proposition 9.13.13. By the Chinese Remainder Theorem, the intersection of the $\mathfrak{m}_i$ coincides with their product. Thus, for some $k > 1$ we have

$$\left(\prod_1^n \mathfrak{m}_i \right)^k = \prod_1^n \mathfrak{m}_i^k = 0.$$

Powers of the $\mathfrak{m}_i$ still satisfy the hypothesis of the Chinese remainder theorem, so the natural map

$$R \longrightarrow \prod_1^n R/\mathfrak{m}_i^k$$

is an isomorphism. Since R is Artinian so is $R/\mathfrak{m}_i^k$. We must show that it is local. A maximal ideal of $R/\mathfrak{m}_i^k$ corresponds to a maximal ideal $\mathfrak{m}$ of R with $\mathfrak{m} \supseteq \mathfrak{m}_i^k$. Hence $\mathfrak{m} \supseteq \mathfrak{m}_i$ because $\mathfrak{m}$ is prime. By maximality, $\mathfrak{m}_i = \mathfrak{m}$. Thus the unique maximal ideal of $R/\mathfrak{m}_i^k$ is $\mathfrak{m}_i/\mathfrak{m}_i^k$. $\qquad\qquad\square$

The Case of Commutative rings

For commutative rings, we do not need to distinguish between left and right ideals, and so $J(R)$ is just the intersection of all maximal ideals. In other words, $J(R)$ is the kernel of the homomorphism

$$R \to \prod_\mathfrak{m} R/\mathfrak{m}.$$

(In general, however, the Jacobson radical need not contain any non-trivial nilpotents.)

If R is a commutative ring then it is clear that for any nilpotent $n \in R$, the element rns is also nilpotent; hence $1 - rns$ is invertible. The collection of nilpotent elements forms an ideal, the nilradical of R, which consequently must lie in $J(R)$.

If R is not commutative, then, in general, the nilpotent elements no longer form an ideal (left or right, let alone two-sided), as the following example shows:

Example 9.13.17. Consider the following two matrices over $\mathbb{Z}$:

$$x = \begin{pmatrix} 0 & 1 \\ 0 & 0 \end{pmatrix} \quad \text{and} \quad y = \begin{pmatrix} 0 & 0 \\ 1 & 0 \end{pmatrix}.$$

An easy calculation shows that both x and y are nilpotent ($x^2 = y^2 = 0$), but their sum

$$x + y = \begin{pmatrix} 0 & 1 \\ 1 & 0 \end{pmatrix}$$

is not, rather it is a unit since $(x + y)^2 = 1$. Hence, the set of nilpotent elements of the non-commutative ring of 2×2 matrices over $\mathbb{Z}$ is not an ideal (it is not even a subgroup).

So, in non-commutative rings we will consider the notion of a nilpotent (left/right) two-sided ideal. Alternatively, we can consider the *nil ideals*, which are ideals all of whose elements are nilpotent. In general going from the commutative setting to a non-commutative one, one switches from "element-wise" to "ideal-wise" conditions.

We say that an ideal (left, right, or two-sided) I is *nil* if it consists only of nilpotent elements and *nilpotent* if $I^n = 0$ for some n.

Proposition 9.13.18. *Let I be a nil left (resp. right) ideal of R. Then $I \subseteq J(R)$.*

Proof. Let $x \in I$. Then rx (resp. xr) is nilpotent for all $r \in R$. Hence $1 - rx$ (resp. $1 - xr$) is left-invertible (resp. right-invertible). $\square$

Proposition 9.13.19. *Let R be left Artinian (or right Artinian). Then $J(R)$ is a nilpotent ideal.*

Proof. The descending chain $J(R) \supseteq J(R)^2 \supseteq J(R)^3 \supseteq \dots$ stabilizes by assumption, and so there exists some k such that $I = J(R)^k = J(R)^{k+1} = \dots$ We claim that $I = 0$.

Suppose otherwise. Among all the ideals U such that $IU \neq 0$, there is a minimal one U_0 by the descending chain condition. Pick $a \in U_0$ such that $Ia \neq 0$. Then, $I(Ia) = Ia \neq (0)$. Hence, by minimality $U_0 = Ia$. It follows that $a = ya$ for some $y \in I \subseteq J(R)$. But $1 - y$ is a unit, so $(1 - y)a = 0$ implies $a = 0$, a contradiction. $\square$

Corollary 9.13.20. *If R is left Artinian, then any nil left or right ideal is nilpotent.*

Corollary 9.13.21. *If R is commutative and Artinian, then $J(R)$ is precisely the nilradical of R.*

It is clear that any semi-simple ring has $J(R) = (0)$. The next result determines under which conditions the converse holds.

Proposition 9.13.22. *R is semi-simple if and only if $J(R) = (0)$ and R is left Artinian.*

Proof. (" $\implies$ ".) By semi-simplicity we have $R = J(R) \oplus I$ as left R-modules for some left ideal I. If I is proper, it is contained in some maximal left ideal which does not contain $J(R)$, a contradiction. Hence $I = R$ and $J(R) = 0$.

(" $\impliedby$ ".) Let $\mathfrak{m}_1$, $\mathfrak{m}_2$, ... be a well-ordering of the maximal left ideals in R. By assumption, the descending chain

$$\mathfrak{m}_1 \supseteq \mathfrak{m}_1 \cap \mathfrak{m}_2 \supseteq \ldots$$

stabilizes. Hence the intersection of all maximal left ideals is an intersection of finitely many maximal left ideals $\mathfrak{m}_1 \cap \ldots \cap \mathfrak{m}_n = J(R) = (0)$, since $J(R) = (0)$. It follows that R embeds into $\bigoplus R/\mathfrak{m}_i$ as a left R-module. This is a direct sum of simple modules, hence semi-simple, so R must also be semi-simple as a left R-module. $\qquad\square$

Since quotients of left Artinian rings are left Artinian, we conclude the following.

Corollary 9.13.23. *Let R be left Artinian. Then $R/J(R)$ is semi-simple.*

9.14   The Hilbert Nullstellensatz Theorem

In this section we prove the important Hilbert Nullstellensatz (i.e. zero places) Theorem. We begin with some lemmas.

Lemma 9.14.1. *Let R be an integral domain and K its quotient field. If $K = R[\alpha_1, \ldots, \alpha_n]$ is a finitely generated ring over R, then $K = R[\alpha]$ for some $\alpha \in K$. In fact, $\alpha = \frac{1}{b}$, where $b \neq 0 \in R$.*

Proof. Let $\alpha_i = \frac{a_i}{b_i}$, where a_i and $b_i \in R$ and $b_i \neq 0$. If $b = b_1 \ldots b_n$, then, since each $b_i \neq 0$ and R is an integral domain, $b \neq 0$. We let $\alpha = \frac{1}{b}$ and show $K = R[\alpha]$.

Each $k \in K$ is a finite sum of terms of the form $r(\frac{a_1}{b_1})^{e_1} \cdots (\frac{a_n}{b_n})^{e_n}$, where r, a_i and $b_i \in R$. Let $e = \max\{e_1, \ldots, e_n\}$ for a fixed multi-index. Then

$$r(a_1)^{e_1} \cdots (a_n)^{e_n} (b_1)^{e-e_1} \cdots (b_n)^{e-e_n} \left(\frac{1}{b_1}\right)^{e_1} \cdots \left(\frac{1}{b_n}\right)^{e_n} = s\left(\frac{1}{b}\right)^e,$$

where $s \in R$ and $e \in \mathbb{Z}$ depend on the multi-index, and so this is a polynomial in α. Thus $K = R[\alpha]$.   □

Definition 9.14.2. Let R be an integral domain and K its quotient field. If $K = R[\alpha_1, \ldots, \alpha_n]$, then we shall call R a $\mathcal{G}$-domain[10]. For example a field is a $\mathcal{G}$-domain.

Lemma 9.14.3. *If R is an integral domain and x is an indeterminate, then R[x] is never a $\mathcal{G}$-domain.*

Proof. Let K be the quotient field of R and assume $R[x]$ is a $\mathcal{G}$-domain. Then the quotient field $R(x)$ of $R[x]$ has the property that $R(x) = R[x][\alpha]$ where $\alpha \in R(x)$ by Lemma 9.14.1. If $\frac{\phi}{\psi} \in K(x)$, then $\frac{b\phi}{d\psi} \in R(x)$ where b and $d \in R$ and $d \neq 0$, and so $\frac{b\phi}{d\psi}$ is a polynomial in α with coefficients in R[x]. But then $\frac{\phi}{\psi}$ is a polynomial in α with coefficients in $K[x]$. This means $K(x) = K[x][\alpha]$.

Claim. $K[x]$ has infinitely many primes.

Proof. To see this just note that since K is a field, $K[x]$ is a UFD (see 5.7.5 of Volume I). If it had only finitely many primes $p_1, \ldots, p_n$, then $q = p_1 \ldots p_n + 1$ could not factor into primes since q is not divisible by any p_i.

Completion of the proof of Lemma 9.14.3.

By Lemma 9.14.1 we may take $\alpha = \frac{1}{f(x)}$, where $f(x) \in R[x] \subseteq K[x]$. This means $K(x)$ consists of elements of the form $\sum_i a_i(x) \left(\frac{1}{f(x)}\right)^i$ for $a_i(x) \in K[x]$. If $p(x)$ is a prime in $K[x]$ which does not divide $f(x)$, then since $\frac{1}{p(x)} \in K(x)$, $f(x)^n = p(x) \sum_i a_i(x) \left((f(x))\right)^{n-i}$ so that $p(x)$ divides $f(x)^n$ and hence $f(x) \in K[x]$, a contradiction.   □

---

[10]According to J. Peter May in [83], Zariski [137] first advocated the desirability of a simpler proof avoiding the use of the Noether normalization theorem. He himself gave two such proofs. A few years later, Oscar Goldman [50] and Wolfgang Krull [74], independently, gave two quite similar elementary proofs of the Nullstellensatz. I. Kaplansky in his book [71] reworked Goldman's argument. He called an integral domain of finite type a $\mathcal{G}$-domain, in honor of Goldman.

Lemma 9.14.4. *Let $R \subseteq S$ be integral domains and S be algebraic over R (the equations need not be monic). Suppose S is finitely generated as a ring over R. If S is a $\mathcal{G}$-domain, then so is R.*

Proof. Let K and L be the quotient fields of R and S, respectively. Then $L = S[v^{-1}]$ by 9.14.1, where $v \in S$ and $S = R[w_1, \ldots, w_k]$, where the $w_i \in S$. Hence $L = R[v-1, w-1, \ldots, w-1]$. Let $\sum_i a_i v_i = 0$ and $\sum_{j=1}^{k} b_j w_j^{n_j} = 0$ be the algebraic equations with coefficients in R. Multiplying the first equation by v^{-n} we see that v^{-1} is algebraic over R. We may assume a_0 is different from 0 since otherwise multiply the first equation by v^{-1} etc. and then proceed as before. Let $R_1 = R[(a_0)^{-1}, (b_1)^{-1}, \ldots, (b_k)^{-1}]$. Then $K \supseteq R_1 \supseteq R$ and so $L = R_1[v^{-1}, w_1, \ldots, w_k]$. But over R_1, $v^{-1}, w_1, \ldots, w_k$ are integral and hence L is integral over R_1. Since L is a field and R_1 is an integral domain it follows that R_1 is a field and hence $K = R_1$. This means $K = R[(a_0)^{-1}, (b_0)^{-1}, \ldots, (b_k)^{-1}]$ and so R is a $\mathcal{G}$-domain. $\quad\square$

We now come to a weak, but useful form of the Nullstellensatz.

Proposition 9.14.5. *Let $k \subseteq K$ be fields with K finitely generated as a k-algebra, i.e. $K = k[x_1, \ldots, x_n]$. Then each x_i is algebraic over k. In particular, if k is algebraically closed then $K = k$.*

Proof. We prove this by induction on n: If $n = 1$ and x is not algebraic/k, then $K = k[x]$ is isomorphic to the polynomial ring in one variable with coefficients in the field k. Since polynomials of positive degree are never units, K can not be a field. Now let $K = k[x_1, \ldots, x_n]$ and L be the quotient field of $k[x_1]$. Then $K = L[x_2, \ldots, x_n]$. Since L is a field each x_i is algebraic over L for $i = 2, \ldots, n$. If x_1 were algebraic over k, then $k[x_1]$ would be isomorphic to $k[x]/(p(x))$ where $p(X)$ is a prime and hence $k[x_1]$ would be a field, that is, L. But $k[x_1]$ would also be algebraic over k so by the transitivity of algebraic dependence (see Chapter 11) K would also be algebraic over k. We may therefore assume that x_1 is transcendental over k. Let $R = k[x_1]$. Then as above $K = R[x_2, \ldots, x_n]$. Since each $x_2, \ldots, x_n$ is algebraic over R, K is a finitely generated R-module. Hence K is finitely generated as a ring over R and each $x_2, \ldots, x_n$ is integral and therefore algebraic over R. Now since as a field K is a $\mathcal{G}$-domain so is R by 9.14.4. This contradicts 9.14.3. $\square$

Taking A/M for the finitely generated k-algebra we have the following corollary.

Corollary 9.14.6. *Let k be a field, A a finitely generated k-algebra and M a maximal ideal in A. Then A/M is a finite algebraic extension of k. In particular, if k is algebraically closed $A/M = k$.*

Regarding a polynomial $f(X) \in k[X] = k[x_1, \ldots, x_n]$ as a function we call a point $p \in k^n$ (called an affine n-space) a zero of f if, upon substitution, $f(p) = 0$.

Corollary 9.14.7. *Let K be algebraically closed and $K[X]$ be the polynomials in n variables with coefficients in K. To each maximal ideal M corresponds a unique point $p \in K^n$ via point evaluations.*

Proof. Clearly to each point $p \in K^n$ corresponds a K-algebra homomorphism $\gamma_p : k[X] \to k$ given by $\gamma_p(f) = f(p)$ which by considering constant functions is clearly onto. Now $M_p = \operatorname{Ker} \gamma_p = \{f : f(p) = 0\}$ is therefore a maximal ideal in $k[X]$. Conversely let M be a maximal ideal in $k[X]$. Then since $K[X]$ is finitely generated as a K-algebra, $K[X]/M = K$. If $\gamma : K[X] \to K$ denotes the corresponding homomorphism with $\operatorname{Ker}(\gamma) = M$, then $\gamma(x_i) = a_i \in k$ and $p = (a_1, \ldots, a_n) \in K^n$. Let $f(X)$ be a polynomial in $K[X]$.

$$f(X) = \sum a_i, \ldots x_1^{e_1} \ldots x_n^{e_n}.$$

Then,

$$\gamma(f) = \sum_i a_{i_1, \ldots, i_n} \gamma(x_1)^{e_1} \ldots \gamma(x_n)^{e_n} = f(p).$$

Hence, $f(p) = 0$ if and only if $\gamma(f) = 0$. This means $M = M_p$. Since the points of an affine n-space can be separated by linear polynomials, p is unique. $\square$

Let S be a subset of $K[X]$, $I(S)$ be the finitely generated ideal generated by S and $V(S)$ the (algebraic variety) set of all common zeros in K^n of elements of S. Then, clearly, $V(S) = V(I(S))$. Hence from the point of view of $V(S)$ we may as well assume that S is an ideal. Conversely, given a variety V in K^n we can consider all $f \in K[X]$ which vanish on V. This is clearly an ideal, usually called $J(V)$. We observe that the bigger the ideal I the smaller the variety V and vice versa. We shall deal with such questions fully in the next section.

Corollary 9.14.8. *If K is algebraically closed and I is an ideal in $K[X]$ with $V(I) = \varnothing$, then $I = K[X]$.*

Proof. Since I has no common zeros it suffices to show that each non-trivial ideal in $K[X]$ has a zero. If I is a proper ideal then $I \subset M$ a maximal ideal 5.4.2 of Volume I. But by the above $M = M_p$ for some $p \in K^n$ and so for $f \in I$, $f(p) = 0$. $\qquad\square$

We now come to one of the many forms of the Hilbert Nullstellensatz:

Theorem 9.14.9. *Let K be an algebraically closed field and $f, f_1, \ldots, f_s \in K[X]$. If f vanishes at all common zeros in K^n of $f_1, \ldots, f_s$, then there is an integer r and $g_1, \ldots, g_s \in K[X]$ so that $f(X)^r = \sum g_i(X) f_i(X)$. Conversely, if $f(X)^r = \sum g_i(X) f_i(X)$, then clearly f vanishes at all common zeros of $f_1, \ldots, f_s$.*

Proof. Let t be a new indeterminate and consider polynomials $f_1(X), \ldots, f_s(X)$, $1 - tf(X)$ in $t, x_1, \ldots, x_n$. These have no common zeros in K^{n+1}. For if (t, X) were such a zero, then X would be a common zero of $f_1, \ldots, f_s$ and $1 = tf(X)$. But $f(X) = 0$ so that $1 = 0$. By the previous corollary the ideal in $K[t, X]$ generated by these $s+1$ polynomials must be $K[t, X]$ itself. Hence there exist $h_1(t, X), \ldots, h_s(t, X)$, $h(t, X)$ such that

$$\sum_i h_i(t, X) f_i(X) + h(t, X)(1 - tf(X)) = 1.$$

Substituting $\frac{1}{f(X)}$ for t throughout gives $\sum_i h_i(\frac{1}{f(X)}, X) f_i(X) + 0 = 1$. Clearing denominators by multiplying by a sufficiently high power of $f(X)$ yields $\sum_i g_i(X) f_i(X) = f(X)^r$. $\qquad\square$

In view of finite generation of ideals in $k[X]$ (the Hilbert basis theorem) we may reformulate the Nullstellensatz as follows:

Theorem 9.14.10. *Let I be an ideal in $K[X]$, where K is an algebraically closed field. Then, the $\mathrm{rad}(I) = \{f : f = 0 \text{ on } V(I)\}$, where $\mathrm{rad}(I)$ stands for $\{f : f^r \in I \text{ for some } r\}$, and $V(I)$ is the algebraic variety determined by I. In other words, $J(V(I)) = \mathrm{rad}(I)$.*

Corollary 9.14.11. *If P is a prime ideal in $K[X]$ where K is an algebraically closed field, then $P = J(V(P))$. In other words, the functions $V(\cdot)$ and $J(\cdot)$ are bijective inverses of one another on the set of prime ideals and its image under V.*

We now list several other variants of the Nullstellensatz. Here again K must be an algebraically closed field. We invite the reader to check these all follow from what we have done.

Nullstellensatz Version 1. Suppose $f_1, ..., f_m \in K[x_1, ..., x_n]$. If the ideal $(f_1, ..., f_m) \neq (1) = K[x_1, ..., x_n]$, then the system of equations $f_1 = \cdots = f_m = 0$ has a solution in K.

Nullstellensatz Version 2. The maximal ideals of $K[x_1, ..., x_n]$ are precisely those maximal ideals which come from points, i.e. those ideals of the form $(x_1 - a_1, ..., x_n - a_n)$ for some $a_1, \ldots , a_n \in K$.

Nullstellensatz Version 3 (The *Weak Nullstellensatz*). If I is a proper ideal in $K[x_1, ..., x_n]$, then $V(I) \neq \varnothing$.

Nullstellensatz Version 4. $I(V(I)) = \sqrt{I}$, which implies radical ideals are in one-to-one correspondence with algebraic sets.

Nullstellensatz Version 5. A radical ideal of $K[x_1, ..., x_n]$ is the intersection of the maximal ideals containing it.

Nullstellensatz Version 6. If $f_1, ..., f_r, g$ are in $K[x_1, ..., x_n]$, and g vanishes wherever $f_1, ..., f_r$ vanish, then some power of g lies in the ideal $(f_1, ..., f_r)$, i.e.

$$g^m = a_1 f_1 + \cdots + a_r f_r$$

for some $m > 0$ and some $a_i \in K[x_1, ..., x_n]$.

Alternatively, if I is a finitely generated ideal of $K[x_1, ..., x_n]$, and g is 0 on the vanishing set of I ($g \in V(I)$), then $g \in \sqrt{I}$.

Exercise 9.14.12. Let $R = K[x_1, \ldots, x_n]$ where K is an algebraically closed field. Show R is semi-simple i.e. $\mathcal{J}(R) = (0)$. In particular, $\mathcal{J}(R) = Nilrad(R)$. Suggestion: the maximal ideals in $K[x_1, \ldots, x_n]$ are given by the Hilbert Nullstellensatz as the point evaluations. That is, if $(a_1, \ldots, a_n)$ is a point in K^n, then the associated maximal ideal is $(x_1 - a_1, \ldots, x_n - a_n)$.

9.15   Algebraic Varieties and the Zariski Topology

Here we make some remarks on Algebraic Varieties and the Zariski[11] topology, particularly over $\mathbb{C}$. Let K be an algebraically closed field, V

---

[11] Oscar Zariski (Ascher Zaritsky) (1899-1986) Jewish mathematician, born in what is now Belarus. During the chaos of World War I and the Russian revolution the family fled to Ukraine where Zariski tried to study in Kiev which was also devastated by war so he decided to go to Italy to continue his studies. In Rome he came under the influence of the great algebraic geometers Castelnuovo, Enriques and Severi and obtained a doctorate (1924) and it was there that Enriques suggested that Ascher Zaritsky italianize his name. Zariski had gone to Italy to escape problems in Belarus and Ukraine. However, the political situation in Italy began to deteriorate rapidly due to Italian fascism which made life for Zariski particularly difficult. The Italians encouraged him to view Lefschetz's topological methods as being the road ahead for algebraic geometry and Lefschetz helped him get to the United States where he got a position at Johns Hopkins University. There he made great strides in making algebraic geometry rigorous by dealing with its algebraic foundations. His work and ideas have had a profound effect on the modern development of the subject and Zariski was appointed to a chair at Harvard where he was to remain until he retired in 1969. From the late 1970s he suffered from Alzheimer's disease and his last few years were difficult ones as his health failed. Zariski contributed greatly both to the AMS and mathematical publishing over many years. He was vice president of the Society between 1960 and 1961 and president from 1969 to 1970. He was an editor of the American Journal Mathematics, the AMS Transactions and the Annals of Mathematics. In 1944 he received the Cole Prize for Algebra, in 1965 the National Medal of Science and in 1981 the Steele Prize for the cumulative influence of his total mathematical research and in the same year, the Wolf prize.

be a vector space over K of dimension n and $K[x_1, ..., x_n] = K[X]$ be the polynomial ring in n variables.

Definition 9.15.1. Let $A \subset K[X]$. We call the *affine algebraic variety* determined by A the space,

$$V(A) = \{x = (x_1, ..., x_n) \in K^n \mid f(x) = 0 \ \forall f \in A\}.$$

The reason $V(A)$ is called affine is because $V(A) \subseteq K^n$. Let $I(A)$ be the ideal in $K[X]$ generated by A, i.e. the set of all polynomials which vanish on A.

Definition 9.15.2. We define the *Zariski topology* on K^n by taking for the closed sets its affine subvarieties, namely sets which are simultaneous zeros in K^n of a family of polynomials in $K[X]$. Of course the affine variety determined by A is the same as the one determined by $I(A)$.

Example 9.15.3.   1. Affine n-space K^n is an algebraic set; it is the zero set of the polynomial 0.

2. The empty set is an algebraic set; it is the zero set of the polynomial (1).

3. A single point $(a_1, ..., a_n)$ in K^n is an algebraic set; it is the zero set of the ideal $(x_1 - a_1, ..., x_n - a_n)$. Since points are closed the Zariski topology is T_1.

4. An affine (or linear) subspace of K^n is an algebraic set since it is affinely homeomorphic with K^m for some $m \leq n$. For more details on this see the use of affine actions in the section on conics in Chapter 4 of Volume I.

5. Evidently in the case of polynomials in one variable, all non-trivial closed sets are finite.

We leave the proof of the following lemma to the reader.

Lemma 9.15.4. *Let A and B be ideals in $K[X]$. Then*

1. $V(A) \cup V(B) = V(AB)$.

2. *If A_i is a family of ideals, then $\cap V(A_i) = V(\sum A_i)$.*

3. *$A \subseteq B$ if and only if $V(B) \subseteq V(A)$.*

From the first two items of the lemma it follows that any intersection of closed sets is closed and a finite union of closed sets is closed. Thus the Zariski topology is indeed a topology. On a subvariety, $V(A)$, we take for its topology the relative Zariski topology and so a closed subset of $V(A)$ is just a Zariski closed set of K^n which is contained in $V(A)$.

Proposition 9.15.5. *Although the Zariski topology is T_1, it is not T_2.*

Proof. K^n is not Hausdorff in the Zariski topology. Take two points a and $b \in K^n$, with $a \neq b$. Consider the line through a and b and choose an affine linear function that maps it Zariski homeomorphically onto K. Thus, it is enough to show the one dimensional space K itself is not Hausdorff. In this case $K[X] = K[x]$ so every ideal, I, in it is principal. Thus $I = (f)$ for some polynomial f in one variable which has only a finite number of roots. Hence, the closed sets of K (other than K itself) are finite. Now suppose U and V are disjoint open sets $a \in U$ and $b \in V$. Then their closed complements U^c and V^c are finite, so $(U \cap V)^c = U^c \cup V^c$ is also finite. In particular, since K must be infinite, $U \cap V$ is non-empty a contradiction.   □

The significance of the Zariski topology not being T_2 tells something about the vanishing of polynomials. Namely, that the behavior of a polynomial at a "few" points can determine a lot about its behavior at seemingly distant points. We shall see a dramatic example of this in the proof of Theorem 9.15.9.

Because $K[X]$ is Noetherian, the last item of the lemma just above shows it is not possible to have an infinitely decreasing chain of affine varieties (closed sets) in K^n. For this reason we shall also say K^n is Noetherian. This in turn tells us that an affine variety X has a finite number of connected components. For if X is not connected, since it can be written as the disjoint union of sets $X_1 \cup X_2$ both of which are closed and if either of these is not connected we can continue this process. But as we just saw this process must stop after a finite number of steps. This leads us to the following proposition.

Proposition 9.15.6. *Any variety has at most a finite number of Zariski connected components.*

An important feature of the Zariski topology is the following: Let A be a closed set in K^n, i.e. the simultaneous zeros of a family of polynomials f_i where i is in some index set. Then A is the simultaneous zeros of the ideal $I(A)$ in $K[X]$ generated by the f_i. By the Hilbert basis theorem, $I(A)$ is generated by a finite number of polynomials and the zeros of these are certainly the same as the zeros of $I(A)$. Thus we may always assume without loss of generality that Zariski closed sets, A in K^n are defined by a *finite number of polynomials*.

$$A = \bigcap_{j=1}^{m} \{x \in K^n : g_j(x) = 0\}.$$

As a final remark along these lines, the Hilbert Nullstellensatz tells us there is a bijective correspondence between algebraic sets in K^n and radical ideals in $K[X]$.

We also mention that if R is an Artinian ring, since, as we saw, $\operatorname{Spec}(R)$ consists of a finite set of points, and points are Zariski closed it follows that here the Zariski topology on $\operatorname{Spec}(R)$ is discrete.

9.15.1   Zariski Density

We now turn to some questions of Zariski density. The importance of this is due to the fact that if two polynomials agree on a Zariski dense set, then they agree identically. As we have already seen, this is very useful tool in proving polynomial identities.

Let $M(n, K)$ be the $n \times n$ matrices over an algebraically closed field K and let $D_n(K)$ denote the set of *diagonalizable* elements of $M(n, K)$. Here we consider $M(n, K) = V$.

Proposition 9.15.7. $D_n(K)$ *contains the non-trivial Zariski open set,* $\{T \in M(n, K) : f_0(T) \neq 0\}$, *where f_0 is a non-trivial polynomial in* $K[x_{i,j}]$.

Proof. For a matrix $T \in M(n, K)$, as usual we denote its characteristic polynomial by $\chi_T = \det(T - xI) \in K[x]$. Since K is an algebraically closed field, χ_T factors completely into linear factors. If χ_T is square-free, then these linear factors are distinct so T has n distinct eigenvalues and so (see 3.4.12 of Volume I) is diagonalizable over K. Set

$$U_n(K) = \{T \in M(n.K) \mid \chi_T \text{ square-free}\}.$$

Then $U_n(K) \subseteq D_n(K)$.

We will show $U_n(K)$ is Zariski open in $M(n, K)$, by finding a non-trivial polynomial $f_0 \in K[x_{i,j}]$, $1 \leq i, j \leq n$, such that

$$U_n(K) = \{T \in M(n.K) \mid f_0(T) \neq 0\}.$$

To do this consider the maps $\sigma_1, ..., \sigma_n : M(n, K) \to K$, defined by

$$\chi_T(x) = x^n + \sum_{i=1}^{n} \sigma_i(A)x^{n-i}.$$

These σ_i are polynomials in the coordinates of T. For example, $\sigma_1(T) = -\operatorname{tr}(T)$ and then the signs alternate ending with $\sigma_n(T) = \pm \det(T)$, according to whether n is even or odd. The discriminant of χ_T (see Chapter 5), $\Delta(\chi_T)$ is a polynomial in the σ_i and $f_0 = \Delta(\chi_T) \neq 0$ if and only if χ_T is square-free. Hence

$$T \in U_n(K) \Longleftrightarrow f_0(T) \neq 0.$$

Thus f_0 is a *non-trivial polynomial* in the n^2 variables, $x_{i,j}$, and

$$U_n(K) = \{T \in M(n, K) : f_0(T) \neq 0\}.$$

$\square$

It is useful to compare the Zariski and Euclidean topologies. To do so we take $K = \mathbb{C}$. The Zariski closed sets in $V = \mathbb{C}^n$ are the simultaneous zeros of any family of polynomials in $\mathbb{C}[X]$, while the Euclidean closed sets certainly include all simultaneous zeros of all continuous complex valued functions on $\mathbb{C}^n$. Hence there are many more Euclidean closed sets than Zariski closed sets. Since the closure of a set D (in either topology) is the intersection of closed sets containing it, it follows that the closures are related by

$$D^{-Euc} \subseteq D^{-Zar}.$$

That is,

Proposition 9.15.8. *If d_{nu} is a net converging to x in the Euclidean topology, then it converges to x in the Zariski topology. In particular, if D is Euclidean dense in $\mathbb{C}(V)$, then it is Zariski dense as well.*

Theorem 9.15.9. *$D_n(\mathbb{C})$ is Euclidean dense and therefore Zariski dense in $M(n, \mathbb{C})$.*

Proof. Let x be a point where $f_0(x) = 0$ and consider Euclidean disks U in $M(n, \mathbb{C})$ about x. If each such disk contains a point y where $f_0(y) \neq 0$, then $D_n(\mathbb{C})$ would be Euclidean dense in $M(n\mathbb{C})$. Otherwise there must exist some disk U so that on it f_0 is identically zero. It would then follow from the identity theorem (see e.g. Chapter 6 of [88]) that f_0 is identically zero on all of $M(n, \mathbb{C})$, a contradiction. This proves $D_n(\mathbb{C})$ is Euclidean dense in $M(n, (\mathbb{C})$. It now follows from the Proposition just above that $D_n(\mathbb{C})$ is Zariski dense in $M(n, (\mathbb{C})$. $\square$

Exercise 9.15.10. Extend Theorem 9.15.9 to get the same conclusion under the assumption that $D_n(\mathbb{C})$ contains a non-trivial Zariski open set.

We conclude this section with some remarks concerning another important example of a Zariski dense set. Let G be a connected real semi-simple Lie group of non-compact type contained in $\mathrm{GL}(n, \mathbb{R}) = \mathrm{GL}(V)$ and H be a closed subgroup of G of cofinite volume. A theorem of A. Borel then tells us that H is Zariski dense in G. That is, both G and H have the same Zariski closure. For example, if $G = \mathrm{SL}(n, \mathbb{R})$, where $n \geq 2$ and $H = \mathrm{SL}(n, \mathbb{Z})$, then H is Zariski dense in G. (We remark that since here H is discrete and therefore Euclidean closed, this shows that although Euclidean convergence implies Zariski convergence, the converse does not hold in general.) For a generalization of Borel's theorem see Chapter 9 of [1].

We now present an application of Borel's Density Theorem. Here G and H are as above.

Corollary 9.15.11. *Let W be an H-invariant subspace of V. Then W is G-invariant.*

Proof. As W is H-invariant choose a basis for V so that in it all the matrices $h \in H$ have a block of zeros in the upper right corner. Since this set is Zariski closed, Borel's theorem tells us the same is true for each $g \in G$. Therefore, W is G-invariant.   □

Corollary 9.15.12. *Let $v_0 \neq 0 \in V$ and suppose for all $h \in H$,*

$$hv_0 = \lambda(h)v_0,$$

where $\lambda(h)$ is real for each $h \in H$. Then $gv_0 = v_0$ for all $g \in G$ and $\lambda(h) = 0$ for all $h \in H$. In particular, if H fixes v_0, then so does G.

Proof. Since here the hypothesis is that each $h \in H$ preserves the line through v_0, by Corollary 9.15.11 the same is true for each $g \in G$. Therefore, for all $g \in G$,

$$gv_0 = \mu(g)v_0,$$

where $\mu(g)$ is real for each $g \in G$. Evidently, μ is a continuous multiplicative function on G. Since G is semi-simple the only such function is identically 1. Thus $gv_0 = v_0$ for all $g \in G$. In particular, the same is true of all $h \in H$. Since $v_0 \neq 0$, $\lambda(h) = 0$ for all $h \in H$.   □

9.16   The Completion of a Ring

Let R be a commutative ring with 1, and

$$I_1 \supset I_2 \supset \cdots$$

be a decreasing sequence of ideals in R with

$$\bigcap_{n=1}^{\infty} I_n = \{0\}.$$

We will use these ideals to define a *metric* on R, by taking $I_0 = R$ and for $k \geq 1$,

$$d(x, y) = e^{-k}, \quad \text{where } x - y \in I_k, \text{ but } x - y \notin I_{k+1}.$$

One checks that $d(x, y)$ is a *distance* on R. Indeed,

1. By definition, $d(x, y) \geq 0$ and $d(x, y) = 0$ if and only if

$$x - y \in \bigcap_n I_n = \{0\},$$

   so that $x = y$.

2. By definition, $d(x, y) = d(y, x)$.

3. If $x - y \in I_j$ and $y - z \in I_k$, then $x - z \in I_{\min(j,k)}$, and hence

$$d(x, z) \leq \max\big(d(x, y), d(y, z)\big),$$

   which is the *strong* form of the *triangle inequality*.

With this metric R becomes a topological ring, that is, R is a topological space and the addition and multiplication maps $R \times R \to R$ are continuous.

We now explain what the completion of such an object is.

Definition 9.16.1. The *completion* of R with respect to the metric d above is called the *d-completion* of R and will be denoted by $\widehat{R_d}$.

As a set $\widehat{R_d}$ is just the usual topological completion of R, that is,

$$\widehat{R_d} = \{\text{equivalence classes of Cauchy sequences of elements of } R\}.$$

In other words, we form all sequences in R so that

$$\lim_{n,m\to\infty} d(x_n, x_m) = 0$$

and we shall say $(x_n)_{n\in\mathbb{N}}$ is equivalent to $(y_n)_{n\in\mathbb{N}}$ $((x_n)_{n\in\mathbb{N}} \sim (y_n)_{n\in\mathbb{N}})$ if $d(x_n, y_n) \to 0$). Then, $\widehat{R_d}$, equipped with the ring operations

$$[x_n] + [y_n] = [x_n + y_n] \text{ and } [x_n] \cdot [y_n] = [x_n \cdot y_n],$$

where $[x_n]$ is the equivalence class of the sequence $(x_n)_{n\in\mathbb{N}}$, becomes a topological ring.

An important example of this is when $R = \mathbb{Z}$ and p is a prime. Then one takes $I_j = (p^j)$ for the decreasing sequence of ideals in $\mathbb{Z}$. The induced topology on $\mathbb{Z}$ is called the p-adic topology and the completion is the ring of p-adic integers. This topology is the subject of Chapter 10.

We now give an alternative, and more algebraic way of defining the completion of R.

Definition 9.16.2. The completion, $\widehat{R}$, of R relative to the ideals I_n is:

$$\widehat{R} = \left\{ (r_1 + I_1, r_2 + I_2, r_3 + I_3, \ldots) \in \prod_n R/I_n \mid r_m - r_n \in I_n \; \forall\, m \geq n \right\}.$$

With this comes a natural completion homomorphism,

$$\phi : R \longrightarrow \widehat{R} \text{ such that } r \mapsto \phi(r) = (r + I_1, r_2 + I_2, \ldots).$$

Then $\widehat{R}$ is a ring under component wise addition and multiplication, with zero element, $(r_1 + I_1, r_2 + I_2, r_3 + I_3, \ldots)$. Here $r_n \in I_n$ for each n.

Henceforth we will restrict ourselves to the case where all the ideals I_n's are equal to some ideal I and its powers. Then we write $\widehat{R_I}$ (resp. $\widehat{R_d}$ for the induced metric d) for the completion of R relative to the ideal I and we will call it the I-adic completion of R. Here the natural homomorphism ϕ is given by

$$\phi : R \longrightarrow \widehat{R_I} \text{ such that } \phi(r) = (r + I, r + I^2, \ldots).$$

One sees immediately that

$$\mathrm{Ker}(\phi) = \bigcap_n I^n,$$

and therefore, since we want ϕ to be injective, we require (in the definition of I-adic topology) $\mathrm{Ker}(\phi) = 0$.

The following theorem shows the connection between inverse limits and completions, and so gives us another way of viewing the completion of a ring.

Theorem 9.16.3. *Let R be a ring and I be an ideal of R such that*

$$\bigcap_{n=1}^{\infty} I^n = \{0\}.$$

Consider the inverse system of rings

$$R/I \longleftarrow R/I^2 \longleftarrow \cdots,$$

where each π_i is a natural surjection. Then

$$\varprojlim_{n \in \mathbb{N}} R/I^n \cong \widehat{R_I},$$

as rings.

Proof. Let

$$x \in \varprojlim_{n \in \mathbb{N}} R/I^n.$$

By definition,

$$x = (a_1 + I,\, a_2 + I^2,\, \ldots)$$

with

$$\pi_n(a_{n+1} + I^{n+1}) = a_n + I^n, \ \forall\, n \in \mathbb{N}.$$

On the other hand, the projections π_i (by definition) satisfy

$$\pi_n(a_{n+1} + I^{n+1}) = a_{n+1} + I^n.$$

Therefore,
$$a_{n+1} - a_n \in I^n.$$

Now, for any $m > n$,

$$a_m - a_n = \sum_{i=n}^{m-1} (a_{i+1} - a_i) \in I^n.$$

This shows that the sequence $(a_n)_{n \in \mathbb{N}}$ is a Cauchy sequence in the I-adic metric.

Define the map

$$\phi : \varprojlim_{n \in \mathbb{N}} R/I^n \longrightarrow \widehat{R_I},$$

as

$$x = (a_1 + I,\, a_2 + I^2,\, \ldots) \mapsto [a_n],$$

where $[a_n]$ is the equivalence class of $(a_n)_{n \in \mathbb{N}}$. We leave to the reader the proof that the map ϕ is well defined, bijective and respects the ring operations.   □

Proposition 9.16.4. *Let R be a ring. Then $R[[x]]$ is the completion of $R[x]$.*

Proof. Let $I = (x)$. We will show that $\widehat{R[x]}_I$, i.e. the I-adic completion of $R[x]$, is isomorphic to $R[[x]]$, via the isomorphism

$$\psi : R[[x]] \longrightarrow \widehat{R[x]}_I,$$

given by

$$p(x) \mapsto \psi(p(x)) = \big(p(x) + (x),\, p(x) + (x^2),\, p(x) + (x^3),\, \ldots\big),$$

where $p(x) = \sum_{i=0}^{\infty} r_i x^i$, with $r_i \in R$ for all i. To prove ψ is an isomorphism we construct its inverse,

$$\psi^{-1} : \widehat{R[x]}_I \longrightarrow R[[x]]$$

in such a manner that

$$\psi^{-1}\big(p(x) + (x),\, p(x) + (x^2),\, \ldots\big) = p(x).$$

To do so, let

$$s(x)_j = \sum_{i=0}^{j-1} r_i x^i, \ \ j \geq 1.$$

Then,

$$\big(s(x)_1 + (x), \ s(x)_2 + (x^2), \dots\big)$$

is $\psi(p(x))$. Indeed,

$$\begin{aligned}
\psi(p(x)) &= \big(p(x) + (x), \ p(x) + (x^2), \ \dots\big) \\
&= \big(r_0 + (x), \ r_0 + r_1 + (x^2), \ \dots\big) \\
&= \big(s(x)_1 + (x), \ s(x)_2 + (x^2), \dots\big).
\end{aligned}$$

Now, using the $s(x)_i$ we construct the inverse ψ^{-1} by

$$\begin{aligned}
\psi^{-1}\big(p(x) &+ (x), \ p(x) + (x^2), \dots\big) \\
&= s(x)_1 + \big(s(x)_2 - s(x)_1\big) + \big(s(x)_3 - s(x)_2\big) + \cdots \\
&= r_0 + \big(r_1 x + s(x)_1 - s(x)_1\big) + \big(r_2 x^2 + s(x)_2 - s(x)_2\big) + \cdots \\
&= p(x).
\end{aligned}$$

The existence of ψ^{-1} shows that ψ is an isomorphism. $\square$

Exercise 9.16.5. Prove that $R[[x_1, ..., x_n]]$ is the completion of $R[x_1, ..., x_n]$.

9.16.1   Exercises

Exercise 9.16.6. Let R be a ring such that each local ring $R_{\mathfrak{p}}$ is Noetherian. Is R necessarily Noetherian?

Exercise 9.16.7. Let R be a ring such that

1. for each maximal ideal $\mathfrak{m}$ of R, the local ring $R_{\mathfrak{m}}$ is Noetherian,

2. for each $x \neq 0$ in R, the set of maximal ideals of R which contain x is finite.

Show that R is Noetherian.

Exercise 9.16.8. Let $I \subseteq R$ be a finitely generated ideal satisfying $I = I^2$. Prove that I is generated by one idempotent element $e \in R$.

Exercise 9.16.9. Let R be a left Artinian ring and M be a left R-module. Prove that there is a unique smallest submodule K of M (called the *radical* of M) such that M/K is a semi-simple R-module.

Exercise 9.16.10. Compute the length of $\mathbb{Z}[x]/(x^2 - 3, 35)$.

Exercise 9.16.11. Let R be a ring of finite length which is an integral domain. Show that R is a field.

Exercise 9.16.12. Prove that:

1. If R is a ring of finite length and $\mathfrak{p} \subset R$ is a prime ideal, then $\mathfrak{p}$ is maximal.

2. Now, if R is any Noetherian ring and $\mathfrak{m} \subset R$ is a minimal prime ideal then, $R_\mathfrak{m}$ is of finite length.

Exercise 9.16.13. Prove that the Jacobson radical contains no nonzero idempotents in each of the following ways:

1. Using the characterization of $J(R)$ as the intersection of maximal ideals.

2. Using the characterization of $J(R)$ as the largest ideal J such that $1 + J$ consists of Units.

3. Using the characterization of $J(R)$ as the intersection of annihilators of all simple modules.

Exercise 9.16.14. Given $\varphi : A \longrightarrow B$ a surjective homomorphism of rings, show that the image by φ of the Jacobson radical of A is contained in the Jacobson radical of B.

Exercise 9.16.15. Let R be a commutative integral domain and let $R[x_1, x_2, \ldots, x_n]$ be the polynomial ring over R in the n variables x_1, $x_2, \ldots, x_n$. If $\mathbf{a} = (a_1, \ldots, a_n)$ is an n-tuple of elements of R, then there is an evaluation homomorphism

$$\phi_{\mathbf{a}} : R[x_1, \ldots, x_n] \longrightarrow R$$

given by $\phi_{\mathbf{a}}(f) = f(a_1, \ldots, a_n)$.

1. If R is the field of complex numbers and if I is a proper ideal of $R[x_1, \ldots, x_n]$, show that there exists an n-tuple $\mathbf{a} = (a_1, \ldots, a_n)$ with $\phi_{\mathbf{a}}(I) \neq R$.

2. Now let R be the ring of integers and let I be the ideal of the polynomial ring $R[x]$ in one variable generated by 3 and $x^2 + 1$. Show that I is a proper ideal of $R[x]$ but that $\phi_{\mathbf{a}}(I) = R$ for all 1-tuples $\mathbf{a} = (a)$ with $a \in R$.

Exercise 9.16.16. Show that $\mathbb{N}^2$ is dense in $\mathbb{C}^2$ in the Zariski topology.

Chapter 10

Valuations and the p-adic Numbers

In this chapter we introduce and study the p-adic integers, where p is a prime, both intuitively by discussing how one represents a p-adic integer and formally using the concept of inverse limits. This latter approach will allow us to show that the p-adic integers form a ring, denoted by $\mathbb{Z}_p$. We will then consider fractions of p-adic integers, that is, p-adic numbers[1], which we will show form a field we denote by $\mathbb{Q}_p$.

10.1   Valuations

Let $(G, +)$ be an Abelian group.

Definition 10.1.1. We say that the Abelian group G is an *ordered* Abelian group, if on G there is a binary relation $\leq$ such that

1. For any two elements g and h of G, either $g \leq h$ or $h \leq g$.

2. $g \leq g$.

3. If $g \leq h$ and $h \leq g$, then $g = h$.

---

[1]p-adic numbers were introduced in 1904 by the German mathematician K Hensel.

4. If $g \leq h$ and $h \leq k$, then $g \leq k$.

5. If $g \leq h$, then $g + k \leq h + k$ for all g, k, $h \in G$.

Let G be an ordered Abelian group and let ∞ be a symbol satisfying

$$\infty = \infty + \infty = g + \infty = \infty + g$$

for all $g \in G$. In this chapter, k will be a field of characteristic zero.

Definition 10.1.2. A *valuation* on the field k is a surjective map

$$v : k \to G \cup \{\infty\}$$

such that for all x, y in k,

1. $v(x) = \infty$ if and only if $x = 0$.

2. $v(xy) = v(x) + v(y)$.

3. $v(x + y) \geq \min\{v(x), v(y)\}$

We call $G_v = G$ the *value group* of the valuation v.

This is often called a *non-Archimedean* valuation. The inequality 3 is easily seen to imply

$$v(x + y) \leq v(x) + v(y).$$

When this weaker inequality holds one calls this an *Archimedean* valuation.

The following exercise illustrates some basic properties of a valuation:

Exercise 10.1.3. Let v be a valuation on a field k. Then

1. $v(1) = 0$.

2. $v(\infty) = 0$.

3. $v(\zeta) = 0$ for all roots of unity $\zeta \in k$.

4. If x, $y \in k$ with $v(x) \neq v(y)$, then

$$v(x + y) = \min\{v(x),\, v(y)\}.$$

5. $v(x^{-1}) = -v(x)$.

6. $v(-x) = v(x)$.

Definition 10.1.4. A valuation v on the field k is called *discrete* if $v(k - \{0\}) = \lambda\mathbb{Z}$ for a real number $\lambda > 0$.

10.2   Absolute Value or Norm

Definition 10.2.1. We say that a function $|\cdot| : k \longrightarrow \mathbb{R}$ is an *absolute value*, or a *norm* on k, if it satisfies the following conditions:

1. $|x| \geq 0$ for any $x \in k$ and $|x| = 0 \Leftrightarrow x = 0$.

2. $|xy| = |x| \cdot |y|$, for any $x, y \in k$.

3. $|x + y| \leq |x| + |y|$ (the *triangle inequality*).

Notice that $|\cdot|$ defined by $|x| = 1$ for any $x \neq 0$, and $|0| = 0$, is an absolute value on the field k called the *trivial absolute value*.

Remark 10.2.2. In the case $k = \mathbb{Q}$, it is sufficient to define the absolute value $|\cdot|$ on $\mathbb{Z}$. Indeed, using condition 2 it follows that for any $\frac{a}{b} \in \mathbb{Q}$, $|a| = |b \cdot \frac{a}{b}| = |b| \cdot |\frac{a}{b}|$. Hence $|\frac{a}{b}| = \frac{|a|}{|b|}$.

Exercise 10.2.3. Let $|\cdot|$ be an absolute value on the field k. Prove the following statements.

1. $|1| = 1$.

2. $|\zeta| = 1$ for all roots of unity $\zeta \in k$.

3. $|x^{-1}| = |x|^{-1}$.

4. $\left||x| - |y|\right| \leq |x - y|$.

Definition 10.2.4. If we replace the condition 3 by the stronger condition (called the *ultra-metric* inequality),

$$|x + y| \leq \max\{|x|, |y|\}, \qquad (3*)$$

then we call $|\cdot|$ a *non-Archimedean* norm.

The standard absolute value is an Archimedean norm on $\mathbb{Q}$.

Proposition 10.2.5. *Let $|\cdot|$ be an absolute value on a field k. Then the following statements are equivalent:*

1. $|\cdot|$ satisfies the ultra-metric inequality.

2. For each $n \in \mathbb{Z}$, $|n \cdot 1| \leq 1$.

Proof. Suppose $|\cdot|$ satisfies the ultra-metric inequality. Then, $|1 + 1| \leq \max\{|1|, |1|\} = 1$. Using induction one sees that the set $\{|n \cdot 1| : n \in \mathbb{Z}\}$ is bounded by 1.

For the converse, suppose $|n \cdot 1| \leq C$. Then,

$$|x + y|^n = |(x + y)^n| \leq \sum_{k=0}^{n} \binom{n}{k} x^k y^{n-k} \leq (n + 1)C(\max(|x|, |y|))^n.$$

Taking the n-th root of both sides we get

$$|x + y| \leq \sqrt[n]{(n + 1)C}(\max(|x|, |y|)).$$

Letting $n \to \infty$, we get the ultra-metric inequality. $\qquad \square$

The above proposition shows the difference between Archimedean and non-Archimedean absolute values. In contrast to the non-Archimedean case, the Archimedean absolute value of an integer can be arbitrarily large since it is equivalent to the *Archimedean property* of a field k which states that given two elements a and b with $a \neq 0$, there is an integer n such that $|na| > |b|$.

A non-Archimedean norm has a surprising property:

Proposition 10.2.6. *If x and y are elements of a non-Archimedean field k and $|x - y| \leq |y|$, then $|x| = |y|$. In other words, all triangles are isosceles.*

Proof. The ultra-metric inequality implies

$$|x| = |x - y + y| \leq \max\{|x - y|, |y|\} = |y|.$$

On the other hand,

$$|y| = |y - x + x| \leq \max\{|y - x|, |x|\}.$$

Now, if $|x - y| > |x|$, then $|y| \leq |x - y|$, a contradiction. So, $|x - y| \leq |x|$ which implies $|y| \leq |x|$. Hence, $|x| = |y|$.   □

Proposition 10.2.7. *For any $x \in k$, $|x| < 1$ if and only if*

$$\lim_{n \mapsto \infty} x^n = 0.$$

Proof. " $\Longrightarrow$ " If $|x| < 1$, then, since $|x^n| = |x|^n$, we get

$$\lim_{n \mapsto \infty} |x^n| = 0, \quad \text{and so} \quad \lim_{n \mapsto \infty} x^n = 0.$$

" $\Longleftarrow$ " Now, if $|x| \geq 1$, then $|x^n| \geq 1$ for any $n \in \mathbb{N}$. Therefore

$$\lim_{n \mapsto \infty} x^n \neq 0.$$

□

Proposition 10.2.8. $|x| = |x|^\alpha$, $\alpha > 0$ *is a norm on $\mathbb{Q}$ if and only if $\alpha \leq 1$.*

Proof. If $\alpha \leq 1$, then we can check easily that conditions 1 and 2 of the definition of a norm are satisfied. For the triangle inequality, assume that $|x| \leq |y|$. Since $|x + y| \leq |x| + |y|$,

$$|x + y|^\alpha \leq (|x| + |y|)^\alpha = |y|^\alpha \left(1 + \frac{|x|}{|y|}\right)^\alpha. \tag{1}$$

But since $\left(1 + \frac{|x|}{|y|}\right) \geq 1$ and $\alpha \leq 1$,

$$\left(1 + \frac{|x|}{|y|}\right)^\alpha \leq \left(1 + \frac{|x|}{|y|}\right). \tag{2}$$

Hence, using (2) in inequality (1), we get

$$|x + y|^\alpha \leq |y|^\alpha \left(1 + \frac{|x|}{|y|}\right). \tag{3}$$

Because $|x| \leq |y|$, $0 \leq \frac{|x|}{|y|} \leq 1$, this implies (since $\alpha \leq 1$)

$$\frac{|x|}{|y|} \leq \frac{|x|^\alpha}{|y|^\alpha}. \tag{4}$$

Thus, using (4) in inequality (3), we have

$$|x + y|^\alpha \leq |y|^\alpha + |x|^\alpha.$$

Finally, if $\alpha > 1$ then the triangle inequality fails. To see this, consider for example

$$|1 + 1|^\alpha = 2^\alpha > 1^\alpha + 1^\alpha = 2.$$

$\square$

Remark 10.2.9. As we shall see later, by Ostrowski's Theorem (see 10.4.4) the above norm is equivalent to the usual absolute value $|\cdot|$ on $\mathbb{Q}$.

Now, let $|\cdot|_1$ and $|\cdot|_2$ be two norms on the field k.

Definition 10.2.10. We say that $|\cdot|_1$ and $|\cdot|_2$ are *equivalent norms*, and we write $|\cdot|_1 \sim |\cdot|_2$, if they define the same topology on k.

It is clear that the equivalence of norms is an equivalence relation on the set of all norms on the field k.

Let $|\cdot|_1$ and $|\cdot|_2$ be two norms on k. Then, they are equivalent if and only if $|\cdot|_1 < 1 \iff |\cdot|_2 < 1$.

Indeed because of the symmetry, it is enough to consider an $a \in k$ with $|a|_1 < 1$ and prove that $|a|_2 < 1$. Let

$$U = \{x \in k \; : \; |x|_2 < 1\}.$$

The set U is a neighborhood of 0 under the topology induced by either norm. Now, $|a^n|_1 = |a|_1^n$ can be made arbitrarily small by choosing n large enough. In particular, we can choose an n such that $a^n \in U$. But then $|a|_2^n < 1$. And so $|a|_2 < 1$.

By taking inverses we deduce that $|a|_1 > 1$ if and only if $|a|_2 > 1$. Hence, $|a|_1 = 1$ if and only if $|a|_2 = 1$. In particular, if one of the norms is trivial, then so is the other (see [102]).

Definition 10.2.11. (**Ring of Integers**) Suppose $|\cdot|$ is a non-Archimedean absolute value on a field k. Then

$$\mathcal{O} = \{a \in k : |a| \leq 1\}$$

is a ring called the *ring of integers* of k with respect to $|\cdot|$.

That $\mathcal{O}$ is a subring of k follows easily from the ultra-metric inequality and the multiplicative property of the non-Archimedean absolute value $|\cdot|$. Indeed, if a, $b \in \mathcal{O}$ then $|ab| = |a| \cdot |b| \leq 1$. Therefore, $ab \in \mathcal{O}$, and also $|a + b| \leq \max\{|a|, |b|\} \leq 1$. Thus, $a + b \in \mathcal{O}$.

Proposition 10.2.12. *Two non-Archimedean norms $|\cdot|_1$ and $|\cdot|_2$ are equivalent if and only if they give the same $\mathcal{O}$.*

Proof. Suppose $|\cdot|_1$ is equivalent to $|\cdot|_2$, and so $|\cdot|_1 = (|\cdot|_2)^c$ for some $c > 0$. Then $|c|_1 \leq 1$ if and only if $(|c|_2)^c \leq 1$, i.e., if $|c|_2 \leq 1^{1/c} = 1$. Thus $\mathcal{O}_1 = \mathcal{O}_2$.

Conversely, suppose $\mathcal{O}_1 = \mathcal{O}_2$. Then, $|a|_1 < |b|_1$ if and only if $a/b \in \mathcal{O}_1$ and $b/a \notin \mathcal{O}_1$. Therefore,

$$|a|_1 < |b|_1 \iff |a|_2 < |b|_2. \tag{10.1}$$

The topology induced by $|\cdot|_1$ has as a neighborhood basis the set of open balls

$$B_{<r}(z) = \{x \in k : |x - z|_1 < r\},$$

where $r > 0$, and likewise for $|\cdot|_2$. Since the absolute values $|b|_1$ get arbitrarily close to 0, the set $\mathcal{U}$ of open balls $B_{<|b|_1}(z)$ also forms a basis of the topology induced by $|\cdot|_1$ (and similarly for $|\cdot|_2$). By (10.1) we have

$$B_{<|b|_1}(z) = B_{<|b|_2}(z).$$

Therefore, the two topologies both have $\mathcal{U}$ as a basis and hence coincide. That the same topologies imply equivalence of the corresponding norms will be proved in Section 10.5.6. □

The set of $a \in \mathcal{O}$ with $|a| < 1$ forms an ideal $\mathfrak{p}$ in $\mathcal{O}$. The ideal $\mathfrak{p}$ is maximal since if $a \in \mathcal{O}$ and $a \notin \mathfrak{p}$, then $|a| = 1$, and so $|1/a| = 1/|a| = 1$. Hence $1/a \in \mathcal{O}$, and therefore a is a unit.

Lemma 10.2.13. *A non-Archimedean norm* $|\cdot|$ *is discrete if and only if* $\mathfrak{p}$ *is a principal ideal.*

Proof. First suppose that $|\cdot|$ is discrete. Choose $x \in \mathfrak{p}$ with $|x|$ maximal, which we can do since

$$S = \{\log |a| : a \in \mathfrak{p}\} \subset (-\infty, 1],$$

and so the discrete set S is bounded above. Suppose $a \in \mathfrak{p}$. Then

$$\left|\frac{a}{x}\right| = \frac{|a|}{|x|} \leq 1,$$

and so $a/x \in \mathcal{O}$. Thus

$$a = x \cdot \frac{a}{x} \in x\mathcal{O}.$$

Conversely, suppose $\mathfrak{p} = (x)$ is principal. For any $a \in \mathfrak{p}$ we have $a = xb$ with $b \in \mathcal{O}$. Thus

$$|a| = |x| \cdot |b| \leq |x| < 1.$$

Thus $\{|a| : |a| < 1\}$ is bounded away from 1, which is exactly the definition of discrete. □

10.3 Non-Archimedean Absolute Values vs. Valuations

Let k be a field. The following theorem gives us a relation between the valuations and the absolute values on k.

Theorem 10.3.1. *Let $|\cdot|$ be a non-Archimedean absolute value on k and $\lambda \in \mathbb{R}$, $\lambda > 0$. Then, the function*

$$v_\lambda : k \longrightarrow \mathbb{R} \cup \{\infty\} \ : \ v_\lambda(x) = \begin{cases} -\lambda \log |x| & \text{if } x \neq 0, \\ \infty & \text{if } x = 0 \end{cases}$$

is a valuation on k. Furthermore, if $\lambda, \lambda' \in \mathbb{R}$, with $\lambda \neq \lambda'$ and $\lambda, \lambda' > 0$, v_λ is equivalent to $v_{\lambda'}$.

Conversely, if v is a valuation on k and $\mu \in \mathbb{R}$, with $\mu > 1$, the function

$$|\cdot|_\mu : k \longrightarrow \mathbb{R} \ : \ |x|_\mu = \begin{cases} \mu^{-v(x)} & \text{if } x \neq 0, \\ 0 & \text{if } x = 0 \end{cases}$$

is an absolute value on k. In addition, if $\mu, \mu' \in \mathbb{R}$, with $\mu, \mu' > 1$ and $\mu \neq \mu'$, $|\cdot|_\mu$ is equivalent to $|\cdot|_{\mu'}$.

Proof. We first show v_λ is a valuation. Indeed, from its definition, $v_\lambda = \infty$ if and only if $x = 0$. Now, let x, y be in k. If one of the two is $= 0$, then $xy = 0$ which implies

$$v_\lambda(xy) = v_\lambda(0) = \infty = v_\lambda(x) + v_\lambda(y).$$

If both $x, y \neq 0$ then,

$$v_\lambda(xy) = -\lambda \log |xy| = -\lambda \log(|x||y|) = -\lambda \log |x| - \lambda \log |y| = v_\lambda(x) + v_\lambda(y).$$

Now, if $x = y = 0$, then

$$v_\lambda(x + y) = v_\lambda(0) = \infty = \min\{v_\lambda(x), v_\lambda(y)\}.$$

If $x = 0$ and $y \neq 0$ (or $y = 0$ and $x \neq 0$), we have either

$$v_\lambda(x + y) = v_\lambda(y) = \min\{v_\lambda(x), v_\lambda(y)\},$$

or

$$v_\lambda(x + y) = v_\lambda(x) = \min\{v_\lambda(x), v_\lambda(y)\}.$$

So, we can assume that both x, $y \neq 0$. We get

$$\begin{aligned}
v_\lambda(x + y) &= -\lambda \log |x + y| \\
&\geq -\lambda \log(\max\{|x|, |y|\}) \\
&= \min\{-\lambda \log |x|, -\lambda \log |y|\} \\
&= \min\{v_\lambda(x), v_\lambda(y)\}.
\end{aligned}$$

Therefore, v_λ is a valuation. To finish, let λ, $\lambda' \in \mathbb{R}$, with $\lambda \neq \lambda'$ and λ, $\lambda' > 0$. Then, for each $x \in k$, $x \neq 0$,

$$v_\lambda(x) = -\lambda \log |x| = \left(\frac{\lambda}{\lambda'}\right)(-\lambda' \log |x|) = \frac{\lambda}{\lambda'} v_{\lambda'}(x)$$

i.e. v_λ and $v_{\lambda'}$ are equivalent.

Now, we will show that $|\cdot|_\mu$ is an absolute value. It is clear that $|x|_\mu = 0$ if and only if $x = 0$, and since $\mu > 1$, $|x|_\mu \geq 0$ for all $x \in k$.

Let x, $y \in k$. If $x = 0$, or $y = 0$, then $xy = 0$. Hence $|xy|_\mu = 0 = |x|_\mu |y|_\mu$. If x, $y \neq 0$, we get

$$|xy|_\mu = \mu^{-v(xy)} = \mu^{-v(x)-v(y)} = \mu^{-v(x)} \mu^{-v(y)} = |x|_\mu |y|_\mu.$$

To prove the ultra-metric inequality, again let x, $y \in k$. If $x = y = 0$, then $|x + y|_\mu = 0 = \max\{|x|_\mu, |y|_\mu\}$. If only one of the x, y is $\neq 0$, say $x \neq 0$, then, $|x + y|_\mu = |x|_\mu = \max\{|x|_\mu, |y|_\mu\}$. Finally, if both x and y are $\neq 0$, we have

$$\begin{aligned}
|x + y|_\mu = \mu^{-v(x+y)} &\leq \mu^{-\min\{v(x), v(y)\}} \\
&= \max\{\mu^{-v(x)}, \mu^{-v(y)}\} = \max\{|x|_\mu, |y|_\mu\}.
\end{aligned}$$

Hence, $|\cdot|_\mu$ is a non-Archimedean absolute value.

In addition, if μ, $\mu' \in \mathbb{R}$, with μ, $\mu' > 1$ and $\mu \neq \mu'$, we put

$$\lambda = \frac{\log \mu}{\log \mu'}.$$

Then, we have

$$|x|_\mu = \mu^{-v(x)} = (\mu')^{-\lambda v(x)} = |x|_\mu^\lambda.$$

In other words, the two absolute values are equivalent. $\qquad\qquad\square$

10.3.1   The Ring of Integers

Now, a subring $\mathcal{O}$ of the field k is called a *valuation ring* if for all $x \in k$, $x \neq 0$, either $x \in \mathcal{O}$ or $x^{-1} \in \mathcal{O}$.

A valuation v on a field k gives rise to the valuation ring:

$$\mathcal{O}_v = \{x \in k, \; : \; v(x) \geq 0\}.$$

The group of units is

$$\mathcal{O}_v^{\times} = \{x \in k \; : \; v(x) = 0\}.$$

Therefore, the set of non-units is

$$\mathfrak{m}_v = \{x \in k \; : \; v(x) > 0\},$$

and is the only maximal ideal of $\mathcal{O}_v$. This shows that $\mathcal{O}_v$ is a local ring.

Definition 10.3.2. We call the *residue field* of v, the quotient

$$k_v = \mathcal{O}_v / \mathfrak{m}_v.$$

Proposition 10.3.3. *Let $\mathcal{O} \subseteq k$ be a valuation ring of k. Then, there exists a valuation v on k such that $\mathcal{O} = \mathcal{O}_v$.*

Proof. The quotient group $G = k^{\times}/\mathcal{O}^{\times}$ is an Abelian group. Denote the cosets $x\mathcal{O}^{\times}$ by $[x]$. We rewrite G additively by setting

$$[x] + [y] = [xy].$$

Define the following binary relation on G

$$[x] \leq [y] \text{ if and only if } \tfrac{y}{x} \in \mathcal{O}.$$

Then G becomes an ordered Abelian group. Define

$$v : k \longrightarrow G \cup \{\infty\} \; : \; v(0) = \infty \text{ and } v(x) = [x], \; x \neq 0.$$

It is straightforward to check that v is a valuation and that $\mathcal{O} = \mathcal{O}^{\times}$. $\square$

Definition 10.3.4. (**Ring of Integers**) Suppose $|\cdot|$ is a non-Archimedean absolute value on a field k. Then

$$\mathcal{O} = \{a \in k : |a| \le 1\}$$

is a ring called the *ring of integers* (or *valuation ring*) of k with respect to $|\cdot|$.

Proposition 10.3.5. *Suppose $|\cdot|$ is a non-Archimedean absolute value on a field k. Let $\mathcal{O}$ be the ring of integers, and $\mathfrak{p}$ the valuation ideal*

$$\mathfrak{p} = \{x \in k \mid |x| < 1\}.$$

Then,

1. *$\mathcal{O}$ is a local ring with unique maximal ideal $\mathfrak{p}$ and fraction field k.*

2. *If $u \in k$, then u is a unit of $\mathcal{O}$ if and only if $|u| = 1$.*

3. *If the trivial valuation is excluded, then $\mathcal{O}$ is not a field.*

Proof. 1. That $\mathcal{O}$ is a subring of k follows easily from the ultra-metric inequality and the multiplicative property of the non-Archimedean absolute value $|\cdot|$. Indeed, if $a,\, b \in \mathcal{O}$ then $|ab| = |a| \cdot |b| \le 1$ and so $ab \in \mathcal{O}$, and also $|a + b| \le \max\{|a|, |b|\} \le 1$, and therefore $a + b \in \mathcal{O}$.

In addition, k is the fraction field of $\mathcal{O}$, since if $x \in k$, $x \ne 0$, then either x or its inverse belongs to $\mathcal{O}$.

By definition $\mathfrak{p}$ is a proper ideal. If $\mathfrak{m}$ is also a proper ideal of $\mathcal{O}$ then we must have $\mathfrak{m} \subset \mathfrak{p}$. This is because $\mathcal{O} - \mathfrak{p} \subseteq \mathcal{O} - \mathfrak{m}$. Indeed, if $a \in \mathcal{O} - \mathfrak{p}$, then $|a| = 1$, and so $|a^{-1}| = 1$, which means that $a \in \mathcal{O}$. Hence, if $a \in \mathfrak{m}$, then $aa^{-1} = 1 \in \mathfrak{m}$, a contradiction.

2. If $u,\, v \in \mathcal{O}$, and $uv = 1$, then $|u| |v| = 1$. But both $|u|$ and $|v|$ are ≤ 1, so they must be $= 1$.

To prove the converse, if $|u| = 1$, then $|u^{-1}| = 1$. But then both u and u^{-1} are in $\mathcal{O}$, and so u is a unit of $\mathcal{O}$.

3. If $a \ne 0$ and $|a| \ne 1$, then either $|a| < 1$ and $|a^{-1}| > 1$, or $|a| > 1$ and $|a^{-1}| < 1$. This shows that in both cases we have an element a of $\mathcal{O}$ whose inverse lies outside $\mathcal{O}$. $\qquad\square$

Definition 10.3.6. The field $\mathcal{O}/\mathfrak{p}$ is called the *residue field* of $\mathcal{O}$ or of k.

10.4 Absolute Values on $\mathbb{Q}$

We consider the absolute values on the field $\mathbb{Q}$ of rational numbers. The usual absolute value on $\mathbb{Q}$ will be denoted by $|\cdot|_\infty$. So

$$|x|_\infty = \begin{cases} x & \text{if } x \geq 0, \\ -x & \text{if } x < 0. \end{cases}$$

Obviously, $|\cdot|_\infty$ is Archimedean.

Now we want to construct a non-Archimedean norm, $|\cdot|_p$, on $\mathbb{Q}$. First we will define it for the natural numbers. To do this, let p be a prime number. We know that every natural number n can be written as a product of prime numbers, $n = 2^{r_2}3^{r_3}\cdots p^{r_p}\cdots$, and so we define

$$|n|_p = p^{-r_p}, \quad \text{and write} \quad |0|_p = 0, \quad \text{and} \quad |-n|_p = |n|_p.$$

We now extend this definition to the rational numbers, by setting

$$\left|\frac{n}{m}\right|_p = \frac{|n|_p}{|m|_p}, \quad \text{where} \quad m \neq 0.$$

In particular, if $x \in \mathbb{Z}$ is relatively prime to p, we have $|x|_p = 1$. More generally, if $x \in \mathbb{Z}$, $|x|_p = p^{-n}$, where n is the number of times p divides x.

We notice that for any integer $x \in \mathbb{Z}$ we get

$$|x|_p \leq 1,$$

and $|x|_p$ will be close to 0 if x is divisible by a high power of p. Therefore, two integers x, $y \in \mathbb{Z}$ will be close with respect to the p-adic norm if $p^n \mid x - y$ for a large n.

Example 10.4.1. We have:
$$|1|_2 = |3|_2 = 1, \quad |4|_2 = |12|_2 = |20|_2 = 2^{-2}, \quad |64|_2 = 2^{-6},$$

$$\left|\tfrac{3}{2}\right|_3 = 3^{-1}, \quad |p^{-1}|_p = |p|_p^{-1} = p,$$

$$|10|_2 = |2 \cdot 5|_2 = \tfrac{1}{2^1} = \tfrac{1}{2},$$

$$|250|_5 = |2 \cdot 5^3|_5 = \tfrac{1}{5^3} = \tfrac{1}{125},$$

$$\left|\tfrac{21}{27}\right|_3 = \tfrac{|21|_3}{|27|_3} = \tfrac{|3 \cdot 7|_3}{|3^3|_3} = \tfrac{3^{-1}}{3^{-3}} = 3^2.$$

Remark 10.4.2. With respect to $|\cdot|_2$, the closed ball $B_{1/2}(0)$ of radius $1/2$ about 0 is simply all rational numbers (in reduced form) with even numerators. Similarly, the closed ball $B_{1/4}(0)$ is simply all rational numbers (in reduced form) whose numerator is congruent to 0 (mod 4).

Proposition 10.4.3. $|\cdot|_p$ *is a non-Archimedean absolute value on* $\mathbb{Q}$.

Proof. We can check easily that properties 1 and 2 of the absolute value are satisfied. Now, we have to prove the property 3*, i.e. $|x + y| \leq \max\{|x|_p, |y|_p\}$ for any x, y in $\mathbb{Q}$. If $x = 0$ or $y = 0$, we are done. So, suppose x, $y \neq 0$. Write $x = \frac{a}{b}$ and $y = \frac{c}{d}$, in lowest terms. We have $|x|_p = \frac{|a|_p}{|b|_p}$, $|y|_p = \frac{|c|_p}{|d|_p}$ and $|x + y|_p = \frac{1}{p^\gamma} = \frac{|ad + bc|_p}{|b|_p \cdot |d|_p}$. Now, let $|a|_p = \frac{1}{p^k}$, $|b|_p = \frac{1}{p^l}$, $|c|_p = \frac{1}{p^m}$ and $|d|_p = \frac{1}{p^n}$. In addition, let $|ad + bc|_p = \frac{1}{p^\nu}$. Now, we have:

$$\gamma = \nu - l - n.$$

Since the highest power of p which divides the sum of two numbers is at least equal to the minimum of the highest power of p which divides the first and the highest power of p which divides the second, we get

$$\gamma \geq \min\{k + n, l + m\} - l - n = \min\{k - l, m - n\}.$$

Therefore,

$$|x + y|_p = \frac{1}{p^\gamma} \leq \max\left\{\frac{1}{p^{k-l}}, \frac{1}{p^{m-n}}\right\} = \max\{|x|_p, |y|_p\}.$$

$\square$

10.4.1 Ostrowski's Theorem

Theorem 10.4.4. *(Ostrowski's Theorem)* *Any possible absolute value on* $\mathbb{Q}$ *is equivalent to one of the following:*

1. *The usual absolute value* $f(x) = |x|^\alpha$, *where* $0 \le \alpha \le 1$.

2. *The p-adic absolute value* $f(x) = |x|_p^\alpha$, $\alpha \ge 0$.

Proof. We distinguish two cases.

Case 1. Assume there are some $n \in \mathbb{Z}$ such that $f(n) < 1$, and let p be the smallest such integer. Then, p must be a prime number. Indeed, if $p = ab$ then $1 > f(p) = f(ab) = f(a)f(b)$, and hence $f(a) < 1$ or $f(b) < 1$, a contradiction. Now, we will prove that f must be the p-adic absolute value $|\cdot|_p$. For this, take an integer m and write it in base p. In other words,

$$m = m_0 + m_1 p + m_2 p^2 + \ldots + m_k p^k, \quad \text{where} \quad 0 \le m_i \le p - 1, \quad \text{and} \quad m_k \ge 1.$$

Therefore, taking into consideration that $f(p) < 1$, $f(m_i) \le m_i \le p - 1$, and also $\log m > k \log p$ (since $m_k \ge 1$), we get

$$f(m) \le f(m_0) + f(m_1) + \cdots + f(m_k) \le (k+1)(p-1) < \left(\frac{\log m}{\log p} + 1 \right)(p-1).$$

This inequality must hold for any $m \in \mathbb{Z}$, and so it holds for m^n for any $n \in \mathbb{N}$. So, we obtain

$$f(m)^n = f(m^n) \le \left(\frac{\log m^n}{\log p} + 1 \right)(p-1) = \left(n\frac{\log m}{\log p} + 1 \right)(p-1).$$

Therefore,

$$f(m) \le \left(n\frac{\log m}{\log p} + 1 \right)^{\frac{1}{n}} (p-1)^{\frac{1}{n}}$$

and taking limits of both sides as $n \to \infty$, we get

$$f(m) \le 1 \quad \text{for all} \quad m \in \mathbb{Z}.$$

To prove that f is the p-adic absolute value, we must prove that $f(m) = 1$ for all m such that $(m, p) = 1$. Indeed, if $(m, p) = 1$ then $(m^n, p^n) = 1$, which means that there are integers λ_n, μ_n such that $m^n \lambda_n + p^n \mu_n = 1$. Therefore, using the triangle inequality,

$$1 = f(1) \le f(m^n \lambda_n) + f(p^n \mu_n) \le f(m)^n + f(p)^n.$$

Since $f(p) < 1$, we have $f(p)^n \to 0$ as $n \to \infty$, which gives us that $f(m) \le 1$. But we know that $f(m) \le 1$, and therefore $f(m) = 1$.

Case 2. Now, we assume that $f(n) \ge 1$ for any $n \in \mathbb{Z}$. Pick up an $\alpha \in \mathbb{N}$, with $\alpha \ge 2$, and use it as a base. Then any integer m can be written as $m = m_0 + m_1\alpha + \cdots + m_k\alpha^k$. Therefore,

$$f(m) \le (\alpha - 1)[1 + f(\alpha) + \cdots + f(\alpha)^k] \le (k + 1)(\alpha - 1)f(\alpha)^k$$

$$< \left(\frac{\log m}{\log \alpha} + 1\right)(\alpha - 1)f(a)^{\frac{\log m}{\log \alpha}}.$$

Now, we replace m by m^n and we get from the above inequality

$$f(m)^n = f(m^n) \le \left(n\frac{\log m}{\log \alpha} + 1\right)(\alpha - 1)f(\alpha)^{n\frac{\log m}{\log \alpha}}$$

and passing to the limit $n \to \infty$ we get

$$f(m) \le f\alpha)^{\frac{\log m}{\log \alpha}}.$$

Repeating the above by interchanging α and m we get

$$f(m)^{\frac{1}{\log m}} = f(\alpha)^{\frac{1}{\log \alpha}}$$

and this for all $\alpha, m \ge 2$. From this last equation it follows that for any integers α and m

$$\frac{\log f(\alpha)}{\log \alpha} = \frac{\log f(m)}{\log m} = \lambda$$

for some λ and so $f(m) = m^\lambda$. $\square$

10.5   Ultra-metric Spaces

Let X be a set and $d : X \times X \longrightarrow \mathbb{R}$ be a map.

Definition 10.5.1. We say that (X, d) is a *metric space* with *metric d* if and only if:

1. $d(x, y) \geq 0$ for any x, $y \in X$, with equality only if $x = y$.

2. $d(x, y) = d(y, x)$ for every x and y.

3. $d(x, y) \leq d(x, z) + d(z, y)$ for any x, y, $z \in X$ (the *triangle inequality*).

Definition 10.5.2. We say that the metric d is an *ultra-metric* if and only if the condition 3 is replaced by the following:
3* - $d(x, y) \leq \max\{d(x, z), d(z, y)\}$, $\forall$ x, y, $z \in X$.
We then say (X, d) is an *ultra-metric* space.

A norm $|\cdot|$ on a field k induces a metric by defining

$$d(x, y) = |x - y|.$$

This in turn gives rise to a topology in which a basis for the neighborhoods of a point are the *open balls*

$$B_{<r}(a) = \{x \in k : |x - a| < r\}$$

for $r > 0$.

Lemma 10.5.3. *Equivalent norms induce the same topology.*

Proof. If $|\cdot|_1 = (|\cdot|_2)^c$, then $|x - a|_1 < r$ if and only if $(|x - a|_2)^c < r$ if and only if $|x - a|_2 < r^{1/c}$ so $B_{<1}(a) = B_{<r^{1/c}}(a)$. Thus the basis of open neighborhoods of a for $|\cdot|_1$ and $|\cdot|_2$ are identical.   □

A norm satisfying the triangle inequality gives a metric for the topology by defining the distance from a to b to be $|a - b|$. Assume for the rest of this section that we only consider norms that satisfy the triangle inequality.

Lemma 10.5.4. *A field whose topology is induced by a norm is a topological field, i.e., the operations sum, product, and reciprocal are continuous.*

Proof. For example, for the product, the triangle inequality implies that

$$|(a + \epsilon)(b + \delta) - ab| \leq |\epsilon||\delta| + |a||\delta| + |b||\epsilon|$$

is small when $|\epsilon|$ and $|\delta|$ are small for fixed a, b, and so it is continuous.
□

Lemma 10.5.5. *Suppose two norms $|\cdot|_1$ and $|\cdot|_2$ on the same field k induce the same topology. Then for any sequence $(x_n)_{n \in \mathbb{N}}$ in k we have*

$$|x_n|_1 \to 0 \iff |x_n|_2 \to 0.$$

Proof. It suffices to prove that if $|x_n|_1 \to 0$ then $|x_n|_2 \to 0$. Let $\epsilon > 0$. The topologies induced by the two norms are the same, so $B_{<\epsilon}(0)$ can be covered by open balls $B_{<r_i}(a_i)$. One of these open balls $B_{<r}(a)$ contains 0. There is $\delta > 0$ such that

$$B_{<\delta}(0) \subset B_{<r}(a) \subset B_{<\epsilon}(0).$$

Since $|x_n|_1 \to 0$, there exists an N such that for $n \geq N$ we have $|x_n|_1 < \delta$. For such n, we have $x_n \in B_{<\delta}(0)$. Therefore, $x_n \in B_{<\epsilon}(0)$, and so $|x_n|_2 < \epsilon$. Thus $|x_n|_2 \to 0$. □

Proposition 10.5.6. *If two norms $|\cdot|_1$ and $|\cdot|_2$ on the same field induce the same topology, then they are equivalent in the sense that there is a positive real α such that $|\cdot|_1 = |\cdot|_2^\alpha$.*

Proof. If $x \in k$ and $i = 1, 2$, then $|x^n|_i \to 0$ if and only if $|x|_i^n \to 0$, which is the case if and only if $|x|_i < 1$. Thus Lemma 10.5.5 implies that $|x|_1 < 1$ if and only if $|x|_2 < 1$. On taking reciprocals we see that $|x|_1 > 1$ if and only if $|x|_2 > 1$, so finally $|x|_1 = 1$ if and only if $|x|_2 = 1$.

Now let $w, z \in k$ be non-zero elements with $|w|_i \neq 1$ and $|z|_i \neq 1$. On applying the foregoing to

$$x = w^m z^n \qquad (m, n \in \mathbb{Z})$$

we see

$$m \log |w|_1 + n \log |z|_1 \geq 0$$

if and only if

$$m \log |w|_2 + n \log |z|_2 \geq 0.$$

Dividing through by $\log |z|_i$ and rearranging, it follows that for every rational number $\alpha = -n/m$,

$$\frac{\log |w|_1}{\log |z|_1} \geq \alpha \iff \frac{\log |w|_2}{\log |z|_2} \geq \alpha.$$

Thus

$$\frac{\log |w|_1}{\log |z|_1} = \frac{\log |w|_2}{\log |z|_2},$$

so

$$\frac{\log |w|_1}{\log |w|_2} = \frac{\log |z|_1}{\log |z|_2}.$$

Since this equality does not depend on the choice of z, we see there is a constant c $(= \log |z|_1 / \log |z|_2)$ such that $\log |w|_1 / \log |w|_2 = c$ for all w. Thus $\log |w|_1 = c \cdot \log |w|_2$, and so $|w|_1 = (|w|_2)^c$, which implies that $| \cdot |_1$ is equivalent to $| \cdot |_2$. $\qquad \square$

Definition 10.5.7. A metric d is called *non-Archimedean* if $d(x, y) \leq \max\{d(x, z), d(z, y)\}$ for any x, y and z.

Remark 10.5.8. If the metric d is induced by a non-Archimedean norm, then it is non-Archimedean, since $d(x, y) = f(x - y) = f(x - z + z - y) \leq \max\{f(x - z), f(z - y)\} = \max\{d(x, z), d(z, y)\}$.

Proposition 10.5.9. *Let k be a field and let $| \cdot |$ be a non-Archimedean norm on it. If x, $y \in k$ and $|x| \neq |y|$, then*

$$|x + y| = max\{|x|, |y|\}.$$

Proof. Without any loss of generality we can assume that $|x| > |y|$. Then, we have

$$|x + y| \leq |x| = \max\{|x|, |y|\}. \qquad (*)$$

On the other hand, writing $x = (x+y) - y$, we get $|x| \leq \max\{|x+y|, |y|\}$. Since by assumption $|x| > |y|$, this can happen only if $\max\{|x+y|, |y|\} = |x+y|$. This gives us $|x| \leq |x+y|$ and so by using $(*)$ we get $|x| = |x+y|$. □

Definition 10.5.10. Let (X, d) be a metric space, $x \in X$ and $r > 0$ be a real number. We call the *open ball* with center x and radius $r > 0$ the set

$$B_{<r}(x) = \{y \in X \ : \ d(x, y) < r\},$$

and the *closed ball* centered at x, of radius r, the set

$$B_{\leq r}(x) = \{y \in X \ : \ d(x, y) \leq r\}.$$

These balls have some strange topological properties from the point of view of the usual Euclidean space. For example,

Proposition 10.5.11. *If (X, d) is an ultra-metric space, any point of a ball can be chosen as its center (so such a ball is not uniquely characterized by its center and radius)!*

Proof. Let the open ball $B_{<r}(x)$ and let $y \in B_{<r}(x)$. We will prove that $B_{<r}(x) = B_{<r}(y)$. To do this, let $a \in B_{<r}(x)$. Then, $d(x, a) < r$, and so $d(a, y) = f(a - y) = f(a - x + x - y) \leq \max\{f(a - x), f(x - y)\} < r$. Hence, $a \in B_{<r}(y)$, i.e. $B_{<r}(x) \subseteq B_{<r}(y)$. Similarly, we prove that $B_{<r}(y) \subseteq B_{<r}(x)$. Thus, $B_{<r}(x) = B_{<r}(y)$. □

Another surprising fact is the following.

Proposition 10.5.12. *In an ultra-metric space (X, d) any two balls have a non-empty intersection if and only if one of them is contained in the other.*

Proof. ("$\Longrightarrow$".) We will show that for any two balls B_1 and B_2 in X, such that $B_1 \cap B_2 \neq \varnothing$, we have $B_1 \subseteq B_2$ or $B_2 \subseteq B_1$. Let $x \in B_1 \cap B_2$. Using proposition 10.5.11, we distinguish the following cases:

1. $B_1 = B_{<\epsilon}(x)$ and $B_2 = B_{<\delta}(x)$.

2. $B_1 = B_{<\epsilon}(x)$ and $B_2 = B_{\leq\delta}(x)$.

3. $B_1 = B_{\leq \epsilon}(x)$ and $B_2 = B_{<\delta}(x)$.

4. $B_1 = B_{\leq \epsilon}(x)$ and $B_2 = B_{\leq \delta}(x)$.

Assuming, without loss of generality, that $\epsilon < \delta$, one can see that in all cases $B_1 \subseteq B_2$, with the remark that in case 3, $B_2 \subseteq B_1$ if $\epsilon = \delta$.
(" $\Longrightarrow$ ".) Obvious. $\square$

Proposition 10.5.13. *Every ball is both open and closed.*

Proof. Let $B_{<r}(a)$ be an open ball, and let x be any point on $\overline{B_{<r}(a)}$. Then, $B_{<r}(a) \cap B_{<s}(x) \neq \varnothing$ for any ball centered at x and of any radius $s > 0$. Without loss of generality we can take $s < r$. Then, since the two balls are not disjoint, one must contain the other. Therefore, since $r > s$, $B_{<s}(x) \subseteq B_{<r}(a)$, and so $B_{<r}(a)$ is closed. $\square$

The above proposition implies the surprising fact that the ball $B_{<r}(a)$ not only has no boundary (!), but the closed ball

$$\overline{B_{<p^n}(a)} = \{x \in \mathbb{Q}_p \;:\; |x - a|_p \leq p^n\}$$
$$= \{x \in \mathbb{Q}_p \;:\; |x - a|_p < p^{n+1}\} = B_{<p^{n+1}}(a)$$

is not the closure of the open ball $B_{<p^n}(a)$, but rather the open ball of radius p^{n+1}! This also implies the topology is *totally disconnected*.

Corollary 10.5.14. *If the ultra-metric space X is connected, then it is empty or consists of a single point.*

Proof. If $X \neq \varnothing$, we pick a point $x \in X$. Proposition 10.5.13 implies $X = B_\epsilon(a)$ for any $\epsilon > 0$, and so $X = \{x\}$. $\square$

The following proposition connects the usual absolute value of $\mathbb{Q}$ with the p-adic absolute values.

Proposition 10.5.15. *If $x \in \mathbb{Q}$, $x \neq 0$,*

$$\prod_p |x|_p = 1,$$

where $p \in \{\infty, 2, 3, 5, \ldots\}$ and $|x|_\infty$ is the real absolute value of x.

Proof. It suffices to prove the statement for a positive integer x from which the general case follows. By the Fundamental theorem of Arithmetic x can be written uniquely as

$$x = p_1^{n_1} p_2^{n_2} \cdots p_j^{n_j}.$$

We have

$$
\begin{aligned}
|x|_q &= 1 && \text{if } q \neq p_i,\ i = 1, 2, ..., j \\
|x|_{p_i} &= p_i^{-n_i} && \text{if } i = 1, 2, ..., j \\
|x|_\infty &= p_1^{n_1} p_2^{n_2} \cdots p_j^{n_j}
\end{aligned}
$$

The result follows. □

Corollary 10.5.16. *For any prime p and any $n \in \mathbb{N}$, $|n|_p \geq \frac{1}{n}$.*

Proof. The product formula in Proposition 10.5.15 and the fact that for any prime q, $|n|_q \leq 1$, imply

$$|n|_\infty \prod_{q \text{ prime}} |x|_q = 1.$$

In other words,

$$n \cdot |n|_p \prod_{\substack{q \text{ prime} \\ q \neq p}} |x|_q = 1$$

and therefore, $n \cdot |n|_p \geq 1$. □

10.6   The Completion of $\mathbb{Q}$

Definition 10.6.1. If (X, d) is a metric (resp. ultra-metric) space, we say that the sequence $(x_n)_{n \in \mathbb{N}}$ is a *Cauchy sequence* if for any $\epsilon > 0$ there is an $N(\epsilon) > 0$ such that $d(x_n, x_m) < \epsilon$ for any $n, m > N(\epsilon)$.

Definition 10.6.2. A metric (resp. ultra-metric) space (X, d) is *complete* provided that every Cauchy sequence (x_i) in X converges to some point $x \in X$.

Definition 10.6.3. Let k be a field equipped with a norm $|\cdot|$. A *completion* of k is a normed field $(\widehat{k}, \|\cdot\|)$, where $\widehat{k}$ is a field extension of k, and $\|\cdot\|$ is a norm on $\widehat{k}$ which extends the norm $|\cdot|$, such that:

1. $\widehat{k}$ is complete.

2. k is dense in $\widehat{k}$.

Theorem 10.6.4. *Let k be a field equipped with a norm $|\cdot|$. Then, it can be completed to a field $\widehat{k} \supseteq k$.*

Proof. Let A be the set of all Cauchy sequences in k. If (x_i), (y_i) are two Cauchy sequences, we know that so is their sum $(x_i) + (y_i) = (x_i + y_i)$ and their product $(x_i) \cdot (y_i) = (x_i y_i)$, and therefore A is a commutative ring. Now, let S be the set of all sequences in k that converge to 0. We know that any convergent sequence is a Cauchy sequence, so $S \subset A$. We can easily check that S is an ideal in A. In fact, it is a maximal ideal. We will prove this by contradiction. Assume S is properly contained in an ideal I. So, there is a Cauchy sequence $(x_i) \in I - S$ which does not converge to 0, i.e. there is an $\epsilon > 0$ and an $N(\epsilon) > 0$ such that $|x_i| \geq \epsilon$ for any $i > N(\epsilon)$. In particular, $x_i \neq 0$ for any $i > N(\epsilon)$. Now, consider the sequence (y_i) defined by $y_i = 0$ if $x_i = 0$ and $y_i = x_i^{-1}$ if $x_i \neq 0$. One sees easily that (y_i) is a Cauchy sequence, and so $(y_i) \in A$. Therefore, $(y_i) \cdot (x_i) \in I$, since I is an ideal. But $(y_i) \cdot (x_i) = (y_i x_i)$ is the constant sequence 1 for all $i > N(\epsilon)$ and 0 for all $i \leq N(\epsilon)$. So, if we consider the sequence (z_i) with $z_i = 1$ for $i \leq N(\epsilon)$ and $z_i = 0$ for $i > N(\epsilon)$, we see that (z_i) converges to 0, and therefore $(z_i) \in S$ and the constant sequence

$$(1, 1, ..., 1, ...) = (x_i y_i) + (z_i) \in I$$

which means that $I = A$, a contradiction. Hence S is maximal.
Now, define $\widehat{k}$ as the quotient ring $\widehat{k} := A/S$. Then, since S is a maximal ideal, $\widehat{k}$ is a field. In addition, we can see that k can be seen as a subfield of $\widehat{k}$ via the injective homomorphism

$$\pi : k \longrightarrow \widehat{k} \quad \text{such that} \quad x \mapsto \pi(x) := (x, x, ...) + S,$$

where $(x, x, ...)$ denotes the constant sequence $x, x, ...$

Now, to define a norm on $\widehat{k}$, we remark that if (x_i) is a Cauchy sequence in k, then $|x_i|$ is a Cauchy sequence in $\mathbb{R}$, and since $\mathbb{R}$ is a complete space the limit

$$\lim_{i \mapsto \infty} |x_i|$$

exists. Since any element of $\widehat{k}$ is of the form $(x_i) + S$, we can define the map

$$\| \cdot \| : \widehat{k} \longrightarrow \mathbb{R}^{\geq 0} \quad : \quad \|(x_i) + S\| = \lim_{i \mapsto \infty} |x_i|.$$

We can see easily that $\| \cdot \|$ is well defined and it is a norm on $\widehat{k}$. In addition, this norm extends the norm on k since the limit of the constant sequence (x) is x.

It remains to show that k is dense in $\widehat{k}$ and $\widehat{k}$ is complete.

Let $\widehat{x} = (x_i) + S$ be an element of $\widehat{k}$, and for any $\epsilon > 0$ let $B_\epsilon^\circ(\widehat{x})$ be the open ball centered at $\widehat{x}$, of radius ϵ. Since (x_i) is a Cauchy sequence we can choose an $n_0 \in \mathbb{N}$ such that $|x_i - x_j| < \frac{\epsilon}{2}$ for all $i > n_0$. Then,

$$\|(x_i) + S - \pi(x_{n_0})\| = \|(x_1 - x_{n_0}, x_2 - x_{n_0}, ...) + S\| = \lim_{i \to \infty} |x_i - x_{n_0}| \leq \frac{\epsilon}{2} < \epsilon$$

and so $\pi(x_{n_0}) \in B(\widehat{x}, \epsilon)$, i.e. $B_\epsilon^\circ(\widehat{x}) \cap k \neq \varnothing$. This means that k is dense in $\widehat{k}$.

Finally, to prove that $\widehat{k}$ is complete, let $\widehat{x}_n$ be a Cauchy sequence in $\widehat{k}$. Since k is dense in $\widehat{k}$, we can find for every $n \in \mathbb{N}$ an element $x_n \in k$ such that $\|\widehat{x}_n - \pi(x_n)\| < \frac{1}{n}$. It follows that the sequence x_n is a Cauchy sequence in k, and so $\widehat{x} = (x_n) + S$ is in $\widehat{k}$. In addition, for each $n \in \mathbb{N}$ we have

$$\|\widehat{x}_n - \widehat{x}\| \leq \|\widehat{x}_n - \pi(x_n)\| + \|\pi(x_n) - \widehat{x}\| < \frac{1}{n} + \lim_{i \to \infty} |x_n - x_i|.$$

The right side of this inequality goes to 0, since (x_n) is a Cauchy sequence, which means that

$$\lim_{n \to \infty} \widehat{x}_n = \widehat{x} \in \widehat{k},$$

i.e. $\widehat{k}$ is complete. $\qquad\qquad\qquad\qquad\qquad\qquad\qquad\qquad\qquad\qquad\square$

We can complete $\mathbb{Q}$ with respect to the p-adic valuation in just the same way. The resulting field is called *the field of p-adic numbers*, and is denoted by $\mathbb{Q}_p$.

What does an element of $\mathbb{Q}_p$ look like? The answer to this question is reminiscent of the theory of meromorphic functions.

Proposition 10.6.5. *Let* $c_{-m} \neq 0$ *and* $0 \leq c_i < p$ *be integers. Then the partial sums of the series*

$$a = c_{-m}p^{-m} + c_{-m+1}p^{-m+1} + \cdots + c_{-1}p^{-1} + c_0 + c_1p + c_2p^2 + \cdots$$

form a Cauchy sequence and therefore a *is an element of* $\mathbb{Q}_p$.

Proof. Let $\epsilon > 0$. Then we can find an $N = N(\epsilon) \in \mathbb{N}$ such that $p^{-N} < \epsilon$, and for $n, k > N$, $k > n$, we have

$$\left| \sum_{i=-m}^{k} c_ip^i - \sum_{i=-m}^{n} c_ip^i \right|_p = \left| \sum_{i=n+1}^{k} c_ip^i \right|_p \leq \max\{|c_{n+1}p^{n+1}|_p, \ldots, |c_kp^k|_p\}$$

$$\leq p^{-N} < \epsilon.$$

$\square$

Proposition 10.6.6. *Let* $x \in \mathbb{Q}$ *with* $|x|_p \leq 1$. *Then for any* i *there is a* unique *integer* $\alpha \in \{0, 1, \ldots, p^i - 1\}$ *so that* $|x - \alpha|_p \leq p^{-i}$.

Proof. Let $x = \frac{a}{b}$ with a and b relatively prime. Since $|x|_p = p^{-\operatorname{ord}p(a)+\operatorname{ord}p(b)} \leq 1$, we get $\operatorname{ord}p(b) = 0$, that is, b and p^i are relatively prime for any i. We can then find integers m and n so that $np^i + mb = 1$. For $\alpha = am$ we get

$$|x-\alpha|_p = \left| am - \frac{a}{b} \right|_p = \left| \frac{a}{b} \right|_p |mb-1|_p \leq |mb-1|_p = |np^i|_p = |n|_p p^{-i} \leq p^{-i}.$$

There is exactly a multiple cp^i of p^i so that $(cp^i + \alpha) \in \{0, 1, \ldots, p^i - 1\}$, and we get

$$|cp^i + \alpha - x|_p \leq \max\{|\alpha - x|_p, |cp^i|_p\} \leq \max\{p^{-i}, p^{-i}\} = p^{-i}.$$

$\square$

This leads us to the following characterization of elements of $\mathbb{Q}_p$.

Theorem 10.6.7. *Let $a \in \mathbb{Q}_p$ with $|a|_p \leq 1$. Then there is exactly one Cauchy sequence $(a_n)_{n \in \mathbb{N}}$ representing a so that for every i*

1. $0 \leq a_i < p^i$.

2. $a_i \equiv a_{i+1} \pmod{p^i}$.

Proof. Let $(c_n)_{n \in \mathbb{N}}$ be a Cauchy sequence representing a. Since $|c_n|_p \to |a|_p \leq 1$, there is an N so that $|c_n|_p \leq 1$ for all $n > N$. By replacing the first N elements we get an equivalent Cauchy sequence so that $|b_n|_p \leq 1$ for all n. Now, for each $j = 1, 2, \ldots$ let $N(j)$ be chosen so that $N(j) \geq j$ and

$$|b_i - b_{i'}|_p \leq p^{-j} \text{ for all } i, i' \geq N(j).$$

From Proposition 10.6.6 we know that for any j there are integers $0 \leq a_j < p^j$ so that

$$|a_j - b_{N(j)}|_p \leq p^{-j}.$$

These a_j also satisfy condition 2. Indeed,

$$|a_{j+1} - a_j|_p = |a_{j+1} - b_{N(j+1)} + b_{N(j+1)} - b_{N(j)} + b_{N(j)} - a_j|_p$$
$$\leq \max\{|a_{j+1} - b_{N(j+1)}|_p, |b_{N(j+1)} - b_{N(j)}|_p, |b_{N(j)} - a_j|_p\}$$
$$\leq \max\{p^{-j-1}, p^{-j}, p^{-j}\} = p^{-j}.$$

This sequence is equivalent to $(b_n)_{n \in \mathbb{N}}$. To see this, for any j take $i \geq N(j)$

$$|a_i - b_i|_p = |a_i - a_j + a_j - b_{N(j)} + b_{N(j)} - b_i|_p$$
$$\leq \max\{|a_i - a_j|_p, |a_j - b_{N(j)}|_p, |b_{N(j)} - b_i|_p\}$$
$$\leq \{p^{-j}, p^{-j}, p^{-j}\} = p^{-j},$$

which proves $|a_i - b_i| \to 0$.

To prove uniqueness, let $(d_n)_{n \in \mathbb{N}}$ be another Cauchy sequence satisfying conditions 1 and 2 and let $(a_n)_{n \in \mathbb{N}} \neq (d_n)_{n \in \mathbb{N}}$. That is, $a_{i_0} \neq d_{i_0}$, for some i_0. Since a_{i_0} and d_{i_0} are between 0 and p^{i_0},

$$a_{i_0} \neq d_{i_0} \pmod{p^{i_0}}.$$

Condition 2 implies that if $i > i_0$, $d_{i_0} = d_i$ (mod p^{i_0}). Thus $a_i \neq d_i$ (mod p^{i_0}) and therefore

$$|a_i - d_i|_p > p^{-i_0}$$

does not converge to 0. This means $(a_n)_{n \in \mathbb{N}}$ and $(d_n)_{n \in \mathbb{N}}$ are not equivalent. $\square$

For $a \in \mathbb{Q}_p$ with $|a|_p \leq 1$ we can write the Cauchy sequence $(a_n)_{n \in \mathbb{N}}$ representing a from the previous theorem as

$$a_i = c_0 + c_1 p + \cdots + c_{i-1} p^{i-1}$$

for $c_i \in \{0, 1, \ldots, p - 1\}$ and a is represented by the convergent series

$$a = \sum_{i=0}^{\infty} c_i p^i,$$

which we can think of as a number written in base p as

$$a = \ldots c_n \ldots c_1 c_0.$$

The numbers $x \in \mathbb{Q}_p$ with $|x|_p \leq 1$ are called *p-adic integers*. The p-adic integers form a ring, denoted by $\mathbb{Z}_p$. For if $x, y \in \mathbb{Z}_p$ then by property 3∗ of the absolute value,

$$|x + y|_p \leq \max(|x|_p, |y|_p) \leq 1,$$

and so $x + y \in \mathbb{Z}_p$. Similarly, by property 1,

$$|xy|_p = |x|_p |y|_p \leq 1,$$

and so $xy \in \mathbb{Z}_p$.
Obviously $\mathbb{Z} \subset \mathbb{Z}_p$. More generally,

$$x = \frac{m}{n} \in \mathbb{Z}_p$$

if $p \nmid n$. (We shall sometimes say that a rational number x of this form is *p-integral*.) In other words,

$$\mathbb{Q} \cap \mathbb{Z}_p = \{\frac{m}{n} : p \nmid n\}.$$

Evidently the p-integral numbers form a subring of $\mathbb{Q}$.

Concretely, each element $x \in \mathbb{Z}_p$ is uniquely expressible in the form

$$x = c_0 + c_1 p + c_2 p^2 + \cdots \quad (0 \le c_i < p).$$

Let $x \in \mathbb{Z}_p$, with $|x|_p = 1$. Then,

$$x = c + yp,$$

where $0 < c < p$ and $y \in \mathbb{Z}_p$. Suppose first that $c = 1$, i.e.

$$x = 1 + yp.$$

Then, x is invertible in $\mathbb{Z}_p$, with

$$x^{-1} = 1 - yp + y^2 p^2 - y^3 p^3 + \cdots.$$

Even if $c \ne 1$, we can find d such that

$$cd \equiv 1 \pmod{p},$$

i.e. that

$$cd = 1 + yp.$$

Then,

$$dx = 1 + dyp,$$

and so x is again invertible in $\mathbb{Z}_p$, with

$$x^{-1} = d \left(1 - dyp + d^2 y^2 p^2 - \cdots \right).$$

Thus the elements $x \in \mathbb{Z}_p$ with $|x|_p = 1$ are all *units* in $\mathbb{Z}_p$ (and all units are of this form). These units form the multiplicative group

$$\mathbb{Z}_p^\times = \{x \in \mathbb{Z}_p : |x|_p = 1\}.$$

10.6.1 The Fields $\mathbb{Q}_p$ are all Non-Isomorphic

Proposition 10.6.8. *If p is a prime and m, n integers not divisible by p, with $n > 1$, then, m has an n-th root in $\mathbb{Q}_p$ if and only if m is an n-th power modulo p.*

Proof. We remark that $m_0 \equiv m \bmod p$ is non-zero. If m is an n-th power we can choose $d_0 \in [0, p-1]$ so that $d_0^n \equiv m \bmod p$, and we must have $d_0 \neq 0$ since $m_0 \neq 0$. Since p does not divide n, we can proceed and compute d_1, d_2,...

To prove the converse, suppose $m = d^n$ for some $d \in \mathbb{Q}_p$. Then $v_p(d^n) = v_p(m) = 0$ so d is a p-adic unit that we can write as $d = 0.d_0 d_1 d_2...$ and we must have $m \equiv (d_0)^n \bmod p$. $\qquad\square$

As a corollary we get

Theorem 10.6.9. *The fields $\mathbb{Q}_p$ are all non-isomorphic and none of them is isomorphic to $\mathbb{R}$.*

Proof. None of the fields $\mathbb{Q}_p$ contains $\sqrt{p}$ because if it did, then it would have a non-integral p-adic valuation which is impossible. On the other hand, $\mathbb{R}$ contains $\sqrt{p}$, so $\mathbb{Q}_p$ cannot be isomorphic to $\mathbb{R}$.

Consider two primes p, q with $p < q$. Clearly q is not $0 \bmod p$, so we can choose $d \in [1, p-1]$ so that $dq \equiv 1^n \bmod p$ is an n-th power modulo p, for all integers n. We have $v_q(dq) = 1$, so dq does not have an n-th root in $\mathbb{Q}_q$ for any $n > 1$. However, by Proposition 10.6.8, dq does have an n-th root in $\mathbb{Q}_p$, for all $n > 1$ and not divisible by p. Therefore $\mathbb{Q}_p \neq \mathbb{Q}_q$. $\qquad\square$

10.7 The Algebraic Definition of $\mathbb{Z}_p$

Let p be a prime number. Denoting by $\mathbb{Z}/p^n\mathbb{Z}$ the ring of residue classes mod p^n for $n \in \mathbb{N}$, we define the sequence of maps

$$\cdots \xrightarrow{\pi_{n+1}} \mathbb{Z}/p^{n+1}\mathbb{Z} \xrightarrow{\pi_n} \mathbb{Z}/p^n\mathbb{Z} \xrightarrow{\pi_{n-1}} \cdots \xrightarrow{\pi_1} \mathbb{Z}/p\mathbb{Z}$$

where, each π_n is the natural projection

$$\pi_n : \mathbb{Z}/p^{n+1}\mathbb{Z} \longrightarrow \mathbb{Z}/p^n\mathbb{Z} \; : \; a \mapsto \pi_n(a) = a \pmod{p^n}.$$

Then, we can define the inverse limit (see section 1.12 of Volume I)

$$\varprojlim \mathbb{Z}/p^n\mathbb{Z} = \Big\{(a_n) \in \prod_{1 \le n} \mathbb{Z}/p^n\mathbb{Z} : \pi_n(a_{n+1}) = a_n\Big\}.$$

The ring of p-adic integers $\mathbb{Z}_p$ can then be defined as the inverse limit,

$$\mathbb{Z}_p = \varprojlim \mathbb{Z}/p^n\mathbb{Z}$$

of the inverse system of rings $\mathbb{Z}/p^n\mathbb{Z}$ with morphisms (π_n) given by reduction modulo p^n, i.e. for a residue class $[x] \in \mathbb{Z}/p^{n+1}\mathbb{Z}$. Simply pick a representative $x \in [x]$ and take its residue class in $\mathbb{Z}/p^n\mathbb{Z}$. The multiplicative identity in $\mathbb{Z}_p$ is $1 = ([1], [1], \ldots)$, where the n-th $[1]$ denotes the residue class of $[1]$ in $\mathbb{Z}/p^n\mathbb{Z}$. The map sending each integer $x \in \mathbb{Z}$ to the sequence $([x], [x], [x], \ldots)$ is a ring homomorphism whose kernel is clearly trivial, since 0 is the only integer congruent to 0 mod p^n for all n. Therefore, the ring $\mathbb{Z}_p$ has characteristic 0 and contains $\mathbb{Z}$ as a subring. But, of course, $\mathbb{Z}_p$ is a much bigger ring than $\mathbb{Z}$.

Theorem 10.7.1. *For each positive integer n, the sequence*

$$0 \longrightarrow \mathbb{Z}_p \xrightarrow{\;[p^n]\;} \mathbb{Z}_p \xrightarrow{\;\pi_n\;} \mathbb{Z}/p^n \longrightarrow 0$$

is exact. Here $[p^n]$ is the map "multiplication" by p^n and π_n is the projection to $\mathbb{Z}/p^n\mathbb{Z}$.

Proof. The map $[p^n]$ shifts the p-adic expansion $(b_0, b_1 \ldots)$ of each element in $\mathbb{Z}_p$ to the right by n digits (filling with zeroes), yielding

$$(c_0, c_1, c_2, \ldots) = (0, \ldots, 0, b_0, b_1, b_2, \ldots),$$

with $c_j = 0$ for $j < n$ and $c_j = b_{j-n}$ for all $j \ge n$. This is clearly an injective operation on p-adic expansions, and hence on $\mathbb{Z}_p$. The image of $[p^n]$ consists of the elements in $\mathbb{Z}_p$ whose p-adic expansion $(c_0, c_1, c_2, \ldots)$ satisfies $c_0 = \ldots = c_{n-1} = 0$.

Conversely, the map π_n sends the p-adic expansion $(b_0, b_1 \ldots)$ to the sum

$$b_0 + b_1 p + b_2 p^2 + \cdots + b_{n-1} p^{n-1}$$

in $\mathbb{Z}/p^n\mathbb{Z}$. Each element of $\mathbb{Z}/p^n\mathbb{Z}$ is uniquely represented by an integer in $[0, p^n - 1]$, each of which can be uniquely represented by a sum as above, with $b_0,\ldots, b_{n-1}$ integers in $[0, p - 1]$. It follows that π_n is surjective, and its kernel consists of the elements in $\mathbb{Z}_p$ whose p-adic expansion $(b_0, b_1 \ldots)$ satisfies $b_0 = \ldots = b_{n-1} = 0$, which is precisely $\mathrm{Im}([p^n])$. $\qquad\square$

Corollary 10.7.2. *For all positive integers* n

$$\mathbb{Z}_p/p^n\mathbb{Z}_p \cong \mathbb{Z}/p^n\mathbb{Z}.$$

In particular, since $\mathbb{Z}_p/p^n\mathbb{Z}_p$ *is a finite Hausdorff space,* $p^n\mathbb{Z}_p$ *is closed (and open) in* $\mathbb{Z}_p$.

Corollary 10.7.3. $\mathbb{Z}_p$ *is an integral domain.*

Proof. If a, b are both non-zero then, $v_p(ab) = v_p(a) + v_p(b) < \infty$, so $ab \neq 0$. $\qquad\square$

We denote by $\mathbb{Z}_p^\times$ the multiplicative group of p-adic units of $\mathbb{Z}_p$.

Theorem 10.7.4.

1. $\mathbb{Z}_p^\times = \mathbb{Z}_p - p\mathbb{Z}_p$ *or equivalently,*

$$\mathbb{Z}_p^\times = \{a \in \mathbb{Z}_p : v_p(a) = 0\}.$$

2. *Every non-zero* $a \in \mathbb{Z}_p$ *can be uniquely written as* $p^n u$, *where* $n \in \mathbb{Z}$, $n \geq 0$, *and* $u \in \mathbb{Z}_p^\times$.

Proof. 1. We first note $v_p(p^n) = n$ for all $n \geq 0$. In particular, $v_p(1) = 0$. If $a \in \mathbb{Z}_p^\times$, then a has a multiplicative inverse a^{-1} and we have

$$v_p(a) + v_p(a^{-1}) = v_p(1) = 0,$$

which implies, since $v_p(a) \geq 0$ for all $a \in \mathbb{Z}_p$,

$$v_p(a) = v_p(a^{-1}) = 0.$$

Conversely, if $a = (a_n)$ with each $a_n \in \mathbb{Z}/p^n\mathbb{Z}$ and $v_p(a) = 0$, then $a_1 \not\equiv 0 \pmod{p}$ is invertible in $\mathbb{Z}/p\mathbb{Z}$, and since $a_n \equiv a_1 \not\equiv 0 \pmod{p}$, each a_n is invertible in $\mathbb{Z}/p^n\mathbb{Z}$. So $a^{-1} = (a_n^{-1}) \in \mathbb{Z}_p$, which proves the first statement.

2. If $a \in \mathbb{Z}_p$, $a \neq 0$, let $v_p(a) = n$. Then $a \in \mathrm{Im}([p^n])$ and therefore $a = p^n u$ for some $u \in \mathbb{Z}_p$. We then have

$$n = v_p(a) = v_p(p^n u) = v_p(p^n) + v_p(u) = n + v_p(u),$$

and so $v_p(u) = 0$. Therefore $u \in \mathbb{Z}_p$. $\qquad\square$

Proposition 10.7.5. *Every non-zero ideal in $\mathbb{Z}_p$ is of the form (p^n) for some integer $n \geq 0$.*

Proof. Let I be a non-zero ideal in $\mathbb{Z}_p$, and let $n = \inf\{v_p(a) : a \in I\}$. Then, since $I \neq (0)$, $n < \infty$, and every $a \in I$ lies in $\mathrm{Im}([p^n]) = (p^n)$. On the other hand, since v_p is discrete, $v_p(a) = n$, and we can write $a = p^n u$ for some unit u. But then since I is closed under multiplication by elements of $\mathbb{Z}_p$,

$$u^{-1}a = p^n \in I,$$

and thus $p^n \in I \subseteq (p^n)$, which implies $I = (p^n)$. $\qquad\square$

Corollary 10.7.6. *The ring $\mathbb{Z}_p$ is a principal ideal domain with a unique maximal ideal. Thus $\mathbb{Z}_p$ is a local ring.*

10.8   The p-adic Numbers, $\mathbb{Q}_p$

Any p-adic number x can be uniquely represented under the form

$$a = p^c(a_0 + a_1 p + a_2 p^2 + \cdots), \tag{1}$$

where $c = c(a) \in \mathbb{Z}$ and a_i are integers such that $0 \leq a_i \leq p-1$, $a_0 > 0$, and $i = 0, 1, \ldots$.

We can see that series (1) converges with respect to the p-adic norm $|\cdot|_p$ because we have

$$|p^c a_i p^i|_p = p^{-(c+i)}, \quad i = 0, 1, \ldots$$

Remark 10.8.1. The representation (1) of a p-adic number is similar to the representation of any real number x in infinite decimal

$$x = \pm 10^c (a_0 + a_1 10^{-1} + a_2 10^{-2} + \ldots), \quad a_i = 0, 1, \ldots, 9, \quad a_0 \neq 0.$$

Remark 10.8.2. The representation of a p-adic number x under the form (1) means that the number x is the limit with respect to the p-adic norm of a sequence $(a_n)_{n \in \mathbb{N}}$ of rational numbers

$$a_n = p^c (a_0 + a_1 p + \ldots + a_n p^n).$$

Remark 10.8.3. One sees easily that

$$|x|_p = p^{-c}.$$

Remark 10.8.4. We can check that the representation (1) gives us a rational number if and only if the numbers a_1, a_2,... beginning from some number form a periodic sequence.

Let p be a prime number. Given an integer $n > 0$, we can write n in base p:

$$n = a_0 + a_1 p + a_2 p^2 + \ldots + a_k p^k$$

with $0 \leq a_i < p$.

Definition 10.8.5. A p-adic integer is a (formal) series

$$a = a_0 + a_1 p + a_2 p^2 + \cdots$$

with $0 \leq a_i < p$. The set of p-adic integers is denoted by $\mathbb{Z}_p$.

10.9   The p-adic Topology

We remind the reader of some notions from topology.

Definition 10.9.1. A topological space X is *Hausdorff* if for all $x \neq y$ in X there are open neighborhoods U_x, U_y of x and y such that $U_x \cap U_y = \varnothing$.

Of course metric spaces are Hausdorff.
We need the following proposition.

Proposition 10.9.2. *Let $(X_n, \phi_n)_{n>0}$ be an inverse system of topological spaces and continuous maps. If the topological spaces, X_n, are all Hausdorff, the inverse limit*

$$(X, \psi_n) := \varprojlim X_n$$

is closed in $\prod_0^\infty X_n$. (Here $\psi_n : X \to X_n$ with $\psi_{n+1} \circ \phi_n) = \psi_n$.)

Proof. We will show that

$$\prod_0^\infty X_n - X$$

is an open set. To do this, let

$$x = (x_n)_{n \in \mathbb{N}} \in \prod_0^\infty X_n - X.$$

Then, there exists an $i \geq 0$ such that $\phi_i(x_{i+1}) \neq x_i$. As each X_i is Hausdorff we can find open neighborhoods, U of $\phi_i(x_{i+1})$, and V of x_i, such that $U \cap V = \varnothing$. Then,

$$\psi_{i+1}^{-1}\left(\phi_i^{-1}(U)\right) \cap \psi_i^{-1}(V) = \left\{ x \in \prod_0^\infty X_n : x_{i+1} \in \phi_i^{-1}(U), x_i \in V \right\}$$

is an open neighborhood of x not intersecting X. This means that $\prod_0^\infty X_n - X$ is open, and so X is closed. $\square$

To see that $\mathbb{Z}_p$ is compact it remains only to know that $\prod_0^\infty X_n$ is compact. Since each $\mathbb{Z}/p^n\mathbb{Z}$ is finite and therefore compact, the result follows from Tychonoff's theorem. Thus,

Proposition 10.9.3. $\mathbb{Z}_p$ *is compact.*

However, since we are working with two different definitions of $\mathbb{Z}_p$, the first as the ring of integers and the second as an inverse limit, we must show they are equivalent.

Theorem 10.9.4. *The map*

$$\varphi : \mathbb{Z}_p \longrightarrow \varprojlim \mathbb{Z}/p^n\mathbb{Z}$$

that sends the p-adic integer

$$x = \sum_{0 \le i} a_i p_i$$

to the sequence $(x_n)_{n \in \mathbb{N}}$ of its partial sums $x_n = \sum_{i<n} a_i p^i \mod p^n$ is a ring isomorphism.

Proof. The coherent sequences in the product $\prod_{1 \le n} \mathbb{Z}/p^n\mathbb{Z}$ are the sequences $(x_n)_{n \in \mathbb{N}}$ of partial sums of a formal series

$$\sum_0^\infty a_i p_i \quad 0 \le a_i \le p-1$$

for all $i \ge 0$. These are precisely the p-adic integers as every $x \in \mathbb{Z}_p$ has a unique representation in canonical form. As the operations in $\prod_1^\infty \mathbb{Z}/p^n\mathbb{Z}$ are defined componentwise, we see that $\varphi(x) = (z_n(x))$ for all $x \in \mathbb{Z}_p$, where $z_n \in \mathbb{Z}/p^n\mathbb{Z}$. Therefore φ is a homomorphism. We show that φ is a bijection. First observe that φ is injective

$$\mathrm{Ker}(\varphi) = \varphi^{-1}(0) = \varphi^{-1}\Big(([0], [0], \ldots)\Big)$$

$$= \Big\{ \sum_0^\infty a_i p_i \in \mathbb{Z}_p \,\Big|\, a_i - 0 \,\forall\, i \ge 0 \Big\} = \{0\}$$

by the uniqueness of the canonical representation. Furthermore, it is easy to see that φ is surjective, since if

$$([x_1], [x_2], \ldots) \in \prod_{1 \leq n} \mathbb{Z}/p^n\mathbb{Z}$$

is a coherent sequence, we can choose $y_i \in [x_i]$ such that $0 \leq y_i < p^i$ for all $i \geq 1$. Then, we can define $b_0 = y_1$ and

$$b_i = \frac{y_{i+1} - y_i}{p^i}$$

for all $i \geq 1$. Therefore,

$$\varphi\left(\sum_0^\infty b_i p_i\right) = ([y_1], [y_2], \ldots) = ([x_1], [x_2], \ldots).$$

In other words, φ is surjective. $\square$

As the map φ in the theorem above is a continuous map from a compact space to a Hausdorff space is a homeomorphism, i.e. it is an *isomorphism of topological rings.*

Proposition 10.9.5. $\mathbb{Z}$ *is dense in* $\mathbb{Z}_p$. *The same holds for* $\mathbb{Q}$ *in* $\mathbb{Q}_p$.

Proof. We must show that for every element $a \in \mathbb{Z}_p$, and every $\varepsilon > 0$, we have $B(a, \varepsilon) \cap \mathbb{Z} \neq \varnothing$. To do this, take an element $a \in \mathbb{Z}_p$ and an $\varepsilon > 0$. There exists an n large enough so that $p^{-n} < \varepsilon$. Denote by $\hat{a} \in \mathbb{Z}$ the integer obtained by cutting the series of a after the term $a_{n-1}p^{k-1}$. Then

$$a - \hat{a} = a_k p^k + a_{k+1} p^{k+1} + \cdots$$

implies

$$|a - \hat{a}|_p \leq p^{-n} < \varepsilon,$$

which means that $\mathbb{Z}$ is dense in $\mathbb{Z}_p$. Similarly, $\mathbb{Q}$ is dense in $\mathbb{Q}_p$. $\square$

Lemma 10.9.6. *As we saw in 10.7.2, the subgroups* $p^n\mathbb{Z}_p$ *are closed (and open) in* $\mathbb{Z}_p$. *In fact, they are the only closed subgroups of* $\mathbb{Z}_p$ *other than* $\{0\}$.

Proof. From Proposition 10.9.5, we know $\mathbb{Z}$ is dense in $\mathbb{Z}_p$. Now suppose S is a closed subgroup of $\mathbb{Z}_p$. Let $s \in S$ be an element of maximal p-adic norm, say

$$\|s\| = p^{-e}.$$

Then

$$s = p^e u,$$

where u is a unit in $\mathbb{Z}_p$, with inverse v. Given any $\epsilon > 0$, we can find $n \in \mathbb{Z}$ such that

$$\|v - n\| < \epsilon.$$

Then

$$ns - p^e = p^e(nu - 1)$$
$$= p^e u(n - v);$$

and so

$$\|ns - p^e\| < \epsilon.$$

Since $ns \in S$ and S is closed, it follows that $p^e \in S$. Hence

$$p^e \overline{\mathbb{Z}} = p^e \mathbb{Z}_p \subset S.$$

Since s is a maximal element in S, it follows that $S = p^e \mathbb{Z}_p$. □

Consider the sphere of center $\alpha \in \mathbb{Q}_p$ and radius $r > 0$

$$S(\alpha, r) = \{x \in \mathbb{Q}_p \, : \, |x - \alpha|_p = r\}.$$

Proposition 10.9.7. *The sphere $S(\alpha, r)$ is an open set in $\mathbb{Q}_p$. In particular, $\mathbb{Z}_p$ is open in $\mathbb{Q}_p$.*

Proof. Let $x \in S(\alpha, r)$, and $\varepsilon > 0$, $r > \varepsilon$. We will show that $B_\varepsilon^\circ(x) \subseteq S(\alpha, r)$. Indeed, let $y \in B_\varepsilon^\circ(x)$. Then $|x - y|_p < |x - \alpha|_p = r$, and by Proposition 10.2.6, $|y - \alpha|_p = |x - \alpha|_p = r$, which shows that $y \in S(\alpha, r)$. □

Remark 10.9.8. A point x of a topological space X is a *boundary point* of a subset $A \subseteq X$ if any open ball centered in x contains points that are in A and points that are not in A, and we can prove that the set A is closed if it contains all its boundary points. It follows from Proposition 10.9.7 that $S(\alpha, r)$ is not the boundary of the open ball $B_r(\alpha)$!

The set of all balls in $\mathbb{R}$ is uncountable (since the set of all positive real numbers is uncountable) but this is not true for the set of all balls in $\mathbb{Q}_p$. Indeed,

Proposition 10.9.9. *The set of all balls in $\mathbb{Q}_p$ is countable.*

Proof. Consider the ball $B_{p^{-l}}(a)$ and write its center a in its canonical form

$$a = \sum_{n=-m}^{\infty} a_n p^n$$

and let

$$a_0 = \sum_{n=-m}^{l} a_n p^n.$$

Then, a_0 is a rational number, and $|a - a_0|_p < p^{-l}$, which shows that $a_0 \in B_{p^{-l}}(a)$. Proposition 10.5.12 implies that the two balls are equal, i.e. $B_{p^{-l}}(a) = B_{p^{-l}}(a_0)$. Since the center and the radius come from countable sets, the set of all pairs (a_0, l) is also countable and so is the set of all balls in $\mathbb{Q}_p$. $\qquad \square$

An important consequence of this is:

Proposition 10.9.10. $\mathbb{Q}_p$ *is a locally compact space.*

Proof. By Proposition 10.9.3 $\mathbb{Z}_p$ is compact. It is also open by Proposition 10.9.7. Hence, any given open set U in $\mathbb{Q}_p$, $U \cap \mathbb{Z}_p$ is open in $\mathbb{Q}_p$ and thus has compact closure. $\qquad \square$

We remind the reader that a sequence of rational numbers $(a_n)_{n \in \mathbb{N}}$ is a Cauchy sequence with respect to the p-adic norm $|\cdot|_p$ if and only if $|a_{n+1} - a_n|_p \longrightarrow 0$ as $n \longrightarrow \infty$.

We now see how much p-adic analysis differs from real analysis.

Proposition 10.9.11. *("**The dream of the bad student**") The series $\sum_1^\infty a_n$ converges in $\mathbb{Q}_p$ if and only if the sequence $\lim_{n\to\infty} a_n = 0$.*

Proof. The difference of partial sums

$$|s_{m+k} - s_m|_p = \left| \sum_{m+1}^{m+k} a_n \right|_p \leq \max\{|a_{m+1}|_p, \ldots, |a_{m+k}|_p\} \to 0.$$

By Cauchy's criteria, the series converges since $\lim_{n\to\infty} a_n = 0$. The converse is always true.   □

More spectacularly, the series

$$\sum_{n=0}^\infty n!$$

is p-adic convergent!

Exercise 10.9.12. Prove that the equation $x^2 + 1 = 0$ has a root in $\mathbb{Q}_p$ if and only if $p \equiv 1 \mod 4$.

In some ways $\mathbb{Q}_p$ and $\mathbb{R}$ share many properties, however in other ways they are quite different. Although neither is algebraically closed, for $\mathbb{R}$, by adjoining $i := \sqrt[2]{-1}$ we get an algebraically closed field. In contrast, the algebraic closure $\overline{\mathbb{Q}}_p$ of $\mathbb{Q}_p$ is of infinite degree over $\mathbb{Q}_p$. Moreover, although $\mathbb{C}$ is complete with respect to the extension of the usual norm of $\mathbb{R}$ while $\overline{\mathbb{Q}}_p$ is not complete with respect to the extension of the p-adic norm. Moreover, after completing it (via the method of Cauchy sequences) one obtains a still larger field, usually denoted by $\mathbb{C}_p$, which turns out to be both algebraically closed and complete. This field is the p-adic analogue of the field of complex numbers. It is algebraically isomorphic to $\mathbb{C}$, but not topologically. Very little more is known about it! (see [127] p. 48.)

10.9.1 $\mathbb{Z}_p$ and $\mathbb{Q}_p$ as Topological Groups and Rings

We note that the product topology on $\mathbb{Z}_p$ is induced by the p-adic metric d_p defined by

$$d_p(x\,,\,y) = p^{-v_p(x-y)}.$$

Lemma 10.9.13. *The map $\mathbb{Z}_p \longrightarrow \mathbb{Z}_p$ defined by $x \mapsto px$ is continuous.*

Proof. Let $x, y \in \mathbb{Z}_p$. Then,

$$\begin{aligned}
d(px\,,\,py) &= p^{-v_p(px-py)} \\
&= p^{-v_p\left(p(x-y)\right)} = p^{-v_p(p)-v_p(x-y)} \\
&= p^{-1}p^{-v_p(x-y)} \\
&= p^{-1}d(x\,,\,y)
\end{aligned}$$

which shows that the multiplication by p in $\mathbb{Z}_p$ is a contraction mapping, therefore continuous. $\square$

Proposition 10.9.14. *$\mathbb{Z}_p$ is a topological ring in the sense that it is a Hausdorff space and the addition and multiplication are continuous. Moreover $\mathbb{Q}_p$ is a topological field.*

Proof. We already know that $\mathbb{Z}_p$ is compact and Hausdorff. First we will show that $\mathbb{Z}_p$ is an (additive) topological group. To do this we shall prove that the two operations

$$\mathbb{Z}_p \times \mathbb{Z}_p \longrightarrow \mathbb{Z}_p \ : \ (x\,,\,y) \mapsto x+y,$$

and

$$\mathbb{Z}_p \longrightarrow \mathbb{Z}_p \ : \ x \mapsto -x$$

are continuous. The space $\mathbb{Z}_p \times \mathbb{Z}_p$ is equipped with the product topology induced by the metric $d_p(x\,,\,a) + d_p(y\,,\,b)$.

Let $(a\,,\,b) \in \mathbb{Z}_p \times \mathbb{Z}_p$. Now assuming that

$$\delta\Big((x\,,\,y)\,,\,(a\,,\,b)\Big) \leq p^{-n},$$

we get $p^{-v_p(x-a)} = d_p(x,a) \leq p^{-n}$, which implies $v_p(x-a) \geq n$. Similarly we obtain $v_p(y-b) \geq n$. But we know that for any z, $w \in \mathbb{Z}_p$ we have

$$v_p(w+z) \geq \min\left(v_p(w), v_p(z)\right) \text{ and } v_p(-w) = v_p(w).$$

Therefore $v_p(x-a-(y-b)) \geq n$, i.e.

$$d_p(x-a, y-b) = p^{-v_p(x-a-(y-b))} \leq p^n.$$

This shows that the map $(x,y) \mapsto x-y$ is continuous at any point (a,b). Hence $\mathbb{Z}_p$ is a topological group.

Now, to prove that $\mathbb{Z}_p$ is a actually a topological ring, i.e. the multiplication $(x,y) \to x \cdot y$ is continuous. To do so, we consider the following neighborhood base at 1

$$\{1 + p^n\mathbb{Z}_p : n \text{ positive integer}\}.$$

Now, let $(a,b) \in \mathbb{Z}_p \times \mathbb{Z}_p$ and suppose that $x \in a + p^n\mathbb{Z}_p$ and $y \in b + p^n\mathbb{Z}_p$. This implies that we can find ℓ and $m \in \mathbb{Z}_p$ such that $x = a + p^n\ell$ and $y = b + p^n m$. Therefore,

$$x \cdot y = a \cdot b + p^n(a + b + p^n\ell m) \in a \cdot b + p^n\mathbb{Z}_p.$$

This shows that the multiplication is continuous at any point (a,b).

The same arguments hold for $\mathbb{Q}_p$ as well.

We also show that $\mathbb{Q}_p$ is actually a topological field. That is the inverse map $x \to x^{-1}$ is continuous on $\mathbb{Q}_p^\times$. Let $a \in \mathbb{Q}_p^\times$ and $x \in a + p^n\mathbb{Z}_p$. Then, we can find ℓ, $m \in \mathbb{Z}_p$ such that $x = a(1 + p^n\ell)$, and

$$(1 + p^n\ell)^{-1} = \sum_{i \geq 0}(-p^n\ell)^i = 1 + p^n m.$$

Hence,

$$x^{-1} = a^{-1}(1 + p^n m) \in a^{-1} + p^n\mathbb{Z}_p,$$

which shows that the inverse map is continuous at a neighborhood of 1 so it is continuous everywhere. $\qquad\square$

Let R be an integral domain, let $a \in R$ $a \neq 0$, and let $S = \{a^k :$ $k \in \mathbb{Z}, k > 0\}$ be a multiplicative set. We define

$$R\left[\frac{1}{a}\right] = S^{-1}R$$

called the *localization of R away from a*.

We now identify the localization of $\mathbb{Z}$ away from p, $\mathbb{Z}[\frac{1}{p}]$, with the set of rational numbers whose denominator is of the form p^k, where k is a non-negative integer. In particular, $\mathbb{Z} \subseteq \mathbb{Z}[\frac{1}{p}]$. Now, if $x \in \mathbb{Q}_p$, write

$$x = \sum_{k \geq v_p(x)} x_k p^k = \sum_{0 > k \geq v_p(x)} x_k p^k + \sum_{k \geq 0} x_k p^k = \{x\}_p + [x]_p.$$

Definition 10.9.15. We call $[x]_p$ the *integral part* of x and $\{x\}_p$ the *fractional part* of x.

We have $[x]_p \in \mathbb{Z}_p$ and $\{x\}_p \in \mathbb{Z}[\frac{1}{p}]$. We define

$$\phi_p : \mathbb{Q}_p \longrightarrow S^1 \text{ such that } \phi_p(x) = e^{2\pi i \{x\}_p},$$

and

$$\phi_\infty : \mathbb{R} \longrightarrow S^1 \text{ such that } \phi_\infty(x) = e^{2\pi i \{x\}},$$

where $[x]$ is the greatest integer $\leq x$ and $\{x\} = x - [x]$. One checks that ϕ_∞ is a homomorphism of topological groups and $\phi_\infty(\mathbb{R}) = S^1$ and $\mathrm{Ker}(\phi_\infty) = \mathbb{Z}$.

Proposition 10.9.16. *If p is a prime number then $\phi_p : \mathbb{Q}_p \longrightarrow S^1$ is a homomorphism of topological groups.*

Proof. Let $x, y \in \mathbb{Q}_p$. We have

$$x - \{x\}_p = [x]_p, \ y - \{y\}_p = [y]_p, \ x + y - \{x + y\}_p = [x + y]_p \in \mathbb{Z}_p.$$

Hence,

$$\{x\}_p + \{y\}_p - \{x + y\}_p = (x - [x]_p) + (y - [y]_p) - (x + y - [x + y]_p)$$
$$= [x + y]_p - [x]_p - [y]_p \in \mathbb{Z}_p.$$

But $\{x\}_p + \{y\}_p - \{x+y\}_p \in \mathbb{Q}$, so the fact that it belongs to $\mathbb{Z}_p$ tells us that p does not divide its denominator. On the other hand, because $\{x\}_p, \{y\}_p, \{x+y\}_p \in \mathbb{Z}[1/p]$, so $\{x\}_p + \{y\}_p - \{x+y\}_p \in \mathbb{Z}[1/p]$ and hence the denominator of $\{x\}_p + \{y\}_p - \{x+y\}_p \in \mathbb{Q}$ is of the form p^k, with k a positive integer . Thus its denominator is 1, showing that $\{x\}_p + \{y\}_p - \{x+y\}_p \in \mathbb{Z}$, say

$$y\{x+y\}_p = \{x\}_p + \{y\}_p + n.$$

Therefore

$$\phi_p(x+y) = e^{2\pi i \{x+y\}_p} = e^{2\pi i \{x\}_p + 2\pi i \{y\}_p + 2\pi i n}$$

$$= e^{2\pi i \{x\}_p} e^{2\pi i \{y\}_p} = \phi_p(x)\phi_p(y),$$

showing that ϕ_p is a homomorphism of groups. To show that it is continuous it suffices to show continuity at $0 \in \mathbb{Q}_p$. For $|x|_p \leq 1 = p^0$, we have $v_p(x) \geq 0$, so $x \in \mathbb{Z}_p$ and hence $\{x\}_p = 0$. Thus, for $|x|_p \leq 1$ we have $\phi_p(x) = 1 = \phi_p(0)$, showing that ϕ_p is continuous at 0 and therefore everywhere.   □

For $x \in \mathbb{Q}_p$ we have $\{x\}_p \in \mathbb{Z}[\frac{1}{p}]$, say $\{x\}_p = \frac{m}{p^k}$, for some $m \in \mathbb{Z}$ and k a positive integer. This implies that

$$\left(\phi_p(x)\right)^{p^k} = (e^{2\pi i m/p^k})^{p^k} = e^{2\pi i m} = 1 \in S^1.$$

Therefore, $\phi_p(x) \in \mathbb{Z}(p^\infty)$, the Prüfer p-group. One checks that $\phi_p(\mathbb{Q}_p) = \mathbb{Z}(p^\infty)$ and that $Ker(\phi_p) = \mathbb{Z}_p$. The group $\mathbb{Z}(p^\infty)$ is a discrete Abelian group and thus is locally compact. (However in the relative topology of S^1, $\mathbb{Z}(p^\infty)$ is not at all discrete, but rather is dense in S^1). $\mathbb{Q}_p$ is locally compact (because $x + \mathbb{Z}_p$ is a compact neighborhood of $x \in \mathbb{Q}_p$) and σ-compact because

$$\mathbb{Q}_p = \bigcup_{n \geq 0} p^{-n} \mathbb{Z}_p.$$

By the open mapping theorem $\phi_p : \mathbb{Q}_p \longrightarrow \mathbb{Z}(p^\infty)$ is open so the first isomorphism theorem gives us a short exact sequence of topological groups

$$(0) \longrightarrow \mathbb{Z}_p \longrightarrow \mathbb{Q}_p \longrightarrow \mathbb{Z}(p^\infty) \longrightarrow (0).$$

Dualizing, by Theorem 2.1 of [89], we get the following short exact sequence,

$$(0) \longrightarrow \widehat{\mathbb{Z}(p^\infty)} = \mathbb{Z}_p \longrightarrow \widehat{\mathbb{Q}_p} \longrightarrow \widehat{\mathbb{Z}_p} = \mathbb{Z}(p^\infty) \longrightarrow (0),$$

where these terms $\widehat{G} = \mathrm{Hom}(G, S^1)$, are the character groups of the various G.

Consider the bilinear map

$$\beta : \mathbb{Q}_p \times \mathbb{Q}_p \to S^1 \; : \; \beta(x, y) = e^{2\pi i x y}.$$

Since $\mathbb{Q}_p$ is a *field* and for fixed y this is a field homomorphism in x. Therefore it is injective and gives rise to a homomorphism $\omega : \mathbb{Q}_p \to \widehat{\mathbb{Q}_p}$. The map ω is continuous, injective and has dense range. Moreover, $\omega(\mathbb{Z}_p)$ is compact and its restriction to $\mathbb{Z}_p$ is a homeomorphism and $\mathbb{Z}_p$ is open in $\widehat{\mathbb{Q}_p}$, showing that $\omega : \mathbb{Q}_p$ is an open subgroup of $\widehat{\mathbb{Q}_p}$ which is therefore also closed. Since it is dense it must be all of $\widehat{\mathbb{Q}_p}$. By the open mapping theorem ω is a topological group isomorphism and so $\mathbb{Q}_p$ is self dual.

10.10   The Geometry of $\mathbb{Q}_p$

We often regard the field $\mathbb{R}$ as a straight line. As we shall see, the field $\mathbb{Q}_p$, is different (for a detailed study of its geometry see [28]). To study the geometry of this field, first, we give some definitions:

1. A *point* is an element of $\mathbb{Q}_p$.

2. A *triangle* is three distinct points x, y, and $z \in \mathbb{Q}_p$.

3. A triangle xyz has side lengths given by $d(x, y)$, $d(y, z)$ and $d(z, x)$.

Note that our definition of a triangle in $\mathbb{Q}_p$ differs from the definition in Euclidean geometry, since here we allow three collinear points to form a triangle. However, we will see that *we never have* three collinear points in p-adic geometry. A well-known characteristic of the p-adic absolute value is most striking when phrased in terms of triangles.

Proposition 10.10.1. *In a metric space* (X, d) *with a non-Archimedean metric, every triangle is isosceles!*

Proof. As we saw in 10.5.9 if two sides of a triangle are unequal, the longer must be equal to the third one. Hence the triangle is isosceles.   □

Proposition 10.10.2. *If a triangle is not equilateral, the unequal side has the largest valuation and hence the shortest length.*

Proof. Consider a triangle xyz with $|x - y|_p = |y - z|_p \neq |x - z|_p$. Then,

$$|x - z|_p = |(x - y) - (z - y)|_p \geq \min\{|x - y|_p, |y - z|_p\},$$

and since $|x - z|_p \neq |x - y|_p = |y - z|_p$ we have $|x - z|_p > |x - y|_p$.   □

In Euclidean geometry with the usual distance d, three points are collinear if and only if $d(x, z) = d(x, y) + d(y, z)$, assuming $d(x, z) > d(x, y)$ and $d(x, z) > d(y, z)$. In $\mathbb{Q}_p$ we see that collinearity is impossible with more than two points.

Corollary 10.10.3. *Given any three distinct points* x, y, *and* $z \in \mathbb{Q}_p$, $d(x, z) < d(x, y) + d(y, z)$, *where* d *is the* p-*adic distance. In other words, no three points in* $\mathbb{Q}_p$ *are collinear.*

We have seen that all triangles in $\mathbb{Q}_p$ are isosceles. Equilateral triangles are easy to construct. Indeed, for any prime $p \geq 3$ the points $2p$, $3p$ and $4p$ form an equilateral triangle since

$$d(2p, 3p) = |p(3 - 2)|_p = p^{-1}, \quad d(3p, 4p) = |p(4 - 3)|_p = p^{-1},$$

and

$$d(2p, 4p) = |p(4 - 2)|_p = p^{-1}.$$

In $\mathbb{Q}_2$, however, *equilateral triangles do not exist*. To see this we first prove the following theorem.

Theorem 10.10.4. *Given a prime p, any subset of $\mathbb{Q}_p$ has at most p equidistant points.*

Proof. Suppose the contrary. That is, there is a set of $p+1$ distinct equidistant points $x_1, x_2, ..., x_{p+1}$ with

$$x_i = \sum_{k=j_i}^{\infty} x_{ik}p^k, \quad x_{ij_i} \neq 0.$$

Since the x_i's are all equidistant, there exists an $m \in \mathbb{Z}$ such that $|x_i - x_j|_p = m$ for all i, j and $x_{ik} = x_{jk}$ for all $k < m$. Thus

$$x_i - x_j = \sum_{k=m}^{\infty} (x_{ik} - x_{jk})p^k$$

$$= \left(p^m \sum_{k=0}^{\infty} x_{i(k+m)}p^k\right) - \left(p^m \sum_{k=0}^{\infty} x_{j(k+m)}p^k\right)$$

$$= p^m \hat{x}_i - p^m \hat{x}_j = p^m(\hat{x}_i - \hat{x}_j)$$

where

$$\hat{x}_i = \sum_{k=0}^{\infty} x_{i(k+m)}p^k \quad \hat{x}_j = \sum_{k=0}^{\infty} x_{j(k+m)}p^k.$$

Therefore

$$|x_i - x_j|_p = m + |\hat{x}_i - \hat{x}_j|_p = m + 0 = m$$

since by assumption $|x_i - x_j|_p = m$ for all $p+1$ points and hence $|\hat{x}_i - \hat{x}_j|_p = 0$. However, since $0 \le \hat{x}_{i0} \le p-1$ for all i and because there are $p+1$ points, there exist distinct i and j so that $\hat{x}_{i0} = \hat{x}_{j0}$. Thus, $|\hat{x}_i - \hat{x}_j| > 0$ for some i and j. □

Corollary 10.10.5. *Equilateral triangles do not exist under the 2-adic metric.*

We next show that in $\mathbb{Q}_p$ right triangles do not exist. A right triangle in $\mathbb{Q}_p$ is a triangle whose side lengths satisfy the Pythagorean theorem.

Theorem 10.10.6. *For any x, y and $z \in \mathbb{Q}_p$ we have $d(x,z)^2 \neq d(x,y)^2 + d(y,z)^2$. In other words, right triangles in $\mathbb{Q}_p$ do not exist.*

Proof. Let xyz be a right triangle with hypotenuse xy, and the two other sides xz, zy. Then, by Proposition 10.10.1, the triangle is isosceles, i.e. $d(z\,,\,x) = d(z\,,\,y)$. Since xyz is a right triangle we have

$$d(x\,,\,y)^2 = d(z\,,\,x)^2 + d(z\,,\,y)^2 = 2d(z\,,\,x)^2.$$

This implies

$$\left(p^{-v(x-y)}\right)^2 = 2\left(p^{-v(z-x)}\right)^2,$$

in other words

$$\frac{1}{2} = \frac{p^{-2v(z-x)}}{p^{-2v(x-y)}} = p^{-2v(z-x)+2v(x-y)}$$
$$= p^{2\frac{v(x-y)}{v(z-x)}}.$$

This implies $\frac{v(x-y)}{v(z-x)} = -\frac{1}{2}$. But this can not be since v takes only integer values. $\square$

10.11   Extensions of Valuations

It is important to be able to extend valuations, that is, for some finite extension field K over k and valuation $|\cdot| : k \to \mathbb{R}$, we wish to extend the valuation to $||\cdot|| : K \to \mathbb{R}$ in such a way that $|x| = ||x||$ for $x \in k$.

Here we show that a valuated field K, which is an extension of a field k, is non-Archimedean if and only if k itself is not Archimedean under the induced norm.

Theorem 10.11.1. (*Chevalley's Theorem*[2]) *Let k be a field, R a subring of k and $\mathfrak{p}$ a prime ideal of R. Then, there is always a valuation ring $\mathcal{O}$ of k such that $R \subseteq \mathcal{O}$ and $\mathfrak{m}_{\mathcal{O}} \cap R = \mathfrak{p}$.*

---

[2]Claude Chevalley (1909-1984) a French mathematician who made important contributions to number theory, algebraic geometry, class field theory and the theory of algebraic groups. By applying algebraic groups over a finite field he contributed to the study of finite simple groups. He was the youngest founding member of the Bourbaki group.

Proof. Let $R_\mathfrak{p}$ be the localization of R at $\mathfrak{p}$, and $\mathfrak{p}R_\mathfrak{p}$ its unique maximal ideal (see Chapter 9). Let

$$\Theta = \{(S, I) : S \text{ a subring of } k, R_\mathfrak{p} \subseteq S, I \text{ ideal of } S: \mathfrak{p}R_\mathfrak{p} \subseteq I\}.$$

We define a partial order $(S, I) \leq (S_1, I_1)$ if and only if $S \subseteq S_1$ and $I \subseteq I_1$. Θ is closed under unions of chains, hence by Zorn's lemma, there is a maximal element $(\mathcal{O}, \mathfrak{m}) \in \Theta$. Then $\mathcal{O}$ is a local ring with maximal ideal $\mathfrak{m}$. We claim that $\mathcal{O}$ is a valuation ring. Suppose it is not, i.e. there is an $x \in k$ such that both x and $x^{-1} \notin \mathcal{O}$. Then,

$$\mathcal{O} \subsetneq \mathcal{O}[x] \text{ and } \mathcal{O} \subsetneq \mathcal{O}[x^{-1}].$$

So $\mathfrak{m}\mathcal{O}[x] = \mathcal{O}[x]$ and $\mathfrak{m}\mathcal{O}[x^{-1}] = \mathcal{O}[x^{-1}]$, i.e.

$$1 = \sum_{i=0}^{n} a_i x^i = \sum_{i=0}^{n} b_i x^{-1} \quad a_i, b_i \in \mathfrak{m}$$

with m, n minimal and $m \leq n$. Then,

$$\sum_{i=1}^{m} b_i x^{-i} = 1 - b_0 \in \mathcal{O}^\times = \mathcal{O} - \mathfrak{m}.$$

Therefore,

$$1 = \sum_{i=1}^{m} \frac{b_i}{1 - b_0} x^{-i} \text{ and so } x^n = \sum_{i=1}^{m} c_i x^{n-i}$$

where $c - i \in \mathfrak{m}$. Then,

$$1 = \sum_{i=0}^{n} a_i x^i = \sum_{i=0}^{n-1} a_i x^i + \sum_{i=1}^{m} a_n c_i x^{n-i},$$

contradicting the minimality of n. Now $\mathfrak{p} \subseteq \mathfrak{m} \cap R$, and

$$R - \mathfrak{p} \subseteq (R_\mathfrak{p})^\times \subseteq \mathcal{O}^\times = \mathcal{O} - \mathfrak{m}$$

which implies $\mathfrak{p} \subseteq \mathfrak{m} \cap R$. $\square$

Corollary 10.11.2. *Let* (k, v) *be a valued field and* K/k *a field extension. Then there is a prolongation* w *of* v *to* K.

Proof. Take $R = \mathcal{O}_v$, $\mathfrak{p} = \mathfrak{m}_v$ and apply Chevalley's Theorem. $\square$

Proposition 10.11.3. *If* $\mathcal{O}$ *is a valuation ring of* k, *then* $\mathcal{O}$ *is integrally closed in* k.

Proof. Suppose $x \in k$ with $x^n + a_{n-1}x^{n-1} + \cdots + a_0 = 0$ with $a_i \in \mathcal{O}$. If $x \notin \mathcal{O}$ then $x^{-1} \in \mathfrak{m}_\mathcal{O}$ and

$$-1 = a_{n-1}x^{-1} + \cdots + a_0 x^{-n} \in \mathfrak{m},$$

a contradiction. $\square$

Corollary 10.11.4. *Let* $R \subseteq k$ *be a subring, and* $\overline{R}$ *be the integral closure of* R *in* k. *Then,*

$$\overline{R} = \bigcap_{R \subseteq \mathcal{O}} \mathcal{O}$$

for all valuation rings $\mathcal{O}$.

Proof. Let $x \in k - \overline{R}$. Then $x \notin R[x^{-1}]$. In particular, $x^{-1} \in \mathfrak{m}$ for some maximal ideal of $R[x^{-1}]$. By Chevalley's theorem, there is a valuation ring $\mathcal{O}$ of k such that $k[x^{-1}] \subseteq \mathcal{O}$ and $x^{-1} \in \mathfrak{m}_\mathcal{O}$. $\square$

10.12   An Application: Monsky's Theorem

Here we ask if it is possible to dissect a square in $\mathbb{R}^2$ into an odd number of triangles of equal area?

The person who first thought about this was Fred Richman (1965) of New Mexico State University. He was preparing a masters exam, and he wanted to include this but could not solve it so he published his question in the Amer. Math. Monthly. Paul Monsky proved the theorem in 1970 ([86]).

Theorem 10.12.1. *(**Monsky's Theorem**) If a square is cut into triangles of equal area, the number of triangles must be even.*

The proof of this theorem utilizes tools from two distinct areas of mathematics:

1. Combinatorics and Topology.

2. Number theory and Algebra.

As of this writing the only published proof of this result is the following.

We start with the topological part.

Let Π be a polygon in the plane, and consider a triangulation of it. Color each vertex of this triangulation by one of three colors: $1 = $ red, $2 = $ blue, and $3 = $ black. An edge is called a *12-edge* if its endpoints are colored 1 and 2, and a triangle is said to be *complete* if each of its vertices has a different color.

Lemma 10.12.2. *(Sperner's Lemma)*[3] *If we have a polygon whose vertices are colored by three colors and a triangulation is given for this polygon, the number of complete triangles (with three different colors) is congruent* mod2 *to the number of* 12-edges *on the boundary of the polygon.*

We will need the following lemma that is a simple corollary of Sperner's Lemma.

Lemma 10.12.3. *Consider a triangulation T of a triangle ABC. Color the outer three vertices A, B and C as 1, 2 and 3, respectively. Use only colors $\{1,2\}$ on the edge AB, only colors $\{1,3\}$ on the edge AC, and only colors $\{2,3\}$ on the edge BC. This triangulation contains a complete triangle.*

Proof. A 12-boundary edge can only appear on the side with endpoints colored with 1 and 2, and so there are an odd number of 12-edges on the boundary. $\square$

---

[3]Emanuel Sperner (1905-1980) was a German mathematician. In 1928 Sperner (who was 23 at the time) produced a simple combinatorial proof (see [118]) from which both Brouwer's fixed point theorem and the invariance of the dimension under bijective continuous maps could be deduced.

The second component of the proof of the theorem entails properties of the 2-adic norm.

Definition 10.12.4. Any rational number $x = p/q$ can be written as $x = 2^n \frac{a}{b}$, where a and b are odd. (Here, n might be negative.) The *2-adic norm* of x is defined as $|x|_2 = (1/2)^n$. Hence, an integer has 2-adic norm less than 1 if and only if it is even.

Proof of Monsky's Theorem
We can color points in $\mathbb{P}^2$ according to the following sets:

$$S_1 = \{(x, y) : |x|_2 < 1, |y|_2 < 1\}$$
$$S_2 = \{(x, y) : |x|_2 \geq 1, |x|_2 \geq |y|_2\}$$
$$S_3 = \{(x, y) : |y|_2 \geq 1, |y|_2 > |x|_2\}.$$

Notice that classes 2 and 3 are translation invariant under points in class 1.

Lemma 10.12.5. *Let a triangle with vertices in $\mathbb{R}^2$ be complete. Then its area satisfies*

$$|area|_2 > 1.$$

This means that it has 2 in the denominator.

Proof. By translation invariance, we can assume that the triangle is at the origin, in which case we can write its area as a determinant. Let (x_2, y_2) and (x_3, y_3) be colored 2 and 3. Then,

$$\frac{1}{2} \begin{vmatrix} x_2 & x_3 \\ y_2 & y_3 \end{vmatrix} = \frac{x_2 y_3 - x_3 y_2}{2}.$$

Furthermore, we have $|x_2|_2 \geq |y_2|_2$ and $|y_3|_2 > |x_3|_2$, where $|x_2|_2, |y_3|_2 > 1$. Therefore, $x_2 y_3$ dominates, having $|x_2 y_3|_2 > |x_3 y_2|_2$. This means that

$$|area|_2 = \left| \frac{1}{2} \right|_2 \cdot |x_2 y_3 - x_3 y_2|_2$$
$$= 2 \cdot \max(|x_2 y_3|_2, |x_3 y_2|_2)$$
$$= 2 \cdot |x_2 y_3|_2 = 2|x_2|_2 |y_3|_2 \geq 2.$$

$\square$

We can now complete the proof of Monsky's theorem.

Theorem 10.12.6. *Given a decomposition of a square S into m triangles of equal area, m is even.*

Proof. Without loss of generality, let the vertices of the square S be $(0,0)$, $(0,1)$, $(1,0)$, $(1,1)$. Consider a triangulation T of S, and just as above, we can color the vertices of T. Observe that on the boundary S, 12-edges can only occur on the edge connecting $(0,0)$ and $(1,0)$. Furthermore, vertices colored 3 cannot occur on this edge, so there must be an odd number of 12-edges.

Therefore, by Sperner's lemma, there is a complete triangle in T. So its area A has a 2-adic norm $|A|_2 > 1$. But the area of S is $mA = 1$, and therefore $|m|_2 < 1$, which means that m is even. □

Remark 10.12.7. We recall the famous Hilbert 5-th Problem which asks what it takes for a locally compact group to be a Lie group. This was resolved in the early fifties by Montgomery-Zippin, Gleason and Yamabe. The result being that a locally compact group is a Lie group if and only if it has no small subgroups, which of course $\mathbb{Z}_p$ has! Actually, the question Hilbert asked was somewhat ambiguous because it could also be interpreted as follows: Let G be a locally compact group acting *faithfully* on a smooth manifold M, when is it a Lie group and the action smooth? This is known as the Hilbert-Smith conjecture and usually M is assumed to be connected. Some time ago a positive answer to this question was reduced to showing that the $\mathbb{Z}_p$, the Abelian group of p-adic integers, can not act faithfully on M and classically, this was verified when M has dimension 1 or 2. In 2013 Pardon [106] using cohomology methods and properties of $\mathbb{Z}_p$, showed that $\mathbb{Z}_p$ can not act on a 3 manifold faithfully because if it did $\mathbb{Z}_p$ would also act non-trivially on H^1 of a submanifold and he proves in dimension 3 this can not happen.

10.12.1 Exercises

Exercise 10.12.8. Let p be a prime and r a rational number. Prove that the p-adic expansion of r is repeating.

Exercise 10.12.9. Let p be a prime. Prove that in $\mathbb{Q}_p$, every bounded infinite sequence has a convergent subsequence (i.e., $\mathbb{Q}_p$ satisfies the Bolzano-Weierstrass property).

Exercise 10.12.10. Find the 5-adic expansion of $3/7$.

Exercise 10.12.11. In $\mathbb{Q}_5$, what rational number has the 5-adic expansion $1 + 3 \cdot 5 + 1 \cdot 5^2 + 3 \cdot 5^3 + \ldots$?

Exercise 10.12.12. Show that the equation $x^2 = 2$ has a solution in $\mathbb{Z}_7$.

Exercise 10.12.13. Write the numbers $\frac{2}{3}$ and $-\frac{2}{3}$ as 5-adic numbers.

Exercise 10.12.14. Let $n \in \mathbb{N}$, and $n = a_0 + a_1 p + \cdots + a_{r-1} p^{r-1}$ its p-adic expansion, with $0 \le a_i < p$, and $s_n = a_0 + a_1 + \cdots + a_{r-1}$. Show that
$$v_p(n!) = \frac{n - s_n}{p - 1}.$$

Exercise 10.12.15. Prove that the sequence $a_n = \frac{1}{10^n}$, $n = 0, 1, \ldots$ does not converge in $\mathbb{Q}_p$ for any p.

Exercise 10.12.16. Show that for every $a \in \mathbb{Z}$ such that $\gcd(a, p) = 1$, the sequence
$$\left(a^{p^n} \right)_{n \in \mathbb{N}}.$$
converges in $\mathbb{Q}_p$.

Exercise 10.12.17. Show that $|z| = (z\bar{z})^{1/2}$ is the only valuation of $\mathbb{C}$ which extends the absolute value $|\cdot|$ of $\mathbb{R}$.

Exercise 10.12.18. Prove that the only automorphism of the field $\mathbb{Q}_p$ is the identity.

Exercise 10.12.19. Prove that:

1. the polynomial $x^2 - 5$ is irreducible in $\mathbb{Q}_7[x]$,

2. the polynomial $x^p - x - 1$ is irreducible in $\mathbb{Q}_p[x]$ for any prime $p > 2$,

3. the polynomial $x^4 + 4x^3 + 2x^2 + x - 6$ is reducible in $\mathbb{Q}_{11}[x]$,

4. the polynomial $x^4 - x^3 - 2x^2 - 3x - 1$ is reducible in $\mathbb{Q}_5[x]$.

Exercise 10.12.20. Find the p-adic expansions for

1. $\frac{2}{3}$ in $\mathbb{Q}_2$,

2. $-\frac{1}{6}$ in $\mathbb{Q}_7$,

3. $\frac{1}{10}$ in $\mathbb{Q}_{11}$,

4. $\frac{1}{120}$ in $\mathbb{Q}_5$.

Chapter 11

Galois Theory

11.1 Field Extensions

Let K be a field and $K \supset k$ be a subfield. We call this a *field extension.* Then, as is easily seen, K is a vector space over k. In this context we shall call $\dim_k K$ the degree of K over k and denote it by $[K : k]$. Here we shall not need to make distinctions between infinite cardinals, only that $[K : k]$ is finite or infinite.

If this degree is finite we shall say K is a *finite extension* of k. For example, $[\mathbb{C} : \mathbb{R}] = 2$, while $[\mathbb{A} : \mathbb{Q}]$ is infinite, where $\mathbb{A}$ is the field of algebraic numbers which is discussed in Section 11.8. This last because if p is any prime, then any n-th root is irrational (see Eisenstein criterion in of Volume I, Chapter 5) and so adjoining the n-th roots of p to $\mathbb{Q}$ is a subfield of $\mathbb{A}$ over $\mathbb{Q}$ of infinite dimension. In fact as we shall see, $\mathbb{A}$ is the algebraic closure of $\mathbb{Q}$. Another example of a field extension is the rational function field $k(x)$. This is the quotient field of the polynomial ring, $k[x]$. Evidently, $[k(x) : k]$ is also infinite since $\frac{1}{x^n}$, $n \in \mathbb{Z}^+$ are linearly independent.

Proposition 11.1.1. *Suppose $K \supset L \supset k$ are three fields, then*

$$[K : k] = [K : L] \cdot [L : k],$$

and similarly for any finite number of fields.

Proof. Let $\{\alpha_i\}$ be linearly independent in K over L and $\{\beta_j\}$ be linearly independent in L over k. We shall show $\{\alpha_i\beta_j\}$ is linearly independent in K over k. For if $\sum_{i,j} a_{i,j}\alpha_i\beta_j = 0$, where $a_{i,j} \in k$, then $0 = \sum_i (\sum_j a_{i,j}\beta_j)\alpha_i$. Since $\sum_i (\sum_j a_{i,j}\beta_j) \in L$ and the α_i are linearly independent over L, $\sum_j a_{i,j}\beta_j = 0$ for each i. Because the β_j are linearly independent over k, each $a_{i,j} = 0$. Hence $[K : k] \geq [L : k] \cdot [K : L]$. Thus if either of the latter two is infinite so is $[K : k]$ and the proposition is proved. Otherwise, we may assume each of these is finite. Then taking the $\{\alpha_i\}$ as a basis for K over L and $\{\beta_j\}$ as a basis for L over k, the previous calculation shows $\{\alpha_i\beta_j\}$ is linearly independent in K over k and therefore is a basis of K over k, proving the proposition. $\qquad\square$

Now, let K be a field containing k and S be a subset of K. We will denote by $k(S)$ the intersection of all subfields of K which contain both k and S. Obviously, $k(S)$ is a field and is the smallest field extension of k containing S. We call $k(S)$ the *adjunction* of S to k. If S is a finite subset of K, say $S = \{s_1, \ldots, s_n\}$, we will write $k(s_1, \ldots, s_n)$ instead of $k(S)$.

Exercise 11.1.2. Let K be a field extension of k and S and T be subsets of K. Prove $k(S \cup T) = k(S)(T) = k(T)(S)$.

11.2   Algebraic Extensions and Splitting Fields

Definition 11.2.1. A field extension $K \supset k$ is called algebraic and we say K is *algebraic over* k if every $\alpha \in K$ satisfies *some* polynomial equation with coefficients from k. That is $f(\alpha) = 0$, where $f \in k[x]$.

Proposition 11.2.2. *If K is a finite extension of k, then K is algebraic over k.*

Proof. If this is not the case, there is an $\alpha \in K$ which satisfies no polynomial equation with coefficients from k. Then $k(\alpha)$ is isomorphic to the field of rational functions in one variable with coefficients from k. Since $[k(\alpha) : k]$ is infinite, it follows from Proposition 11.1.1 that $[K : k]$ is also infinite, a contradiction. $\qquad\square$

Before proceeding it would be well to give an important example of an algebraic extension.

Example 11.2.3. Consider the polynomial $x^2 + 1$ (which is actually monic with coefficients in $\mathbb{Q}$) in $\mathbb{R}[x]$. It obviously has no roots in $\mathbb{R}$ and so is irreducible in $\mathbb{R}[x]$. As we learned in Chapter 5, the principal ideal $(x^2 + 1)$ is maximal and so $\mathbb{R}[x]/(x^2 + 1)$ is a *field*. Let $f(x) \in \mathbb{R}[x]$. By the Euclidean algorithm for polynomials $f(x) = g(x)(x^2 + 1) + r(x)$, where $\deg r \leq 1$. Hence $r(x) = a + bx$, where $a, b \in \mathbb{R}$. Obviously, this field has dimension 2 over $\mathbb{R}$ and

$$\mathbb{R}[x] \to \mathbb{R}[x]/(x^2 + 1) = \{a + bx : a, b \in \mathbb{R}\}.$$

Let $\alpha(= i) = x + (x^2 + 1) \in \mathbb{R}[x]/(x^2 + 1)$. Evidently, $\alpha^2 + 1 = 0$. So we have extended the field $\mathbb{R}$ by adjoining a root of this polynomial. Since $-i$ is also in the field which consists of all $a + bi$, where $a, b \in \mathbb{R}$, we see this field is $\mathbb{C}$. Thus $\mathbb{C}$ is an algebraic extension of $\mathbb{R}$ of degree 2. Since it contains all the roots of $x^2 + 1 = 0$, it is the *splitting field* of $x^2 + 1$ over $\mathbb{R}$.

We will now show all of this is rather general.

Proposition 11.2.4. *Let k be a field and $p(x)$ be an irreducible polynomial in $k[x]$ of degree n. Then there is an algebraic extension K of k of degree n which contains a root α of $p(x)$. Indeed, $K = k(\alpha)$.*

Proof. By the same reasoning as above $K = k[x]/(p(x))$ is a field and $[K : k] = n$. Taking $\alpha = x + (p(x))$, we see $p(\alpha) \in (p(x))$ and so $p(\alpha) = 0$. $\square$

We now want to deal with the remainder of the roots of $p(x)$. Factor $p(x)$ over $K(\alpha)$. One of the factors is certainly $x - \alpha$. Throw it and any other linear factors (such as $x + \alpha$ if this is different from $x - \alpha$ in case $\operatorname{char} k \neq 2$) away and consider any irreducible subfactor q of p (over $k(\alpha)$) if any. If there are none, then all the roots of p lie in $k(\alpha)$ and we are done. Otherwise, we apply the same procedure to q and $k(\alpha)$ that we did to p and k. Continuing in this way we arrive at a finite algebraic extension K over k containing *all* the roots of p.

Now let k be a field and $f(x)$ be any polynomial of degree ≥ 2. Factor it into irreducibles $f(x) = p_1(x)^{e_1} \cdots p_s(x)^{e_s}$ and successively apply the procedure just above to the various p_i. This results in,

Theorem 11.2.5. *Let k be a field and $f(x) \in k[x]$ be any polynomial of degree ≥ 2. Then, there is a field K which contains k as a subfield, unique up to an isomorphism leaving k fixed and containing all the roots of f. K is a finite algebraic extension of k obtained by adjoining all the roots of f, as above.*

Definition 11.2.6. We call K the *splitting field* of f over k.

We recall (see Corollary 5.7.21 of Volume I) that if k has characteristic zero, a polynomial $f(x) \in k[x]$ has only simple roots in its splitting field.

We now turn to the very useful theorem of the primitive element. This tells us among other things, that when k has characteristic zero, the splitting field can be gotten from the base field by a *single* (algebraic) field extension, $K = k(\theta)$. Such a θ is called a primitive element.

Theorem 11.2.7. *Let k be a field of characteristic zero and $K = k(a_1, ..., a_n)$, with each a_i algebraic over k. Then K is actually a simple extension of k, i.e., $K = k(\theta)$ for some $\theta \in K$.*

Proof. Using induction we may assume $n = 2$. We will prove "most" linear combinations of a_1 and a_2 result in a primitive element. Let $\lambda \in k^{\times}$ and $\theta = a_1 + \lambda a_2$. We will show θ is primitive for all but finitely many choices of λ. For a "good" choice of λ it suffices to show that $k(\theta)$ contains a_2, because then $k(\theta)$ would also contain $a_1 = \theta - \lambda a_2$ and therefore $k(\theta) \supseteq k(a_1, a_2)$ so that $k(\theta) = K$.

To do this, it would be sufficient to prove that, except for finitely many λ, the minimal polynomial Φ of a_2 over $k(\theta)$ can not have degree > 2. For $i = 1, 2$ let f_i be the minimal polynomial of a_i over k and let $g(x) = f_1(\theta - \lambda x)$. Then g is a polynomial in $k(\theta)[x]$ and $g(a_2) = f_1(\theta - \lambda a_2) = f_1(a_1) = 0$. Therefore f_2 divides g. Since Φ divides f_2, Φ divides both f_2 and g (all in $k(\theta)[x]$). If we can show $\gcd(f_2, g)$ can not have degree ≥ 2 we would be done. For suppose it did. Then f_2

and g must have a common root, $b_2 \in k(\theta)$, which by Corollary 5.7.21 of Volume I, since k has characteristic 0, is $\neq a_2$. Thus $f_1(\theta - \lambda b_2) = 0$ and $\theta - \lambda b_2 = b_1$ for some root b_1 of f_1 in $k(\theta)$. Since $\theta = a_1 + \lambda a_2 = b_1 + \lambda b_2$, it follows that
$$\lambda = \frac{b_1 - a_1}{a_2 - b_2}.$$

As these bad values of $\lambda \in k^\times$ are finite in number and k is infinite since it has characteristic zero, they can be avoided.

Let k and l be fields and $\alpha : k \to l$ be a field isomorphism between them. Then α gives rise to a ring isomorphism $k[x] \to l[x]$ by mapping the coefficients through α. By abuse of notation we also call this isomorphism α. Now suppose $p(x)$ is irreducible in $k[x]$. Then since α is a ring isomorphism, $\alpha(p)(x)$ is irreducible in $l[x]$. It induces an isomorphism $k[x]/(p(x)) \to l[x]/\alpha(p)(x)$ by

$$\alpha(f(x) + (p(x))) = \alpha(f)(x) + (\alpha(p)(x)).$$

In particular, $\alpha(x + (p(x))) = x + (\alpha(p)(x))$. By the proof above this gives us an isomorphism from $k(a) = k[x]/(p(x))$ to $l(\alpha(a)) = l[x]/(\alpha(p)(x))$. The case of an automorphism of k is the main use of this.

As a final result of this section we have,

Corollary 11.2.8. *Let k and l be fields and $\alpha : k \to l$ be a field isomorphism between them. Let f be a polynomial over k, $\alpha(f)$ the corresponding polynomial over l, and K and L be the respective splitting fields. Then α extends to an isomorphism $K \to L$.*

Proof. Since $k(a) \to l(\alpha(a))$ is an isomorphism, the result follows by induction. $\square$

11.3   Finite Groups of Automorphisms of a Field

The purpose of this section is to prove that if K is a field, G a finite subgroup of $\mathrm{Aut}(K)$ and k is the subfield of K consisting of the points fixed under G. Then $[K : k] = |G|$.

Definition 11.3.1. Let G be a group and K be a field. We call a homomorphism $\lambda : G \to K^\times$ a K-character of G.

Lemma 11.3.2. Let $\lambda_1, \ldots, \lambda_n$ be distinct K-characters of G. Then they are K-linearly independent in the space $\mathcal{F}(G, K)$ of K-valued functions on G.

Proof. Suppose some linear combination $\sum_{i=1}^{n} c_i \lambda_i = 0$, where $c_i \in K$. Select such a linear combination of shortest length r, where all the $c_i \neq 0$. First $r > 1$, because if $c_1 \lambda_1 = 0$, then since $c_1 \neq 0$, it would follow that $\lambda_1 = 0$, a contradiction since $\lambda_1(G) \subseteq K^\times$. Thus $r \geq 2$ and

$$\sum_{i=1}^{r} c_i \lambda_i(x) = 0. \tag{11.1}$$

Replace x by ax in equation 11.1, where $a \in G$. Then since these are characters, $\sum_{i=1}^{r} c_i \lambda_i(a)\lambda_i(x) = 0$. Multiplying 11.1 by $\lambda_r(a)$ and subtracting this from this last equation giving, $\sum_{i=1}^{r-1} c_i (\lambda_r(a) - \lambda_i(a))\lambda_i(x) = 0$, a shorter relation. Therefore all the coefficients must be zero. For if any of these coefficients were $\neq 0$, then just taking these would give an even shorter relation (which must have more than one term). Since the c_i are all $\neq 0$ we conclude $\lambda_r(a) = \lambda_i(a)$ for all $a \in G$ and $i = 1, \ldots, r - 1$. Since the $a \in G$ is arbitrary this means $\lambda_r = \lambda_{r-1}$, where $r - 1 \geq 1$, contradicting the hypothesis that the λ_i are distinct. $\square$

Corollary 11.3.3. Let K be a field and $\sigma_1, \ldots, \sigma_n$ be distinct automorphisms of K. Then these are K-linearly independent in $\mathcal{F}(K, K)$.

Proof. Take $G = K^\times$. If σ is an automorphism of K it takes $K^\times$ to itself. Thus each σ_i is a K-character. $\square$

Lemma 11.3.4. Let K be a field and S be a set of automorphisms of K. Let $k = \{x \in K : \sigma(x) = x, \text{ for all } \sigma \in S\}$. Then k is a subfield of K.

Proof. For $x \in k$ and $\sigma \in S$ we have $\sigma(x \pm y) = \sigma(x) \pm \sigma(y) = x \pm y$ and $\sigma(xy) = \sigma(x)\sigma(y) = xy$. Also, $\sigma(1) = 1$ and so $1 \in k$ and $\sigma(x^{-1}) = \sigma(x)^{-1} = x^{-1}$. $\square$

Proposition 11.3.5. *Let K be a field, $S = \{\sigma_1, \ldots, \sigma_n\}$ be a set of automorphisms of K and*

$$k = \{x \in K : \sigma_i(x) = x \text{ for all } \sigma_i \in S\}.$$

Then $[K : k] \geq n$.

Proof. Suppose the contrary, namely that $[K : k] = r < n$ and $\alpha_1, \ldots \alpha_r$ is a basis of K over k. Consider the system of homogeneous linear equations,

$$x_1\sigma_1(\alpha_1) + x_2\sigma_2(\alpha_1) + \cdots + x_n\sigma_n(\alpha_1) = 0$$
$$x_1\sigma_1(\alpha_2) + x_2\sigma_2(\alpha_2) + \cdots + x_n\sigma_n(\alpha_2) = 0$$
$$\cdots\cdots\cdots\cdots\cdots\cdots\cdots\cdots\cdots\cdots\cdots\cdots\cdots\cdots\cdots$$
$$x_1\sigma_1(\alpha_r) + x_2\sigma_2(\alpha_r) + \cdots + x_n\sigma_n(\alpha_r) = 0$$

where the unknowns $x_i \in K$. Since this is a system of r equations and n unknowns and $r < n$, then, as we know, there must be a non-trivial solution $(x_1, \ldots, x_n) \in K^n$. Now multiply the i-th equation by $c_i \in k$, $i = 1 \ldots r$ and make use of the fact that each σ_i is an automorphism of K fixing k pointwise.

$$x_1\sigma_1(c_1\alpha_1) + x_2\sigma_2(c_1\alpha_1) + \cdots + x_n\sigma_n(c_1\alpha_1) = 0$$
$$x_1\sigma_1(c_2\alpha_2) + x_2\sigma_2(c_2\alpha_2) + \cdots + x_n\sigma_n(c_2\alpha_2) = 0$$
$$\cdots\cdots\cdots\cdots\cdots\cdots\cdots\cdots\cdots\cdots\cdots\cdots\cdots\cdots\cdots$$
$$x_1\sigma_1(c_r\alpha_r) + x_2\sigma_2(c_r\alpha_r) + \cdots + x_n\sigma_n(c_r\alpha_r) = 0$$

Adding these we get,

$$x_1\sigma_1\left(\sum_{i=1}^{r} c_i\alpha_i\right) + \ldots x_n\sigma_n\left(\sum_{i=1}^{r} c_i\alpha_i\right) = 0,$$

and since $\alpha_1, \ldots \alpha_r$ is a basis of K over k and the c_i are arbitrary,

$$x_1\sigma_1(x) + \ldots x_n\sigma_n(x) = 0,$$

where $x \in K$ is arbitrary. By Corollary 11.3.3 all x_i must be zero, a contradiction.   □

Definition 11.3.6. Let K be a field and k be a subfield. We shall write $\mathrm{Aut}(K)|k$ for the group of field *automorphisms on K leaving k pointwise fixed.*

Taking $G = S$, we get as an immediate corollary of Proposition 11.3.5:

Corollary 11.3.7. *Let K be a field, k be a subfield and G be a subgroup of $\mathrm{Aut}(K)|k$ generated by $T = \{\sigma_1, \ldots, \sigma_n\}$. Then of course, the field fixed by T is the same as the field fixed by G, namely k. Moreover, $[K : k] \geq |G|$.*

Example 11.3.8. Let $K = k(x)$, the field of rational functions over k, where k is a field of characteristic zero and $\sigma f(x) = f(x+1)$ is the shift operator on rational functions. Then, the fixed field of σ (or the cyclic group $(\mathbb{Z}, +)$ generated by σ) is evidently just k. Hence $[K : k] = \infty$.

On the other hand, with the same k, if $\sigma f(x) = f(\frac{1}{x})$, then $\sigma^2 f = f$ and so σ has order 2 and generates a group isomorphic with $\mathbb{Z}_2$. If $f \in K$ is σ-fixed, $f(x) = f(\frac{1}{x})$ for all x. Here the fixed field, F, consists of all rational functions of the form

$$a_n x^{-n} + a_{n-1} x^{1-n} + \ldots a_0 + a_1 x + \ldots a_n x^n,$$

and here too $[K, F] = \infty$.

However, it is not these (infinite) transcendental extensions that we are really interested in, but rather those in the theorem below.

Lemma 11.3.9. *Let K be a field, $G = \{\sigma_1, \ldots \sigma_n\}$ be a finite subgroup of $\mathrm{Aut}(K)$, and k be subfield of K consisting of points fixed under G. Then*

$$k = \Big\{ x \in K : x = \sum_{i=1}^{n} \sigma_i(\alpha), \ \ \alpha \in K \Big\}.$$

Proof. We have to show

1. $\sum_{i=1}^{n} \sigma_i(\alpha) \in k$ for all $\alpha \in K$.

2. Given $a \in k$, there is an $\alpha \in K$ so that $\sum_{i=1}^{n} \sigma_i(\alpha) = a$.

Let $\sigma \in G$ and $\alpha \in K$. Then $\sigma(\sum_{i=1}^{n} \sigma_i(\alpha)) = \sum_{i=1}^{n} \sigma\sigma_i(\alpha) = \sum_{i=1}^{n} \sigma_i(\alpha)$. Since $\sum_{i=1}^{n} \sigma_i(\alpha)$ is G-fixed it lies in k. Conversely, let $a \in k$. If $\sum_{i=1}^{n} \sigma_i(\alpha)$ were zero for all $\alpha \in K$, this would violate the linear independence of $\{\sigma_1, \ldots \sigma_n\}$, that is Corollary 11.3.3. Hence there must be some $\alpha \in K$ where $\sum_{i=1}^{n} \sigma_i(\alpha) = c \neq 0$. By the first equation $c \in k$ and since k is a field, $\frac{a}{c} \in k$. Therefore,

$$\sum_{i=1}^{n} \sigma_i\left(\frac{a}{c}\alpha\right) = \sum_{i=1}^{n} \sigma_i\left(\frac{a}{c}\right)\sigma_i(\alpha) = \frac{a}{c}\sum_{i=1}^{n} \sigma_i(\alpha) = \frac{a}{c}c = a.$$

□

We are now in a position to address the main point of this section.

Theorem 11.3.10. *Let K be a field, $G = \{\sigma_1, \ldots \sigma_n\}$ be a finite subgroup of $\mathrm{Aut}(K)$ and k be the subfield of K consisting of points fixed under G. Then $[K : k] = |G|$.*

Proof. Since by Proposition 11.3.5 we already know $[K : k] \geq |G|$, it suffices to show that $[K : k] \leq |G|$. That is, if $\alpha_1, \ldots, \alpha_m$ are any m elements of K with $m > |G| = n$, then $\alpha_1, \ldots, \alpha_m$ are linearly dependent over k.

Consider the system of homogeneous linear equations,

$$x_1\sigma_1(\alpha_1) + x_2\sigma_1(\alpha_1) + \cdots + x_m\sigma_1(\alpha_1) = 0$$
$$x_1\sigma_2(\alpha_1) + x_2\sigma_2(\alpha_2) + \cdots + x_m\sigma_2(\alpha_2) = 0$$
$$\cdots\cdots\cdots\cdots\cdots\cdots\cdots\cdots\cdots\cdots\cdots\cdots\cdots\cdots\cdots\cdots$$
$$x_1\sigma_n(\alpha_1) + x_2\sigma_n(\alpha_n) + \cdots + x_m\sigma_n(\alpha_n) = 0.$$

These are n equations in m unknowns $x_1, \ldots x_m$ in K and since $m > n$, there is a non-trivial solution. By renumbering the α_i we can assume $x_1 \neq 0$. Let $a \neq 0$ be in k and as in Lemma 11.3.9 choose $\alpha \in K$ so that $\sum_{i=1}^{n} \sigma_i(\alpha) = a$. Choose $\beta \in K$ so that $\beta x_1 = \alpha$. Since $(x_1, \ldots, x_m)$ is a solution so is $(\beta x_1, \ldots, \beta x_m)$. Thus there is a solution where the

x_1 is α and $\sum_{i=1}^{n} \sigma_i(\alpha) = a$, where a can be any non-zero element of k. Apply σ_i to the system of equations and get

$$\sigma_i(x_1)\sigma_i\sigma_1(\alpha_1) + \sigma_i(x_2)\sigma_i\sigma_1(\alpha_2) + \cdots + \sigma_i(x_m)\sigma_i\sigma_1(\alpha_m) = 0$$
$$\sigma_i(x_1)\sigma_i\sigma_2(\alpha_1) + \sigma_i(x_2)\sigma_i\sigma_2(\alpha_2) + \cdots + \sigma_i(x_m)\sigma_i\sigma_2(\alpha_m) = 0$$
$$\cdots\cdots\cdots\cdots\cdots\cdots\cdots\cdots\cdots\cdots\cdots\cdots\cdots\cdots\cdots\cdots\cdots\cdots\cdots\cdots\cdots$$
$$\sigma_i(x_1)\sigma_i\sigma_n(\alpha_1) + \sigma_i(x_2)\sigma_i\sigma_n(\alpha_2) + \cdots + \sigma_i(x_m)\sigma_i\sigma_n(\alpha_m) = 0.$$

This is just a permutation of equations with $\sigma_i(x_1), \ldots, \sigma_i(x_m)$ as solutions. Thus, if $x_1, \ldots, x_m$ is a solution so is $\sigma(x_1), \ldots, \sigma(x_m)$ for any $\sigma \in G$. Now the sum of solutions is again a solution. Hence,

$$\left(\sum_{i=1}^{n} \sigma_i(x_1), \ldots, \sum_{i=1}^{n} \sigma_i(x_m) \right)$$

is a simultaneous solution in k^m. But when $x_1 = \alpha$, $\sum_{i=1}^{n} \sigma_i(\alpha) = a$, where $a \neq 0$ is the first coordinate of a simultaneous solution. Since one of the equations in our system is (taking $\sigma_i = I$) is $x_1\alpha_1 + \ldots + x_m\alpha_m = 0$, the α_i are linearly dependent over k. $\qquad\square$

As a consequence,

Corollary 11.3.11. *Let K be a field, $G = \{\sigma_1, \ldots \sigma_n\}$ be a finite subgroup of $\mathrm{Aut}(K)$ and k be subfield of K consisting of points fixed under G. Then every automorphism σ of K leaving k fixed is already in G. That is, $\mathrm{Aut}(K)|k \subseteq G$.*

Proof. Otherwise let σ be an automorphism of K leaving k fixed which is not in G and $S = G \cup \{\sigma\}$. Then $|S| = |G| + 1$. By Proposition 11.3.5 $[K : k] \geq |G| + 1$, contradicting Theorem 11.3.10. $\qquad\square$

Not only is k determined by G, but G is determined by k (and not merely the order of G).

Corollary 11.3.12. *Let G_1 and G_2 be finite subgroups of automorphisms of a field K and k_1 and k_2 the respective fixed fields. If $k_1 = k_2$, then $G_1 = G_2$.*

Proof. Suppose σ is an automorphism of K leaving k_1 fixed. Then it leaves k_2 fixed and therefore $G_1 \subseteq G_2$. Similarly, $G_2 \subseteq G_1$. $\qquad\square$

11.3.1   Exercises

Exercise 11.3.13. Conditions 1 and 2 in the proof of Lemma 11.3.9 enable us to consider $P : K \to k$ defined by $P(\alpha) = \sum_{\sigma \in G} \sigma(\alpha)$ to be a k-linear *projection* operator. Show P is k-linear, surjective and $P^2 = P$.

Exercise 11.3.14. Let K be the splitting field over $\mathbb{Q}$ of the following polynomials.

1. $x^2 - 2 = 0$

2. $x^2 - 2x + 1 = 0$

3. $x^2 - 3x + 1 = 0$

4. $x^3 - 1 = 0$

5. $x^3 - 2 = 0$

6. $x^4 - 1 = 0$

7. $x^5 - 1 = 0$

Calculate K as $\mathbb{Q}$ with appropriate elements adjoined. What is $[K : \mathbb{Q}]$? What is the group of automorphisms of K leaving $\mathbb{Q}$ fixed? What is the subfield of K fixed by all these automorphisms?

11.4   Normal Separable Extensions and the Galois Group

Here we try to understand the elements of K in terms of roots of polynomial equations with coefficients in the subfield k.

Definition 11.4.1. Let k be a field. A polynomial in $k[x]$ is called *separable* if each irreducible component has no multiple roots (in any extension field).

Theorem 11.4.2. *Let K be a field, $G = \{\sigma_1, \ldots, \sigma_n\}$ be a finite subgroup of $\mathrm{Aut}(K)$ and k be the subfield of K consisting of points fixed under G. Then each $\alpha \in K$ is a root of a separable irreducible polynomial in $k[x]$.*

Proof. Let $\alpha \in K$ and $S = \{\sigma_i(\alpha) : i = 1, \ldots, r\}$ be the set of all distinct G-transforms of α. Here $r \leq n$ and depends on α. Since one of the σ_i is I, $\alpha \in S$. Let $\sigma \in G$. Then $\{\sigma\sigma_i(\alpha) : i = 1, \ldots, r\}$ are all distinct so there are r of them and since this is just a rearrangement of the elements of G, $\{\sigma\sigma_i(\alpha) : i = 1, \ldots, r\} = S$. Thus $G(S) = S$.

Let $p(x) = \prod_{i=1}^{r}(x - \sigma_i(\alpha))$. Then $p \in K[x]$ and has distinct roots and degree r. For each $\sigma \in G$, $\sigma(p(x)) = \prod_{i=1}^{r}(x - \sigma(\sigma_i(\alpha))) = p(x)$. Thus p is a G-fixed polynomial. This means the coefficients are G-fixed and so they must actually lie in k. The set of all roots of p (in any extension field) is S. Hence p is separable. Evidently α is a root of p. Since $p \in k[x]$, it only remains to see that p is irreducible.

Let $f = \sum_{i=1}^{m} c_i x^i \in k[x]$ and have α as a root. Then each $\sigma \in G$ fixes each coefficient of f and because $\sum_{i=1}^{m} c_i \alpha^i = 0$ we get $\sum_{i=1}^{m} c_i \sigma(\alpha)^i = 0$ for all $\sigma \in G$. Thus $f(\sigma(\alpha)) = 0$ for every $\sigma \in G$. Hence every element of S is a root of f. In particular, $\deg f \geq r$. Thus p is the polynomial of least degree in $k[x]$ which has α as a root. Write $f(x) = q(x)p(x) + r(x)$, where $\deg r < \deg p$. Then $f(\alpha) = q(\alpha)p(\alpha) + r(\alpha)$ and so $r(\alpha) = 0$. This contradiction shows $p|f$ in $k[x]$. But then if $p(x) = a(x)b(x)$, since $p(\alpha) = 0$, either $a(\alpha) = 0$ or $b(\alpha) = 0$, and so $p|a$, or $p|b$. This means $p = a$ (up to units) and b is a unit (or vice versa) and so p is irreducible. $\square$

Definition 11.4.3. Let $k \subseteq K$ be a field extension. We define $\mathrm{Gal}(K|k)$ to be the group of all automorphisms of K leaving k fixed. $\mathrm{Gal}(K|k)$ is called the *Galois*[1] *group* of the extension $K|k$.

---

[1]Évariste Galois (1811-1832) French mathematician enrolled in his first mathematics class at Louis-le-Grand at age 17. His teacher wrote "It is the passion for mathematics which dominates him, I think it would be best for him if his parents would allow him to study nothing but this, he is wasting his time here and does nothing but torment his teachers and overwhelm himself with punishments". His school reports began to describe him as singular, bizarre, original and closed. Here

we have a situation where perhaps the most original mathematician who ever lived is criticised for being original. In 1828 Galois took the examination of the École Polytechnique and failed. (A secondary motivation for his interest may have been the strong political views of its students, since Galois was an ardent Republican). Back at school Galois enrolled in the mathematics class of Louis Richard. However he worked more and more on his own research and less and less on his schoolwork. He studied Legendre and Lagrange. As Richard reports, "This student works only in the highest realms of mathematics". In 1829 he published a paper on continued fractions and then submitted an article on the algebraic solution of equations to the Académie des Sciences with Cauchy appointed as referee. Shortly after that his father committed suicide and of course Galois was deeply affected by this. Then he took the examination for entry to the École Polytechnique for the second time and again failed. Galois therefore resigned himself to enter the École Normale and received his degree that year. His examiner in mathematics reported:"This pupil is sometimes obscure in expressing his ideas, but he is intelligent and shows a remarkable spirit of research". His literature examiner reported: "This is the only student who has answered me poorly, he knows absolutely nothing. I was told that this student has an extraordinary capacity for mathematics. This astonishes me greatly, for after his examination, I believed him to have but little intelligence".

Galois sent Cauchy further work on the theory of equations, but then learned from the Bulletin de Férussac of a posthumous article by Abel which overlapped with a part of his work. Galois then took Cauchy's advice and submitted a new article On the condition that an equation be solvable by radicals in 1830. The paper was sent to Fourier, the secretary of the Paris Academy, to be considered for the Grand Prize in mathematics. Fourier died in April 1830 and Galois' paper was never subsequently found and so never considered for the prize. Galois, after reading Abel and Jacobi's work, worked on the theory of elliptic functions and Abelian integrals. With support from Sturm, he published three papers in Bulletin de Férussac in April 1830. However, he learnt in June that the prize of the Academy would be awarded jointly to Abel (posthumously) and to Jacobi, his own work never having been considered.

In 1830 a revolution caused Charles 10-th to flee France. There was rioting in the streets of Paris and the director of École Normale, M. Guigniault, locked the students in to avoid them taking part and then wrote newspaper articles attacking the students. Galois wrote a reply in the Gazette des Écoles, attacking M. Guigniault for locking the students up. For this Galois was expelled and he joined the Artillery of the National Guard, a Republican branch of the militia. This was abolished by Royal Decree because the new King, Louis-Phillipe, felt it was a threat to the throne.

Galois had two minor publications in his short life. He was then invited by Poisson to submit another version of his memoir on equations to the Academy. He did so while deeply depressed, without money and having been sent to prison for threats on the King. Then a cholera epidemic swept Paris and prisoners, including Galois, were transferred to the pension Sieur Faultrier. There he fell in love with Stephanie-

Evidently, $\mathrm{Gal}(K|k)$ is a group. However, it may not be finite. Exercise 11.3.14 calculates $\mathrm{Gal}(K|k)$ in a few examples (also see example above where we have transcendental extensions). Actually, even for algebraic extensions the Galois group may not be finite. This is the case when $K = \mathbb{A}$ and $k = \mathbb{Q}$ (here as we noted K is the algebraic closure of k and $[K : k]$ is infinite). Then $|G|$ is infinite, for if it were finite, by Theorem 11.3.10 $[K : k] = |G|$. Since $[K : k] = \infty$ so is $|G|$.

Definition 11.4.4. We call a field extension $K \supseteq k$ *normal* if

1. $\mathrm{Gal}(K|k)$ has fixed field exactly k.

2. $\mathrm{Gal}(K|k)$ is finite.

Theorem 11.4.5. $K \supseteq k$ *is a normal extension if and only if K is the splitting field of a separable polynomial in $k[x]$.*

Proof. Suppose $K \supseteq k$ is normal. Then $[K : k] = |\mathrm{Gal}(K|k)|$ is finite. Let $\alpha_1, \ldots, \alpha_s$ be a basis of K over k and for $i = 1, \ldots, s$, let $f_i(x) \in k[x]$ be a separable irreducible polynomial which has α_i as a root. Then $f(x) = \prod_{i=1}^{s} f_i(x)$ is a polynomial in $k[x]$ which has $\alpha_1, \ldots, \alpha_s$ as roots (and perhaps some others). Hence the splitting field of f over k contains

---

Felice du Motel, the daughter of the resident physician. After he was released Galois exchanged letters with Stephanie who tried to distance herself from the affair. Galois then fought a duel with Perscheux d'Herbinville (the exact reason being unclear, but certainly linked with Stephanie). In his manuscript he writes "There is something to complete in this demonstration. I do not have the time". This led to the exaggerated legend that he spent his last night writing out all he knew about group theory. Galois was wounded in the duel and was abandoned by d'Herbinville and his own seconds and found by a peasant and died the next day. His funeral was the focus for a Republican rally and riots followed which lasted for several days. Galois' brother and his friend Chevalier copied his mathematical papers and sent them to Gauss, Jacobi and others. It had been Galois' wish that Jacobi and Gauss should give their opinions on his work. No record exists of any comment these men made. However the papers reached Liouville who, in September 1843, announced to the Academy that he had found in Galois' papers a concise solution...as correct as it is deep of this lovely problem: "Given an irreducible equation of prime degree, decide whether or not it is soluble by radicals". Liouville published these papers of Galois in his Journal in 1846.

K. Now actually each f_i has all its roots in K because the roots are $\{\sigma(\alpha_i) : \sigma \in \text{Gal}(K|k)\}$ and since $\alpha_i \in K$ so is $\sigma(\alpha_i)$ for each $\sigma \in \text{Gal}(K|k)$. Hence all roots of f are in K so the splitting field of f over k is actually K.

Conversely, suppose K is the splitting field of a separable polynomial $f(x) \in k[x]$. We show K is normal over k. If all roots of $f = 0$ lie in k, then $K = k$, $\text{Gal}(K|k) = (I)$ and the conclusion is trivially true. Now suppose f has $n \geq 1$ roots in K, but not in k. We will prove our result by induction on n. Let $f(x) = p_1(x) \cdots p_r(x)$ be the factorization of f into primes in $k[x]$. We may assume at least one of these primes, say p_1, has degree ≥ 2. Let α_1 be a root of p_1. Then $[k(\alpha_1) : k] = \deg p_1 \geq 2$. Consider $k(\alpha_1)$ as the new ground field. Then fewer roots of p_1 (and therefore fewer roots of f) lie outside $k(\alpha_1)$. Since $f \in k(\alpha_1)[x]$ and K is the splitting field of f over $k(\alpha_1)$, we know by induction that K is a normal extension of $k(\alpha_1)$. Hence $\text{Gal}(K|k(\alpha_1))$ is finite and the fixed field of K under this group is $k(\alpha_1)$ (and not bigger). Therefore every element of $K - k(\alpha_1)$ is moved by some $\sigma \in \text{Gal}(K|k(\alpha_1))$. For otherwise there would be some $\alpha \in K - k(\alpha_1)$ for which $\sigma(\alpha) = \alpha$ for all $\sigma \in \text{Gal}(K|k(\alpha_1))$. That would mean $K \supseteq k(\alpha_1)(\alpha) > k(\alpha_1)$ and $k(\alpha_1)(\alpha)$ is fixed under $\text{Gal}(K|k(\alpha_1))$, a contradiction to our assumptions.

Since f is separable, the roots $\alpha_1, \ldots, \alpha_s$ of p_1 are distinct elements of K. Choose field isomorphisms $\sigma_1, \ldots, \sigma_s$, where $\sigma_i : k(\alpha_1) \to k(\alpha_i)$ which each fix k, and $\sigma_i(\alpha_1) = (\alpha_i)$, $i = 1, \ldots, s$. Extend these σ_i to automorphisms of K, again calling them σ_i. (see Corollary 11.2.8) Then, $\sigma_1, \ldots, \sigma_s \in \text{Gal}(K|k)$. Let $\theta \in K$ be fixed under $\text{Gal}(K|k)$. We will show $\theta \in k$. Now θ is fixed by $\text{Gal}(K|k(\alpha_1))$. Hence by inductive hypothesis, $\theta \in k(\alpha_1)$. Therefore for $c_i \in k$,

$$\theta = c_0 + c_1\alpha_1 + \ldots + c_{s-1}\alpha_1^{s-1}. \tag{11.2}$$

Hence,

$$\sigma_i(\theta) = c_0 + c_1\alpha_i + \ldots + c_{s-1}\alpha_i^{s-1}. \tag{11.3}$$

Now the polynomial $q(x) = c_{s-1}x^{s-1} + \ldots + c_1 x + c_0 - \theta$ has $\alpha_1, \ldots \alpha_s$ as its roots. This is because α_1 is a root by (5.2) and for $i \geq 2$, α_i is a

root by (5.3) and the fact that $\sigma_i(\theta) = \theta$. Since q has degree $\leq s-1$ and
has s distinct roots, it must be identically zero. In particular, $c_0 = \theta$
and so $\theta \in k$. Finally, $[K : k] \geq \mathrm{Gal}(K|k)$ and since $[K : k] < \infty$
because K is the splitting field of a polynomial in $k[x]$, $\mathrm{Gal}(K|k)$ is also
finite. $\square$

Now let K be a finite extension of k and suppose $\mathrm{char}(k) = 0$. Then,
conditions (5.2) and (5.3) are equivalent to (5.4). *Every* irreducible
polynomial in $k[x]$ which has a root in K factors completely in $K[x]$.

The proof of this uses the formal derivative defined in Chapter 5
of Volume I. Namely, a polynomial $f \in k[x]$ is separable if and only if
$(f, f') = 1$. In particular, here the irreducible polynomial $p \in k[x]$ is
separable if and only if $p' \neq 0$. When $\mathrm{char}(k) = 0$ and p has degree ≥ 2
this is automatic and holds for all irreducible p.

This results in the following corollary.

Corollary 11.4.6. *Let K be a finite extension of k, where $\mathrm{char} k = 0$.
Then the following are equivalent*

1. *Every irreducible polynomial in $k[x]$ which has a root in K factors
   completely in $K[x]$.*

2. *K is the splitting field of a separable polynomial $f(x) \in k[x]$.*

3. *K is a normal extension.*

Proof. 2. and 3. are equivalent by Theorem 11.5.3. Assume 1. Since K
is a finite separable extension of k we have $K = k(\theta)$ by the Theorem
of the primitive element. Let f be the minimal polynomial of θ. Since
f is irreducible and has a root, $\theta \in K$, f factors completely over K.
Therefore $k(\theta) = K$ is the splitting field of f over k. This proves 2.

Conversely, suppose K is a normal extension of k; that is assume
3. Then each $\alpha \in K$ is a root of a separable irreducible polynomial in
$k[x]$. Therefore, K is separable over k. Now assume 2. Let p be an
irreducible polynomial in $k[x]$ with some root $\alpha \in K$. Let β be another
root of p. Then, there is an isomorphism $\sigma : k(\alpha) \to k(\beta)$ fixing k and
taking α to β which extends to an automorphism $\tau : K(\alpha) \to K(\beta)$ (see
Corollary 11.2.8). But since $\alpha \in K$, $K(\alpha) = K$. On the other hand,

τ being an extension of σ fixes k pointwise and so permutes the roots. Hence they all lie in K as K is the splitting field. Therefore $\tau(K) = K$. But $\tau(K) = K(\beta)$. Since $K = K(\beta)$, $\beta \in K$, this proves 1. $\qquad\square$

When the char$(k) = 0$, we also get the following two results.

Proposition 11.4.7. *If $K \mid k$ is a Galois extension and L is an intermediate field, then $K \mid L$ is also a Galois extension.*

Proof. As we know K is the splitting field of a separable polynomial $f \in k[x]$. Hence K is also the splitting field of a separable polynomial $f \in L[x]$. But is f separable over L? Since $f \in k[x]$ has no multiple roots and all roots lie in K, then the same is true when we regard $f \in L[x]$. By 2 K is a Galois extension of L. $\qquad\square$

Corollary 11.4.8. *Suppose $[K : k] < \infty$ and char$(k) = 0$. Then, the three previous conditions are equivalent to*
4. $|\operatorname{Gal}(K|k)| = [K : k]$.

Proof. Assume 1, 2 and 3. Then by 1 k is the fixed field of K under $\operatorname{Gal}(K|k)$. By Theorem 11.3.10 $|\operatorname{Gal}(K|k)| = [K : k]$, proving 4.

Assume 4. Then since $[K : k] < \infty$, $\operatorname{Gal}(K|k)$ is finite. We show the fixed field under $\operatorname{Gal}(K|k)$ is exactly k. Suppose L is the fixed field. Then $K \supseteq L \supseteq k$. But we know by Proposition 11.4.7 that $|\operatorname{Gal}(K|L)| = [K : L]$ and since $[K : k] = [K : L][L : k]$, $[L : k] = 1$. $\quad\square$

To sum up: let K be a finite extension of k, a field of characteristic zero. Then the following are equivalent.

1. K is a normal extension of k.

2. K is the splitting field of a separable polynomial in $k[x]$.

3. Every irreducible polynomial in $k[x]$ which has a root in K factors completely in $K[x]$.

4. $|\operatorname{Gal}(K|k)| = [K : k]$.

As this is the core of the subject it deserves a name. We call such an extension a *Galois extension*.

11.5   The Fundamental Theorem of Galois Theory

We now come to the *Fundamental Theorem of Galois Theory* which enables us to correlate the intermediate fields $k \subseteq L \subseteq K$ with subgroups H of $\mathrm{Gal}(K|k)$, from which we shall derive some important and not always obvious consequences. For example, there are at most finitely many intermediate fields. This is because a given finite group has only finitely many subgroups since it has only finitely many (2^n) subsets.

Theorem 11.5.1. *Let K be a Galois extension of k and assume $char(k) = 0$. Then*

1. *$L \mapsto \mathrm{Gal}(K|L)$ maps intermediate fields $k \subseteq L \subseteq K$ to subgroups $\mathrm{Gal}(K|L)$ of $\mathrm{Gal}(K|k)$. Such a map is called a* functor. *We call this functor $\mathcal{G}$. Thus, $\mathcal{G}(L) = \mathrm{Gal}(K|L)$.*

2. *To a subgroup H of $\mathrm{Gal}(K|k)$ corresponds an intermediate field $k \subseteq L_H \subseteq K$ given by $L_H = \{\alpha \in K : H(\alpha) = \alpha\}$. This gives us another functor $\mathcal{F}$. That is, $\mathcal{F}(H) = L_H$.*

3. *These functors invert one another so that we have a bijective correspondence between intermediate fields and subgroups of $\mathrm{Gal}(K|k)$.*

Proof. We already know that if L is an intermediate field between $k \subseteq K$, then $\mathrm{Gal}(K|L)$ is a subgroup of $\mathrm{Gal}(K|k)$. Now let H be an arbitrary subgroup of $\mathrm{Gal}(K|k)$ and $L_H = \{\alpha \in K : H(\alpha) = \alpha\}$. Then $L_H \subseteq K$ and since k is the set of things in K fixed under $\mathrm{Gal}(K|k)$, in particular $k \subseteq L_H$. Moreover, since

$$L_H = \bigcap_{h \in H} \{\alpha \in K : h(\alpha) = \alpha\},$$

each $h \in H$ is in $\mathrm{Aut}(K)$ and since the intersection of subfields is a subfield, L_H is a subfield of K. This proves 2.

To see that these functors invert one another, let L be an intermediate field and $H = \mathcal{G}(L) = \mathrm{Gal}(K|L)$. Then $\mathcal{F}(H) = L$. This follows since $K|L$ is also a Galois extension of L. Hence

$$\mathcal{F}(H) = \bigcap_{\sigma \in \mathrm{Gal}(K|L)} \{\alpha \in K : \sigma(\alpha) = \alpha\} = L.$$

Going the other way, let H be a subgroup of $\mathrm{Gal}(K|k)$ and $L = \mathcal{F}(H)$. Then $K|L$ is also a Galois extension since it is finite and normal. Also, $\mathrm{Gal}(K|L)$ consists of all the automorphisms of K leaving L fixed. But if $\sigma \in H$, then by the definition of $\mathcal{F}(H)$, σ leaves L fixed so that $H \subseteq \mathrm{Gal}(K|L)$. But since $K|L$ is a normal extension actually $H = \mathrm{Gal}(K|L)$. Thus

$$\mathcal{G}(\mathcal{F}(H)) = H.$$

In particular, if L is any intermediate field $\mathcal{G}(\mathcal{F}\mathcal{G}(L)) = \mathcal{G}(L)$. Hence $[K : \mathcal{F}(\mathcal{G}(L))] = |\mathcal{G}(L)| = [K : L]$. On the other hand, since $L \subseteq \mathcal{F}(\mathcal{G}(L))$ this forces $[\mathcal{F}(\mathcal{G}(L)) : L] = 1$ and so, finally,

$$\mathcal{F}(\mathcal{G}(L)) = L.$$

$\square$

When functors, such as $\mathcal{G}$ and $\mathcal{F}$, invert one another they are called *adjoint* functors.

We now turn to inclusion and intersection relations coming from the Fundamental Theorem of Galois Theory.

Corollary 11.5.2. *Let $K|k$ be a Galois extension, H_1 and H_2 subgroups of $\mathrm{Gal}(K|k)$ and L_1 and L_2 be the corresponding intermediate fields. Let L be the subfield of K generated by L_1 and L_2 and H be the subgroup generated by H_1 and H_2. Then*

1. $H_1 \subseteq H_2$ *if and only if* $L_1 \supseteq L_2$.

2. L *corresponds to* $H_1 \cap H_2$.

3. H *corresponds to* $L_1 \cap L_2$.

Proof. 1. Evidently, if $H_1 \subseteq H_2$, then $L_1 \supseteq L_2$. Conversely, if $L_1 \supseteq L_2$, that is if $\mathcal{F}(H_1) \supseteq \mathcal{F}(H_2)$, then by what was just said, $\mathcal{G}(\mathcal{F}(H_1)) \subseteq \mathcal{G}(\mathcal{F})(H_2)$. Thus $H_1 \subseteq H_2$.

2. Since $H_1 \cap H_2$ is the largest subgroup of $\mathrm{Gal}(K|k)$ contained in both H_1 and H_2, then by 1, $\mathcal{F}(H_1 \cap H_2)$ is the smallest intermediate field containing $\mathcal{F}(H_1) = L_1$ and $\mathcal{F}(H_2) = L_2$. This is L.

3. Since H is the smallest subgroup containing H_1 and H_2, the corresponding field is the largest subfield contained in L_1 and L_2 which is L. $\square$

Now we turn to the question of normality.

Corollary 11.5.3. *Let $K|k$ be a Galois extension with Galois group $\mathrm{Gal}(K|k)$. Then*

1. *L is a normal extension of k if and only if $\mathrm{Gal}(K|L)$ is a normal subgroup of $\mathrm{Gal}(K|k)$.*

2. *If H is a normal subgroup, then $L|k$ is a Galois extension and $\mathrm{Gal}(L|k) \cong \mathrm{Gal}(K|k)/H$ (which is $\mathrm{Gal}(K|k)/\mathrm{Gal}(K|L)$).*

Proof. Let $\sigma \in \mathrm{Gal}(K|k)$ and L be any intermediate field. We first show that
$$\mathcal{G}(\sigma(L)) = \sigma L \sigma^{-1}.$$
To see this, let $\tau \in \mathcal{G}(\sigma(L))$. Then $\sigma\tau\sigma^{-1}$ is a k automorphism of $\sigma(L)$. Hence $\mathcal{G}(\sigma(L)) \supseteq \sigma L \sigma^{-1}$. On the other hand, if $\rho \in \mathcal{G}(\sigma(L))$, then $\sigma\rho\sigma^{-1}$ is a k automorphism of L. Thus $\sigma\rho\sigma^{-1} \in \mathcal{G}(L)$. Hence, $\rho \in \sigma\mathcal{G}(L)\sigma^{-1}$, proving the other inclusion.

1. Now observe, that $L|k$ is a normal extension if and only if $\sigma(L) = L$ for all $\sigma \in \mathrm{Gal}(K|k)$. This occurs if and only if taking $\mathcal{G}$ of both sides are equal: $\mathcal{G}(\sigma(L)) = \mathcal{G}(L) = H$. But the left side of this is $\sigma L \sigma^{-1}$. Thus $L|k$ is a normal extension if and only if H is normal.

2. As we just saw if H is normal, then L is a normal extension of k. Since $[L : k] \leq [K : k] < \infty$, $L|k$ is a Galois extension. The map $\sigma \mapsto \sigma|L$ is clearly a group homomorphism. Since any k automorphism of L extends to a k automorphism of K, this map is surjective. Its kernel is $\{\sigma \in \mathrm{Gal}(K|k) : \sigma|L = Id\} = H$. The isomorphism theorem completes the proof. $\square$

11.6 Consequences of Galois Theory

Here we shall give solutions to a number of classical problems.

11.6.1 Non Solvability of Equations of Degree ≥ 5

When we speak of solving a polynomial equation $f(x) = 0$ by formula we mean a formula which is a function of the coefficients of f involving the four arithmetical operations together with extracting various r-th roots of arithmetical combinations of these coefficients. When $\deg f = 1$ this is completely trivial and the result lies in the ground field, k. When $\deg f = 2$ this problem was solved in antiquity and results in a field extension of degree 2 (unless the discriminant happens to be a perfect square) and the roots lie in an algebraic number field of degree 2. When $\deg f = 3$ or 4 this problem was solved in the 16-th century by Cardano and Tartaglia when $\deg f = 3$ and by Ferrari when $\deg f = 4$. These not so trivial formulas are derived in the last section of the chapter (see Section 11.7).

This brings us to the question of solving polynomial equations of degree ≥ 5, or higher. Although great effort has been expended since the 16th century to derive such formulas, we shall now explain why there can not be formulas similar to the ones in Section 11.7 for solutions to the general polynomial equation $f(x) = 0$ of degree ≥ 5 with coefficients from a field k of characteristic zero. This fact, due independently to Abel and Galois in the early 19th century, was a radical change in the way mathematics is done and is *understood*. Before this, one could either prove, or fail to prove something, whereas here we are saying *it is impossible to prove something* and not a question of whether one is clever enough or not. Our objective here is to demonstrate the following theorem.

Theorem 11.6.1. *Let k be a field of characteristic zero containing all the r-th roots of unity and $f \in k[x]$ have degree ≥ 5. There can be no general formula involving the four arithmetic operations of the field k, together with extracting r-th roots that gives the solutions of the equation $f(x) = 0$ in terms of the coefficients.*

To prove this we shall use the method of Galois. In order to do so we must make precise what it means to have such a formula. The formulas in 11.7 involve the four arithmetic operations on the field k of coefficients, together with the extracting of various r-th roots for an arbitrary integer $r \geq 2$. Of course, since k is a field, using any of the four field operations on the coefficients of f will keep us in k. The only issue is what happens when one takes r-th roots and where does Galois theory work?

We illustrate the situation with the familiar quadratic formula. If $ax^2 + bx + c = 0$, then

$$x = \frac{-b \pm \sqrt{b^2 - 4ac}}{2a}.$$

Thus when the coefficients lie in k, the roots lie in $k(\sqrt{b^2 - 4ac})$. Similar results, using more complex formulas and higher roots, hold for equations of degree 3 and 4.

Since we seek such a formula for the roots of the polynomial $f \in k[x]$ and such a formula will involve various r-th roots we may as well assume our ground field k contains all r-th roots of 1. For technical reasons we shall also assume the characteristic of k is zero. (All this is certainly true for $\mathbb{A}$, the field of algebraic numbers or $\mathbb{C}$, the complex numbers.)

Lemma 11.6.2. *Let k be a field of characteristic zero containing all the r-th roots of unity. Then the Galois group of $x^n - a \in k[x]$ is isomorphic to a subgroup of the additive group $\mathbb{Z}_n$. In particular, it is Abelian.*

Proof. Let E be the splitting field of $x^r - a$ over k which we can assume is $\neq k$. Then there is a $b \in E - k$ and $b^r = a$. Let ω be a primitive r-th root of unity. Then the roots of $x^r - a$ are $b, b\omega, \ldots b\omega^{r-1}$. Thus $E = k(b)$ and the elements of the Galois group are determined by their effect on b. Let τ_i and $\tau_j \in \text{Gal}(E/k)$. Then $\tau_i(b) = b\omega^i$ and $\tau_j(b) = b\omega^j$, where i, j run from 0 to $r - 1$. Hence,

$$\tau_i \tau_j(b) = \tau_i(b\omega^j) = \tau_i(b)\tau_i(\omega^j) = b\omega^j\omega^j = b\omega^{i+j}.$$

This means the map $\tau_j \mapsto j$ is a monomorphism from $\text{Gal}(E/k) \to \mathbb{Z}_n$. $\square$

Such a field extension (for any r-th root) is called an *extension by radicals* of k. Our objective is to find out what it means for the roots of f to lie in a field gotten from the base field, k, by a sequence of radical extensions.

Definition 11.6.3. Let $f \in k[x]$. We shall say $f(x) = 0$ is *solvable by radicals* if there is a finite sequence of field extensions

$$k = k_0 \subseteq k_1 \subseteq \ldots \subseteq k_m$$

where for each $i \geq 1$, $k_{i+1} = k_i(a_i)$, $a_i^{r_i} \in k_i$ and $k_m = E$ the splitting field of f over k.

Evidently, if f is not solvable by radicals, there can be no general formula in terms of the coefficients for the roots of $f(x) = 0$. Conversely, if $f(x) = 0$ is solvable by radicals. Then for each i, $k_{i+1} = k_i(a_i)$, where $a_i^{r_i} \in k_i$ These extensions are normal. Hence the various Galois groups are normal subgroups. Since by Lemma 11.6.2 each of these extensions is Abelian, if f is solvable by radicals the Fundamental Theorem of Galois Theory, together with 2.2.8 of Volume I, tells us $\mathrm{Gal}(E/k)$ must be solvable.

Since our general formula is supposed to apply to *all* polynomials in $k[x]$ of degree ≥ 5, to prove our theorem, we need only construct an example where it fails. To do so choose a transcendental number $y_1 \in \mathbb{R} - \mathbb{Q}$. Then choose $y_2 \in \mathbb{R} - \mathbb{Q}(y_1)$ and so on. This can be done by a countability argument since the function field with rational coefficients is countable and $\mathbb{R}$ is not. Continuing in this way, we get $y_n \in \mathbb{R} - \mathbb{Q}(y_1, \ldots, y_{n-1})$. In other words, $y_1, \ldots, y_n$ are algebraically independent transcendental numbers. In particular, the y_i are distinct. Let $E = \mathbb{Q}(y_1, \ldots, y_n)$ and

$$f(x) = \prod_{i=1}^{n} (x - y_i),$$

which has only simple roots in E, the splitting field of f over $\mathbb{Q}$. Now let σ_i be the i-th symmetric polynomial in $y_1, \ldots, y_n$, for $i = 1, \ldots, n$ and $F = \mathbb{Q}(\sigma_1, \ldots, \sigma_n) \subseteq E$. By Vieta's formula 5.5.11 of Volume I the coefficients of f are $\pm$ the various σ_i. Every permutation of the σ_i in S_n induces a *distinct* automorphism of E leaving $\mathbb{Q}$ fixed. Moreover, these automorphisms also leave f fixed, so they are in $\text{Gal}(E/F)$.

Since $|\text{Gal}(E/F)| \leq n!$ (see 11.3.10), we conclude $\text{Gal}(E/F) = S_n!$ (This is what is meant when some authors say that most Galois groups are the full symmetric group.) This holds for any n, not just those ≥ 5. Thus, as above, on the one hand $\text{Gal}(E/F)$ must be solvable and on the other, in this case, it is isomorphic to S_n. But by Theorem 2.4.12 of Volume I when $n \geq 5$, S_n is definitely not solvable, because A_n is simple. This contradiction proves there can be no such general formula.

11.6.2 Classical Ruler and Compass Constructions

Since the material here follows roughly the same pattern as the solution to the question of polynomial equations of degree ≥ 5 and is actually simpler, we shall be briefer.

As is well known, the ancients were able to bisect any angle using only a straight edge and compass. This naturally gave rise to the question, as to whether one can trisect any angle using only a straight edge and compass? They did not succeed in answering this question because, as above, they were really only in a position to attempt to answer such a question if the answer was yes. We will apply the ideas of Galois theory to this and another problem of antiquity, namely duplicating the cube. That is, constructing the length of a cube whose volume is twice that of a given cube. As we shall see the answer to both these questions is no. Similarly, one can consider last famous problem of antiquity, namely squaring the circle. In other words, construct, using ruler and compass, a square equal in area to a given circle. We deal with this first. Let the circle have radius 1. Then the side x of the square must satisfy $x^2 = \pi$. Since then $x = \sqrt{\pi}$ if x were constructible would have to be an algebraic number (see below). But then its square would also be an algebraic number which is impossible since π is transcendental by a famous theorem of Lindermann[2]. The proof of the transcendence of π

---

[2]Carl Louis Ferdinand von Lindermann (1852-1939) was a German mathematician who gave the first proof of the transcendence of π in 1882. He was student of Felix

appears in the Appendix.

Before attempting to answer these questions we must define precisely what "constructible" means.

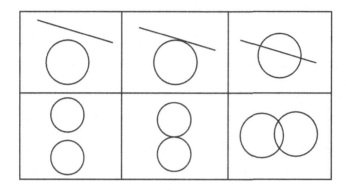

We will show the constructible numbers form a field, Const, lying between $\mathbb{Q}$ and $\mathbb{R}$ and that each construction lies in a subfield k_r of degree some power of 2 over $\mathbb{Q}$. Now, a construction is done by a finite iteration of constructions i.e. of extensions of the ground field $\mathbb{Q} = k_0$, of degree 2, by intersecting two lines, or intersecting a line with a circle (or an arc of a circle), or intersecting two circles, or duplicating a distance. Suppose we are in k_i, a previously constructed extension field. Evidently, in the case of intersecting lines we stay within k_i (check!). If we intersect a line $y = mx+b$ with a circle $(x-c)^2+(y-d)^2 = r^2$, solving this system by substituting the linear equation into the second degree equation gives a polynomial $ax^2 + bx + c = 0$, where $a, b, c \in k_i$ (check!). Here $a = 1 + m^2$ so $a \neq 0$ (since the constructible numbers are inside $\mathbb{R}$) and we indeed have a quadratic function. If f is irreducible over k_i, we have an extension of degree 2. If not, we remain in k_i. Similarly, if we have the intersection of two circles we get the same conclusion since this intersection is the same as the intersection of a line and a circle (no matter whether there are one or two points of intersection, or the intersection is empty). The reader should check this. If the construction required duplicating a length, this is again intersecting with a circle.

Klein and doctoral advisor of D. Hilbert, H. Minkowski, O. Perron, A. Sommerfeld and about 45 others!

Thus a construction corresponds to a finite sequence of field extensions, $k_0 \subseteq k_1 \dots \subseteq k_r$ all of degree 2. Taking into account trivial extensions, Proposition 11.1.1 tells us that after r steps $[k_r : k_0] = 2^\nu$, where $\nu \leq r$.

We now show that one can not trisect, for example, a 60 degree angle. That is, the angle 20 degrees can not be constructed. Here is a useful trigonometric identity $\cos(3\theta) = 4\cos^3(\theta) - 3\cos(\theta)$ (check!). Letting $\theta = 20$ degrees, we see $\frac{1}{2} = 4x^3 - 3x$, where $x = \cos(20)$. The equation for $x = \cos(20)$ is $8x^3 - 6x - 1 = 0$. Next we show the polynomial $f(x) = 8x^3 - 6x - 1$ is irreducible over $\mathbb{Q}$. Notice here the Eisenstein criterion does not help because the constant term is a unit. Suppose $\frac{a}{b} = x$ were a rational root of $f = 0$ and assume $\gcd(a, b) = 1$. Since $8\frac{a^3}{b} - 6\frac{a}{b} - 1 = 0$. we see that $8a^3 - 6ab^2 = b^3$. Let p be a prime divisor of a. Then p divides b^3 and therefore p divides b. This is a contradiction unless a is a unit. On the other hand $x \in \mathbb{Q} \cap \mathbb{A} = \mathbb{Z}$ (see Theorem 11.8.3 of Chapter 11) so b is also a unit. Thus $x = \pm 1$. But then $8x^3 - 6x - 1$ is either 1 or -3 and not zero, a contradiction. Since $f(x) = 8x^3 - 6x - 1$ has degree 3 and no rational root, it is irreducible over $\mathbb{Q}$ and consequently, the splitting field has degree 3 over $\mathbb{Q}$. By uniqueness of prime factorization, x is not constructible.

Now let us consider a cube say of side 1. Then the side of the cube to be constructed satisfies $x^3 = 2$. As we saw in Chapter 5 of Volume I, the polynomial $f(x) = x^3 - 2$ is irreducible over $\mathbb{Q}$ by the Eisenstein irreducibility criterion and Gauss' lemma. Hence, its splitting field has degree 3 over $\mathbb{Q}$ and so the cube root of 2 is not constructible.

11.6.3   The Fundamental Theorem of Algebra

As we saw above $\mathbb{C}$ is the splitting field of $x^2 + 1$ over $\mathbb{R}$. So it contains the roots $\pm i$. But actually it contains the roots of all polynomial equations with coefficients from $\mathbb{C}$ of all degrees. This is the statement of the Fundamental Theorem of Algebra. Interestingly, there seems to be no purely algebraic proof of this important fact[3]. The following largely

---

[3] A statement of the Fundamental Theorem of Algebra first appeared early in the 17th century in the writings of several mathematicians, including Peter Roth, Albert Girard, and René Descartes. Carl Friedrich Gauss is credited with the first correct

algebraic proof is due to L. Horowitz [63]. In the Appendix the reader will find an alternative proof of this.

Here we assume the following facts. The first has to do with connectedness of the real line in the form of the intermediate value theorem, while the second follows from the quadratic formula.

1. Every polynomial of odd degree in $\mathbb{R}[x]$ has a root in $\mathbb{R}$.

2. Every polynomial of degree 2 in $\mathbb{C}[x]$ has a root in $\mathbb{C}$.

Lemma 11.6.4. *Let k be a field. Suppose every polynomial of odd degree in $k[x]$ has a root in k. Let $L \mid k$ be a finite Galois extension. Then the Galois group G of $L \mid k$ is a 2-group.*

Proof. We can assume that $L \neq k$. By the theorem of the primitive element, there exists a θ with $L = k(\theta)$. By the assumption of the Lemma we may take the degree of the minimal polynomial of θ to be even. Hence so is $|G|$. Write $|G| = 2^r m$, where m is odd. Now we recall the Sylow theory of Chapter 2 Volume I and let P be the Sylow 2-subgroup of G and M be the subfield fixed by P. Since $[M : k] = m$ is odd, $m = 1$ by reasoning similar to the above. Hence $|G| = 2^r$.   □

Theorem 11.6.5. *(The Fundamental Theorem of Algebra) The field of complex numbers $\mathbb{C}$ is algebraically closed.*

Proof. Let $f(x)$ in $\mathbb{R}[x]$ be a non-constant polynomial. It suffices to prove that $f(x)$ splits in $\mathbb{C}$. Let $L \mid \mathbb{C}$ be a splitting field of $f(x)$. Since $L \mid \mathbb{R}$ is a splitting field of $(x^2 + 1)f(x)$, $L \mid \mathbb{R}$ is Galois. Let G be the Galois group of $L \mid \mathbb{R}$ and H be the Galois group of $L \mid \mathbb{C}$. By our first assumption together with the lemma, G is a 2-group. Hence H is also a 2-group. Therefore H is nilpotent. Evidently, we may assume $|H| > 1$. As a nilpotent group of order 2^n, H has a normal subgroup N of index $[H : N] = 2$. Let F be the subfield fixed by N. Then $[F : \mathbb{C}] = 2$, a contradiction by our second assumption. Thus $H = \{1\}$ and this means $L = \mathbb{C}$.   □

proof in his doctoral dissertation of 1799. Today there are many known proofs of the Fundamental Theorem of Algebra, usually via complex analysis, but also via Fixed Point theorems of Topology.

Exercise 11.6.6. Prove that a nilpotent group H of order 2^n has a normal subgroup N of index $[H : N] = 2$. Hint: Consider the non-trivial center $\mathcal{Z}(H)$ and reduce to the Abelian case. Then apply the Fundamental Theorem of Abelain Groups.

11.7   Solution of the General Cubic and Quartic Equations

Here we solve the general cubic and quartic equations as they were first done in the sixteenth century by Tartaglia-Cardano and Ferrari, respectively. We shall assume for simplicity that the coefficients are real or complex although the arguments work for any field of characteristic zero. We denote the field by k.

In general we consider the equation,

$$a_0 x^n + a_1 x^{n-1} + \ldots a_{n-1} x + a_n = 0,$$

where $a_j \in k$ for all $j = 0 \ldots n$ and $n = 3$ or 4. Since we are in a field and the lead coefficient is non-zero, by dividing it we may assume $a_0 = 1$.

Thus in the case of the cubic, the equation we want to solve is

$$x^3 + a_1 x^2 + a_2 x + a_3 = 0,$$

while for the quartic it is

$$x^4 + a_1 x^3 + a_2 x^2 + a_3 x + a_4 = 0.$$

Lemma 11.7.1. *We may always assume $a_1 = 0$. One then says the equation is in reduced form.*

Proof. Our equation is $x^n + a_1 x^{n-1} + \ldots a_{n-1} x + a_n = 0$. Perform a change of variables by translation $x \mapsto x + h$, where h is to be chosen so that in the new coordinates $a_1 = 0$. We leave it as an exercise to the reader to show that $h = -\frac{a_1}{n}$ will do this. $\qquad\square$

We now turn specifically to the general cubic. The equation is $x^3 + px = q$, where $p = a_2 - \frac{a_1^2}{3}$ and $q = -(\frac{2a_1^3}{27} - \frac{a_1 a_2}{3} + a_3)$. Thus p and $q \in k$. Evidently, we may assume $p \neq 0$. For if $p = 0$ the equation reduces to $x^3 = q$ and x is one of the three cube roots of q.

To solve the equation $x^3 + px = q$ we note the identity

$$(z - w)^3 + 3zw(z - w) = z^3 - w^3.$$

This follows immediately from the Binomial Theorem 0.4.2 of Volume I since,

$$(z - w)^3 = z^3 - 3z^2 w + 3zw^2 - w^3.$$

Lemma 11.7.2. *We can find z and $w \in k$ satisfying $z^3 - w^3 = q$ and $zw = \frac{p}{3}$.*

Proof. Solve the second equation for $w = \frac{p}{3z}$ and substitute this into the first getting $z^3 - \frac{p^3}{27z^3} = q$. Thus $z^6 - qz^3 - \frac{p^3}{27} = 0$, which is a quadratic equation in z^3. To solve it we use the quadratic formula:

$$z^3 = \frac{q \pm \sqrt{q^2 + 4\frac{p^3}{27}}}{2}$$

Then z is any cube root of z^3 and w is determined by the second equation. $\qquad\square$

Notice that if $k = \mathbb{R}$, then to be sure z^3 and hence z is real we must have $q^2 + 4\frac{p^3}{27} \geq 0$. That is, $\Delta = \frac{q^2}{4} + \frac{p^3}{27} \geq 0$. In any case this quantity, Δ, is the discriminant (see Chapter 5) of the reduced equation.

Now squaring the first of these equations and cubing the second yields,

$$z^6 - 2z^3 w^3 + w^6 = q^2$$

and

$$4z^3 w^3 = \frac{4p^3}{27}.$$

Adding the two yields, $(z^3 + w^3)^2 = q^2 + \frac{4p^3}{27}$. Hence $z^3 + w^3 = \sqrt{q^2 + \frac{4p^3}{27}}$. Of course we also have $z^3 - w^3 = q$.

Adding and subtracting these two equations tells us

$z^3 = \frac{q}{2} + \sqrt{\frac{q^2}{4} + \frac{p^3}{27}}$ and $w^3 = -\frac{q}{2} + \sqrt{\frac{q^2}{4} + \frac{p^3}{27}}$.

Thus $z = (\frac{q}{2} + \sqrt{\Delta})^{\frac{1}{3}}$ and $w = (-\frac{q}{2} + \sqrt{\Delta})^{\frac{1}{3}}$. So finally since $x = z - w$,

$$x = \left(\frac{q}{2} + \sqrt{\frac{q^2}{4} + \frac{p^3}{27}} \right)^{\frac{1}{3}} + \left(\frac{q}{2} - \sqrt{\frac{q^2}{4} + \frac{p^3}{27}} \right)^{\frac{1}{3}}.$$

Observe that although if $\Delta < 0$ we get conjugate complex numbers when we take the square root, this may be undone when we finally take cube roots. As an example and an exercise, the reader should find the (three real) roots of $x^3 = 15x + 4$.

We now turn to the quartic equation. Just as we have reduced the solution of the cubic to a quadratic, we shall reduce the solution of the quartic to that of a cubic (and a quadratic).

The equation we want to solve is $x^4 + px^2 + qx + r = 0$, or equivalently, $x^4 + px^2 = -qx - r$. Completing the square we get

$$(x^2 + p)^2 = px^2 + p^2 - qx - r.$$

Now let $z \in k$ be a new variable to be chosen later. Then

$$(x^2 + p + z)^2 = px^2 + p^2 - qx - r + 2(x^2 + p)z + z^2,$$

which as a polynomial in x is $(p + 2z)x^2 - qx + (p^2 - r + 2pz + z^2)$.

We want to choose z so that this quadratic polynomial in x is a perfect square. That means its (quadratic) discriminant is zero. Thus we want $q^2 - 4(p + 2z)(p^2 - r + 2pz + z^2) = 0$. Since p, q and r are fixed in k, this is a cubic equation in z which *we have just solved above*. Hence we can find such a z_0. Thus

$$(x^2 + p + z_0)^2 = (\alpha x + \beta)^2,$$

where α and β are computable constants in k. Then $x^2 + p + z_0 = \pm(\alpha x + \beta)$ and hence, $x^2 = \pm(\alpha x + \beta) - p - z_0$. This means our solution is

$$x = \pm\sqrt{\pm(\alpha x + \beta) - p - z_0}.$$

Exercise 11.7.3. Find α and β in terms of p, q, r and z_0.

11.8 Some Preliminaries from Algebraic Number Theory

Definition 11.8.1. We say that $z \in \mathbb{C}$ is an *algebraic number* (respectively an *algebraic integer*), if z is a root of a monic polynomial with rational (respectively integer) coefficients.

Definition 11.8.2. We denote the set of algebraic numbers by $\mathbb{A}$ and the set of algebraic integers by $\mathbb{A}_0$.

Every algebraic number a has a minimal polynomial $p(x)$, which is the monic polynomial with rational coefficients of the smallest degree such that $p(a) = 0$. Any other polynomial $q(x)$ with rational coefficients such that $q(a) = 0$ is divisible by $p(x)$. Roots of $p(x)$ are called the *algebraic conjugates* of a; they are roots of any polynomial $q(x)$ with rational coefficients such that $q(a) = 0$.

Since an algebraic conjugate of an algebraic integer is obviously also an algebraic integer it follows from Vieta's Theorem (see Chapter 5), that the minimal polynomial of an algebraic integer has integer coefficients.

Theorem 11.8.3. $\mathbb{A}_0 \cap \mathbb{Q} = \mathbb{Z}$.

Proof. Let $z \in \mathbb{A}_0 \cap \mathbb{Q}$. Since $z \in \mathbb{A}_0$, there is a monic polynomial $p(x) = x^n + a_1 x^{n-1} + \ldots + a_{n-1} x + a_n$ such that $p(z) = 0$, and, since $z \in \mathbb{Q}$, $z = \frac{p}{q}$ with $p, q \in \mathbb{Z}$ and $\gcd(p,q) = 1$. Therefore,

$$\frac{p^n}{q^n} + a_1 \frac{p^{n-1}}{q^{n-1}} + \ldots + a_{n-1} \frac{p}{q} = -a_n.$$

In other words,

$$p^n + a_1 q p^{n-1} + \ldots + a_{n-1} q^{n-1} p = -a_n q^n.$$

Now, since q divides $-a_n q^n$, it must divide the left side of the above equation. But q divides all terms except p^n, since p, q are relatively prime. Therefore $q = \pm 1$, and so $z \in \mathbb{Z}$.

Now, if $z \in \mathbb{Z}$, then it is a root of the monic polynomial $p(x) = x - z$ and so $z \in \mathbb{A}_0 \cap \mathbb{Q}$. $\qquad\square$

Proposition 11.8.4. *The following two statements are equivalent:*

1. *z is an algebraic integer (respectively an algebraic number).*

2. *z is an eigenvalue of a matrix with integer (respectively rational) entries.*

Proof. If λ is an eigenvalue of $A \in M_n(\mathbb{Z})$, then $\chi_A(x)$ is a monic polynomial in $\mathbb{Z}[x]$, and $\chi_A(\lambda) = 0$. Hence λ is an algebraic integer. To prove the converse, assume that $f(\lambda) = 0$, for some monic polynomial $f(x) = x^n + a_1 x^{n-1} \cdots a_{n_1} x + a_n \in \mathbb{Z}[x]$. Let A be the companion matrix of $f(x)$, i.e.

$$
\begin{pmatrix}
0 & 1 & 0 & \cdots & 0 \\
0 & 0 & 1 & \cdots & 0 \\
\vdots & \vdots & \ddots & \ddots & \vdots \\
0 & 0 & \cdots & 0 & 1 \\
-a_n & -a_{n-1} & \cdots & -a_2 & -a_1
\end{pmatrix}
$$

Expanding $\chi_A(x) = \det(A - xI)$ by the first column and using induction, it follows that $\chi_A(x) = f(x)$. Therefore λ is an eigenvalue of $A \in M_n(\mathbb{Z})$. A similar argument works for algebraic numbers, monic polynomials in $\mathbb{Q}[x]$ and matrices in $M_n(\mathbb{Q})$. $\qquad\square$

Theorem 11.8.5. *The set $\mathbb{A}_0$ is a commutative ring with identity and $\mathbb{A}$ is a field.*

Proof. We have to prove that if α and β are algebraic integers (respectively algebraic numbers), then so are $\alpha \pm \beta$, $\alpha\beta$, and β/α (if $\alpha \neq 0$). By Lemma 11.8.4 α and β are eigenvalues of the matrices $A \in M_n(\mathbb{Z})$ (respectively $M_n(\mathbb{Q})$) and $B \in M_m(\mathbb{Z})$ (respectively $M_m(\mathbb{Q})$). Then, there exist vectors $x \in \mathbb{C}^n$ and $y \in \mathbb{C}^m$ such that

$$
Ax = \alpha x \quad \text{and} \quad By = \beta y.
$$

It follows immediately that

$$
(A \otimes B)(x \otimes y) = \alpha\beta \, (x \otimes y)
$$

and

$$(A \otimes I_n \pm I_m \otimes B)(x \otimes y) = (\alpha \pm \beta)((x \otimes y),$$

i.e. $\alpha\beta$ and $\alpha \pm \beta$ are also algebraic integers (respectively algebraic numbers) which means that both $\mathbb{A}_0$ and $\mathbb{A}$ are rings. Clearly 1 is an identity of both. By the Fundamental Theorem of Algebra $\mathbb{A} \subseteq \mathbb{C}$. Hence $\mathbb{A}$ and of course also $\mathbb{A}_0$ are commutative. Thus $\mathbb{A}$ and $\mathbb{A}_0$ are commutative rings with identity. To see that $\mathbb{A}$ is a field, let $\alpha \neq 0$ be an element of $\mathbb{A}$. Since $\mathbb{A}$ is a (countable) union of splitting fields for polynomials in $\mathbb{Q}[x]$, α must be in one of these splitting fields. Hence α^{-1} is also in this splitting field and therefore $\alpha^{-1} \in \mathbb{A}$, i.e. $\mathbb{A}$ is a field. $\qquad\square$

Exercise 11.8.6. The field $\mathbb{A}$ is the algebraic closure of $\mathbb{Q}$. In particular, $\mathbb{A}$ is algebraically closed.

Proposition 11.8.7. *Let M be an $n \times n$ matrix with entries from $\mathbb{A}$ and λ be an eigenvalue of M. Then $\lambda \in \mathbb{A}$.*

Proof. The characteristic polynomial, $\chi_M(x) = \det(M - xI)$, is a monic polynomial of the form,

$$x^n - \mathrm{tr}(M)x^{n-1} + \sigma_2(M)x^{n-2} - \sigma_3(M)x^{n-3} + \ldots + (-1)^n \det(M),$$

where each σ_i is a homogeneous polynomial in $\mathbb{Z}[m_{i,j}]$, $i, j = 1 \ldots n$ of degree i. Since the entries $m_{i,j}$ of M lie in $\mathbb{A}$, $\mathbb{A} \supseteq \mathbb{Z}$ and is a ring, $\sigma_i(M) \in \mathbb{A}$ for all i. Hence $\chi_M(x) \in \mathbb{A}[x]$. By Exercise 11.8.6, $\mathbb{A}$ is algebraically closed. Hence, all of the eigenvalues of M are in $\mathbb{A}$. $\qquad\square$

Our last lemma will be needed in the Appendix.

Lemma 11.8.8. *Let α be an algebraic number with minimum polynomial $g(x) \in \mathbb{Z}[x]$ and leading coefficient b. Then $b\alpha$ is an algebraic integer.*

Proof. Let $g(x) = bx^n + a_{n-1}x^{n-1} + \cdots + a_0 = 0$. Then, $b\alpha$ satisfies the polynomial equation

$$x^n + ba_{n-1}x^{n-1} + \cdots + b^{n-1}a_1 x + b^n a_0 = 0,$$

and so $b\alpha \in \mathbb{Z}$. Thus each algebraic number α can be written in the form

$$\alpha = \frac{b\alpha}{m}$$

where m is an integer.   □

11.8.1   Exercises

Exercise 11.8.9. Let K/k be a field extension and α and $\beta \in K$. Show that if $[k(\alpha) : k]$ and $[k(\beta) : k]$ are relatively prime, then

$$[k(\alpha, \beta) : k] = [k(\alpha) : k] \cdot [k(\beta) : k].$$

Exercise 11.8.10. Let ζ be a (complex) primitive n-th root of unity. Show that $\mathbb{Q}(\zeta)$ is a splitting field of the polynomial $x^n - 1$ over $\mathbb{Q}$.

Exercise 11.8.11. Is it possible to have a nontrivial field extension K/k with isomorphic fields K and k?

Exercise 11.8.12. Let K/k be a field extension, $\alpha, \beta \in K$. Prove that the extension

$$k(\alpha, \beta)/k(\alpha + \beta, \alpha\beta)$$

is algebraic.

Exercise 11.8.13. Let $F \subseteq E$ be a field extension of characteristic $\neq 2$ and assume that the degree $[E : F] = 4$.

1. Show that $E = F(\alpha)$ for some α.

2. If $E = F(\alpha)$, where α is a root of a polynomial of the form $x^4 + ax^2 + b \in F[x]$, prove that there exists an intermediate field properly between E and F.

3. Now let $E = F(\beta)$ with no assumption on β and let $L \supseteq E$ be a splitting field for the minimal polynomial of β over F. If $\mathrm{Gal}(L/F)$ is isomorphic to the symmetric group Sym_4, show that there is no intermediate field properly between E and F.

Exercise 11.8.14. Let $\mathbb{Q}$ be the field of rational numbers and let $K = \mathbb{Q}[\sqrt{2}]$. Suppose $f(x) \in \mathbb{Q}[x]$ is a monic irreducible polynomial of odd degree $n \geq 1$ and notice that $f(x + \sqrt{2})$ is a monic polynomial of degree n in $K[x]$. Then,

1. Show that the coefficient of x^{n-1} in $f(x + \sqrt{2})$ is not rational.

2. Show that the polynomial $f(x + \sqrt{2})$ is irreducible in $K[x]$.

3. Prove that the polynomial $g(x) = f(x+\sqrt{2})f(x-\sqrt{2})$ is irreducible in $\mathbb{Q}[x]$.

Exercise 11.8.15. Let α be a real root of the polynomial $x^3 + 3x + 1$. Prove that α can not be constructed by ruler and compass.

Exercise 11.8.16. Let α, β be complex numbers. Prove that if $\alpha + \beta$, $\alpha\beta$ are algebraic numbers, then α, β are algebraic.

Chapter 12

Group Representations

12.1 Introduction

In this chapter we deal with the classical representation theory of a finite group G. Actually, many of the results, except for the last few sections, work just as well (but with complications) when G is compact and the representations are continuous. However, here we will restrict our attention to finite groups. The case of an Abelian group is, of course, of special interest.

The central object of study here is the set of all finite dimensional, irreducible, complex representations. As we shall see, actually every such representation is equivalent to a unitary one. In section 12.1 we introduce the players and prove the Peter-Weyl theorem in our case. In section 12.2 we do the Schur orthogonality relations. Section 12.3 deals with characters and class functions. In section 12.4 we study induced representations and the Frobenius reciprocity theorem. In our final sections we will find the degrees of the irreducible representations of a finite group have important divisibility properties and also prove Burnside's p, q theorem.

Representation theory has a wide spectrum of applications, going from Theoretical Physics (Quantum Mechanics) and Chemistry to Number theory and Probability. The subject was born in 1896, in the work of F. Frobenius. Major contributions were made by Burnside, Schur.

Let $\rho : G \to \mathrm{GL}(V_\rho)$ be a *finite dimensional complex representation* of G, by which we mean a homomorphism from G to $\mathrm{GL}(V_\rho)$, where V_ρ is a finite dimensional complex vector space. We call V_ρ the representation space of ρ and $d_\rho = \dim V_\rho$ its degree. Of course, $\mathrm{GL}(V_\rho) = \mathrm{GL}(d_\rho, \mathbb{C})$. Other accounts of representation theory are done over more general fields.

Here we list the basic definitions of the subject.

Definition 12.1.1.

- Given two such representations ρ and σ of G we shall call a linear operator $T : V_\rho \to V_\sigma$ an *intertwining* operator if we have a commutative diagram $T\rho_g = \sigma_g T$, $g \in G$.

- ρ and σ are said to be *equivalent* if there exists an *invertible* intertwining operator between them. Of course, equivalent representations have the same degree.

- We shall say ρ is a *unitary* representation if $\rho(G) \subseteq \mathrm{U}(n, \mathbb{C})$. As we shall see every representation of G is equivalent to a unitary representation.

- An *Hermitian inner product* $\langle , \rangle$ on V_ρ is called *invariant* if $\langle \rho_g(v), \rho_g(w) \rangle = \langle v, w \rangle$ for all $g \in G$ and $v, w \in V_\rho$.

- A subspace W of V_ρ is called an *invariant subspace* if $\rho_g(W) \subseteq W$ for all $g \in G$.

- ρ is called *irreducible* if it has no non-trivial invariant subspaces. Notice that if $d_\rho = 1$, then ρ is automatically irreducible.

Now suppose we have a representation ρ of G on V_ρ and V_ρ is a direct sum, $V_1 \oplus \cdots V_r$, of invariant irreducible subspaces. Then we say ρ is *completely reducible* and we write $\rho = \rho_1 \oplus \cdots \rho_r$, where $\rho_i = \rho | V_i$. Thus ρ is the direct sum of a finite number of irreducible representations.

This is implied (and is actually equivalent) to ρ having the property that every invariant subspace W has a complementary invariant subspace W'. This clearly follows by induction on $\dim(V_\rho)$ since both W and W' have lower dimension.

We denote by $\mathcal{R}(G)$ the *equivalence classes of finite dimensional, continuous, irreducible unitary representations* of G.

We leave as an exercise to the reader to verify that if ρ and σ are equivalent and ρ is irreducible, respectively completely reducible, then so is σ.

We now make the following important observation of I. Schur: G has an invariant integral. That is, given a function $f : G \to \mathbb{C}$ on the group, we get a complex number, $\int_G f$ (or $\int_G f\,dx$), in such a way that when we translate f on the left $f_g(x) = f(g^{-1}x)$, or on the right $f^g(x) = f(xg)$, where $g, x \in G$, the integral is unchanged. Moreover, the constant function 1 has integral 1. We do this by simply averaging the values of f and normalizing. Since G is finite this is simply,

$$\int_G f = \frac{1}{|G|} \sum_{x \in G} f(x),$$

where $|G|$ is the order of G. If there is no possibility of confusion we may sometimes just write $\int f$. This integral is both left and right invariant since each of these translations is just a permutation of the values of f. A good notation for this space of functions on G is, $\mathcal{F}(G)$. It is a vector space over $\mathbb{C}$ and has $\dim \mathcal{F}(G) = |G|$ since each $f \in \mathcal{F}(G)$ is completely determined by its values on the $|G|$ elements of G. As G has order n, we shall often write $\mathbb{C}^n$ for this vector space. Given a complex function f on G, $\mathrm{Supp}(f)$ stands for the places $x \in G$ where $f(x) \neq 0$. On $\mathcal{F}(G)$ we have the positive definite Hermitian inner product,

$$\langle \phi, \psi \rangle = \frac{1}{|G|} \sum_{g \in G} \phi(g)\overline{\psi(g)}.$$

Proposition 12.1.2. *Any finite dimensional representation ρ of a finite group G has a G-invariant Hermitian inner product on V_ρ and hence is equivalent to a unitary representation.*

Proof. Let $\langle,\rangle$ be any positive definite Hermitian inner product on V_ρ and for $v, w \in V_\rho$ consider $f(g) = \langle \rho_g(v), \rho_g(w) \rangle$, $g \in G$. Then $\langle v, w \rangle = \int_G \langle \rho_x(v), \rho_x(w) \rangle$. The reader can easily check that this is again a positive definite Hermitian inner product on V_ρ. But in addition, this new Hermitian inner product is G-invariant because for $g, x \in G$,

$$\langle \rho_g v, \rho_g w \rangle = \int_G \langle \rho_x(\rho_g v), \rho_x(\rho_g w) \rangle dx = \int_G \langle \rho_{xg}(v), \rho_{xg}(w) \rangle dx$$

$$= \int_G \langle \rho_x(v), \rho_x(w) \rangle dx = \langle v, w \rangle.$$

$\square$

Henceforth we shall, without loss of generality, assume all our representations are unitary. In particular,

Corollary 12.1.3. *Any finite dimensional representation of a finite group G is completely reducible.*

Proof. Since ρ is equivalent to a unitary representation and complete reducibility is invariant under equivalence we may assume ρ is actually unitary. Let W be an invariant subspace of V. Then $V = W \oplus W^\perp$. Moreover, $W^\perp$ is G invariant. Let $w^\perp \in W^\perp$ and $g \in G$. Then $\rho_g(w^\perp)$ is also in $W^\perp$ because if $w \in W$ then $\langle \rho_g(w^\perp), w \rangle = \langle w^\perp, \rho_{g^{-1}} w \rangle$, which is zero since W is G-invariant. Thus V is the direct sum of the invariant subspaces W and $W^\perp$. $\square$

12.1.1   The Regular Representation

For a finite group G the representation space of ρ_{reg} is $\mathbb{C}^n$. Here the action is given by left translation, $\rho_{\text{reg}_g}(f)(x) = f(g^{-1}x)$, where $g, x \in G$ and $f \in \mathbb{C}^n$. If we fix a basis of $\mathbb{C}^n$ by taking the characteristic function of each point $g \in G$ and consider the matrix describing $\rho_{\text{reg}}(g)$ with respect to this basis. Clearly this is a permutation matrix and $\rho_{\text{reg}}(g)$ is a linear representation. It is actually unitary, that is each ρ_{reg_g} is a unitary operator, since $\langle \rho_{\text{reg}_g}(f_1), \rho_{\text{reg}_g}(f_2) \rangle = \langle f_1, f_2 \rangle$ for all $g \in G$, and $f_1, f_2 \in \mathbb{C}^n$. This is because of the left invariance of Haar measure. We remark that the proof of both this and the previous result requires both left and right invariance of Haar measure. But as we saw, this is so.

In order to prove the Peter-Weyl theorem (in this context) it will be necessary to study the left regular representation. Now ρ_{reg} is also faithful. For if $\rho_{\text{reg}_g} = I$, then $\rho_{\text{reg}_g}(f) = f$ for every $f \in \mathbb{C}^n$. But $\text{Supp}(f_g) = g^{-1} \text{Supp}(f)$. Taking f to be the characteristic function of a point g yields a contradiction unless $g = 1$. Thus ρ_{reg} is faithful. Now decompose the finite dimensional representation ρ_{reg} into irreducibles as above. These must separate the points because if they did not, then ρ_{reg} would not be faithful. In particular, the set of all irreducibles representations, $\mathcal{R}(G)$, of G also separates the points of G. This is one form of the Peter-Weyl theorem.

Corollary 12.1.4. *For a finite group G, $\mathcal{R}(G)$ separates the points of G.*

Definition 12.1.5.

1. If ρ is a finite dimensional irreducible representation of the finite group G, we denote by $R(\rho)$ the $\mathbb{C}$ linear span of the coefficients of ρ, i.e. the *representative functions* associated with ρ.

2. $R(G)$ stands for all the *representative functions* on G. This means the $\mathbb{C}$ linear span of the subspace $R(\rho)$. The reader will notice that both $R(\rho)$ and $R(G)$ are *independent of any choice of coordinates and therefore of equivalent representations.*

Another version of the Peter-Weyl theorem is the following.

Corollary 12.1.6. *For a finite group G, $R(G) = \mathbb{C}^n$.*

This is because as we saw,

$$\text{lin. sp.}_{\mathbb{C}}\{R(\rho) : \rho \text{ contained in reg}\} = \mathbb{C}^n.$$

In section 12.2 we shall see the exact nature of the decomposition into irreducibles of ρ_{reg} acting on $R(G)$.

1. Prove equivalently $R(G)$ is the $\mathbb{C}$ linear span of all coefficient functions of all representations of G.

2. Show $R(\rho)$, and therefore also $R(G)$, is invariant under both left and right translations and hence also under conjugation, that is, by inner automorphisms.

Here is the proof of 2.

Proof. We prove left invariance; right invariance is done similarly. Let $r(x) = \sum c_{i,j}\rho_{i,j}(x)$, where $c_{i,j} \in \mathbb{C}$ is a generic element of $R(\rho)$. Since ρ_{reg} is a linear representation it is sufficient to show $\rho_{\mathrm{reg}}(g)\rho_{i,j} \in R(\rho)$. But $\rho(g^{-1}x) = \rho(g^{-1})\rho(x)$ and so $\rho_{i,j}(g^{-1}x) = \sum_k \rho_{i,k}(g^{-1})\rho_{k,j}(x) \in R(\rho)$. $\qquad\qquad\square$

We leave the following important series of exercises to the reader.

Exercise 12.1.7.

- Show that equivalence of representations is an equivalence relation.

- Given a single representation ρ, show the set of intertwining operators forms a subalgebra of $\mathrm{End}(V_\rho)$.

- Give an example to show that the proposition and corollary above is false if G is not finite, e.g. consider a unipotent representation of $\mathbb{R}$.

- Show that two representations are equivalent if and only if the modules (G, V_ρ, ρ) and (G, V_σ, σ) are isomorphic. (See Chapter 6 of Volume I. Here the ring is the group algebra $\mathbb{C}(G)$). Thus, they share all module theoretic properties such as, a composition series for one corresponds to such a series for the other, etc.

- Show that a finite dimensional representation ρ of G is completely reducible if and only if the corresponding module is semi-simple.

- Show that a finite dimensional representation ρ is irreducible if and only if the corresponding module is simple.

12.1.2   The Character of a Representation

Let $\rho : G \longrightarrow \mathrm{GL}(V)$ be a representation of the group G.

Definition 12.1.8. The *character* χ_ρ of the representation ρ is the function $\chi_\rho : G \longrightarrow \mathbb{C}$ defined by

$$g \mapsto \chi_\rho(g) := \mathrm{tr}(\rho(g)).$$

Equivalent representations have the same character because $\mathrm{tr}(ABA^{-1}) = \mathrm{tr}(B)$.

We shall use $\mathcal{X}(G)$ to mean the set of all characters of all *irreducible* representations of G.

Definition 12.1.9. A function $f : G \longrightarrow \mathbb{C}$ is called *central* means $f(hgh^{-1}) = f(g)$ for any g, h in G. That is, f is constant on conjugacy classes.

Exercise 12.1.10. Show f is a central function if and only if $f(xy) = f(yx)$ for all $x, y \in G$.

The character χ_ρ of the representation ρ is constant on conjugacy classes since $\mathrm{tr}(ABA^{-1}) = \mathrm{tr}(B)$. Hence so is any linear combination of characters of various representations.

Let $(F)^G$ be the vector subspace of $\mathcal{F}(G)$ consisting of central functions. On it we shall also have occasion to use the restriction of the Hermitian inner product.

12.2   The Schur Orthogonality Relations

The Schur orthogonality relations are the following.

Theorem 12.2.1. *Let G be a finite group, $\int dg$ be normalized invariant measure and ρ and σ finite dimensional continuous irreducible unitary representations of G. Then*

1. *If ρ and σ are inequivalent, then $\int_G \rho_{i,j}(g)\sigma_{l,k}(\bar{g})dg = 0$ for all $i, j = 1 \ldots d_\rho$ and $k, l = 1 \ldots d_\sigma$.*

2. $\int_G \rho_{i,j}(g)\rho_{k,l}(\bar{g})dg = \frac{\delta_{i,k}\delta_{j,l}}{d_\rho}$.

The first statement means that, in the language of representative functions, if $\rho \neq \sigma$, then $R(\rho) \perp R(\sigma)$. The second tells us that $\dim_{\mathbb{C}} R(\rho) = d_\rho^2$. In fact, as we shall see, $\{d_\rho^{\frac{1}{2}}\rho_{i,j} : i,j = 1 \ldots d_\rho\}$ is a complete orthonormal basis of $R(\rho)$ in $\mathbb{C}^n$.

Proof. Let V_ρ and V_σ be the respective representation spaces and $B(V_\sigma, V_\rho)$ be the $\mathbb{C}$ vector space of linear operators between them. Let $T \in B(V_\sigma, V_\rho)$ and consider the map $G \to B(V_\sigma, V_\rho)$ given by $g \mapsto \rho_g T \sigma_g^{-1}$. This is an operator valued function on G and so $\int_G \rho_g T \sigma_g^{-1} dg$ is also an operator in $B(V_\sigma, V_\rho)$. For $h \in G$ we have

$$\rho(h)\int_G \rho_g T \sigma_g^{-1} dg\sigma(h)^{-1} = \int_G \rho(h)\rho_g T \sigma_g^{-1}\sigma(h)^{-1} dg$$

$$= \int_G \rho(hg)T\sigma(hg)^{-1} dg = \int_G \rho(g)T\sigma(g)^{-1} dg.$$

Letting $T_0 = \int_G \rho(g)T\sigma(g)^{-1} dg$ we see $\rho_g T_0 = T_0 \sigma_g$ for every $g \in G$. That is, T_0 is an intertwining operator. By Schur's Lemma (see 6.9.1 of Volume I) there are only two possibilities. Either ρ and σ are equivalent and T_0 is invertible (and implements the equivalence), or ρ and σ are inequivalent and $T_0 = 0$. In the latter case $\int_G \rho(g)T\sigma(g)^{-1} dg = 0$, where T is arbitrary. Let $T = (t_{j,k})$. Then $\rho(g)T\sigma(g)^{-1}_{i,l} = \sum_{j,k} \rho_{i,j}(g)t_{j,k}\sigma_{k,l}(g)^{-1}$. Since $(t_{j,k})$ are arbitrary we get $\int_G \rho_{i,j}(g)\sigma_{k,l}(g)^{-1} dg = 0$ for all $i,j = 1 \ldots d_\rho$ and $k,l = 1 \ldots d_\sigma$. Because σ is a unitary representation $\sigma(g)^{-1} = \sigma(g)^* = \sigma(g)^{-t}$. Thus $\int_G \rho_{i,j}(g)\overline{\sigma_{l,k}(g)}dg = 0$ for all $i,j = 1 \ldots d_\rho$ and $k,l = 1 \ldots, d_\sigma$.

We now consider the case when we have equivalence. Here we may as well just take σ to be ρ. In this case Schur's Lemma tells us T_0 is a scalar multiple of the identity. Thus $\int_G \rho(g)T\rho(g)^{-1} dg = \lambda(T)I$. Taking the trace of each side yields

$$\text{tr}\left(\int_G \rho(g)T\rho(g)^{-1} dg\right) = \int_G \text{tr}\left(\rho(g)T\rho(g)^{-1}\right)dg = \int_G \text{tr}(T)dg = \text{tr}(T),$$

while $\mathrm{tr}(\lambda(T)I) = \lambda(T)d_\rho$. We conclude $\lambda(T) = \frac{\mathrm{tr}(T)}{d_\rho}$ and so $\int_G \rho(g)T\rho(g)^{-1}dg = \frac{\mathrm{tr}(T)}{d_\rho}I$. Using reasoning similar to the earlier case one finds $\int_G \rho_{i,j}(g)\overline{\rho_{l,k}(g)}dg = 0$ for all $i,j,k,l = 1\ldots,d_\rho$ whenever $i \neq l$, or $j \neq k$. Now we consider the case when $i = l$ and $j = k$. By taking T to be diagonal with all zero entries except for one we get $\int_G \rho_{i,j}(g)\rho_{i,j}(\bar{g})dg = \frac{1}{d_\rho}$, $i,j = 1\ldots,d_\rho$. Hence in general one has $\int_G \rho_{i,j}(g)\overline{\rho_{k,l}(g)}dg = \frac{\delta_{i,k}\delta_{j,l}}{d_\rho}$. $\qquad\qquad\qquad\qquad\qquad\qquad\qquad\qquad\qquad\qquad\square$

As a corollary of the proof we get,

Corollary 12.2.2. *If G is Abelian, all its irreducible representations are 1 dimensional.*

Proof. Let ρ be an irreducible representation of G and $a, x \in G$. Then $\rho(ax) = \rho(a)\rho(x)$. Since G is Abelian, $\rho(a)\rho(x) = \rho(x)\rho(a)$. Holding a fixed and letting x vary over G, it follows that $\rho(a)$ is a scalar multiple of the identity for each $a \in G$. Since ρ is irreducible this means $d_\rho = 1$. $\quad\square$

Notice that $R(G)$ is stable under conjugation since if ρ is a finite dimensional unitary representation of G so is $\bar{\rho}$, its conjugate. If ρ is irreducible so is $\bar{\rho}$.

Exercise 12.2.3. The reader should verify these statements, particularly the irreducibility of $\bar{\rho}$, by using Schur's lemma.

Returning to $R(\rho)$ and $R(G)$, we already know $\dim R(\rho) = d_\rho^2$. If ρ and σ are distinct in $\mathcal{R}(G)$, then $R(\rho)$ and $R(\sigma)$ are orthogonal and hence linearly independent. Now the linear span of all $R(\rho)$, as $\rho \in \mathcal{R}(G)$, is $R(G)$. Hence by the Peter-Weyl theorem we have,

Corollary 12.2.4. $\mathbb{C}^n = \bigoplus_{\rho \in \mathcal{R}(G)} R(\rho)$ *(orthogonal direct sum).*

Thus we have decomposed $\mathbb{C}^n$ into the orthogonal direct sum of perhaps a large number of finite dimensional *left invariant* subspaces. In order to completely analyze ρ_{reg} we merely need to know which irreducibles occur in each of the $R(\rho)$.

Now if τ is a finite dimensional representation of G on V and ρ is an irreducible sub-representation of G, then $[\tau : \rho]$, the multiplicity that ρ occurs in τ, is given in Corollary 12.3.3 below by $\langle \chi_\tau, \chi_\rho \rangle = \int_G \chi_\tau(x)\overline{\chi_\rho(x)}dx$. In our case, $\tau = \rho_{\text{reg}} \,|\, \{R(\rho)\}$. As we saw earlier $\chi_{\rho_{\text{reg}}} \,|\, \{R(\rho)\}(g) = d_{\bar\rho}\chi_{\bar\rho}$. We also saw $\bar\rho \in \mathcal{R}(G)$, if ρ is. Thus the multiplicity of $\bar\rho$ in $\rho_{\text{reg}} \,|\, \{R(\rho)\}$ is $d_{\bar\rho} = d(\rho)$. Since $\dim_{\mathbb{C}} R(\rho) = d(\rho)^2$ it follows that $\rho_{\text{reg}} \,|\, \{R(\rho)\}$ contains only $\bar\rho$ with multiplicity $d(\bar\rho)$ and nothing else. Since ρ then occurs in $\rho_{\text{reg}} \,|\, \{R(\bar\rho)\}$ with multiplicity $d_{\bar\rho} = d_\rho$ and every representation is its double bar, we arrive at the following corollary.

Corollary 12.2.5. *Each irreducible of G occurs in ρ_{reg} with a multiplicity equal to its degree.*

Corollary 12.2.6. *A finite group has a finite number of inequivalent finite dimensional irreducible representation $\rho_1, \ldots \rho_r$. These are constrained by the requirement that $\sum_{i=1}^r d_{\rho_i}^2 = |G|$.*

Example 12.2.7. Let $G = S_3$, the symmetric group on three letters. This group has two 1-dimensional characters. These are the two characters of $S_3/A_3 = Z_2$ lifted to G. It has no others since $[S_3, S_3] = A_3$. By the relation just above, S_3 must have an irreducible of degree $d > 1$ for otherwise it would be Abelian (see Corollary 12.2.9). Since $1^2 + 1^2 + 2^2 = 6$, the order of S_3, we see $|\mathcal{R}(S_3)| = 3$ and the higher dimensional representation has $d = 2$.

Corollary 12.2.8. *For a finite group G,*

$$\chi_{\rho_{\text{reg}}}(x) = \sum_{\rho \in \mathcal{R}(G)} d_\rho \chi_\rho(x).$$

If we consider the two sided regular representation of G on $\mathbb{C}^n$ (Haar measure is both left and right invariant and left and right translations commute), then a similar analysis shows that this representation on $R(\rho)$ is *actually irreducible* and equivalent to $\bar\rho \otimes \rho$. This can be done by calculating the character of this representation (see beginning of the next section). Here $R(\rho)$ is identified with $V_{\bar\rho} \otimes V_\rho$ and $\chi_{\bar\rho \otimes \rho}(g, h) = \chi_{\bar\rho}(g)\chi_\rho(h)$, $g, h \in G$. Applying Proposition 12.3.2 shows these representations are equivalent and Corollary 12.3.5 that they are irreducible. We leave this verification to the reader as an exercise.

We make a few final remarks concerning the Abelian case. Recall here the irreducibles are all 1 dimensional. That is, they are multiplicative characters $\chi : G \to \mathbb{T}$. This enables us to sharpen the conclusions of the Peter-Weyl theorem in this case.

The next corollary shows that if a finite group has only 1 dimensional irreducible representations it must be Abelian since it can be imbedded in an Abelian group.

Corollary 12.2.9. *For a finite Abelian group G, the characters separate the points and the linear span of the characters is all of $\mathbb{C}^n$.*

We now turn to the Plancherel theorem. Let $\rho \in \mathcal{R}(G)$ and ϕ be a function on G. We define $T_\phi(\rho) = \int_G \phi(g)\rho(g)dg$. Thus $T_\phi(\rho)$ is a linear operator on V_ρ. It is called the *Fourier transform* of ϕ at ρ and so each fixed ϕ gives an operator valued function on $\mathcal{R}(G)$ (but always takes its value in a different space of operators).

Now let ϕ and ψ be functions. We want to calculate $\langle \phi, \psi \rangle$ by Fourier analysis. This is what the Plancherel theorem does.

$$\langle \phi, \psi \rangle = \sum_{\rho \in \mathcal{R}(G)} d_\rho \operatorname{tr}(T_\phi(\rho)T_\psi(\rho))^\star,$$

where $\star$ means the adjoint operator.

To prove this using polarization we may take $\psi = \phi$. Then we get the following formula involving the (Hilbert Schmidt) norm of an operator.

$$\|\phi\|_2^2 = \sum_{\rho \in \mathcal{R}(G)} d_\rho \operatorname{tr}(T_\phi(\rho)T_\phi(\rho))^\star.$$

Matrix calculations similar to those involved in the orthogonality relations themselves yield,

$$d_\rho \operatorname{tr}(T_\phi(\rho)T_\phi(\rho))^\star = \sum_{i,j=1}^{d_\rho} |d_\rho^{\frac{1}{2}} \int_G \phi(g)\rho_{i,j}(g)dg|^2.$$

The Parseval equation together with the orthogonality relations and the decomposition of ρ_{reg} completes the proof.

12.3   Characters and Central Functions

Corollary 12.3.1. *Let ρ and σ be finite dimensional irreducible, unitary representations of G. Then $\langle \chi_\rho, \chi_\sigma \rangle = 0$ if ρ and σ are inequivalent and $\langle \chi_\rho, \chi_\rho \rangle = 1$.*

Thus the set $\mathcal{X}(G)$ consisting of the characters of the irreducible unitary representations of G form an orthonormal family of functions in $\mathbb{C}^n$. In particular, they are linearly independent.

Proof.

$$\sum_{i=1}^{d_\rho} \rho_{i,i}(g) \left(\sum_{j=1}^{d_\sigma} \overline{\sigma_{j,j}(g)} \right) = \sum_{i=1}^{d_\rho} \sum_{j=1}^{d_\sigma} \rho_{i,i}(g) \overline{\sigma_{j,j}(g)}.$$

Hence

$$\int_G \chi_\rho \overline{\chi_\sigma} dg = \sum_{i=1}^{d_\rho} \sum_{j=1}^{d_\sigma} \int_G \rho_{i,i}(g) \overline{\sigma_{j,j}(g)} dg.$$

This is clearly 0 if ρ and σ are inequivalent. Now if $\sigma = \rho$, then

$$\|\chi_\rho\|_2^2 = \sigma_{i=1}^{d_\rho} \sum_{j=1}^{d_\rho} \int_G \rho_{i,i}(g) \overline{\rho_{j,j}(g)} dg.$$

If $i \neq j$ we get 0. Hence

$$\|\chi_\rho\|_2^2 = \sum_{i=1}^{d_\rho} \int_G \rho_{i,i}(g) \overline{\rho_{i,i}(g)} dg = 1,$$

by what we learned in Section 12.2. $\qquad\qquad\qquad\qquad\qquad\qquad \square$

Proposition 12.3.2. *Let ρ and σ be unitary representations of G. Then ρ and σ are equivalent if and only if they have the same character.*

Proof. Evidently, equivalent representations have the same character. We now suppose $\chi_\rho \equiv \chi_\sigma$. Decompose ρ and σ into irreducibles,

$$\rho = \sum_{i=1}^{r} n_i \rho_i, \quad n_i > 0 \quad \text{and} \quad \sigma = \sum_{i=k}^{s} m_i \rho_i, \quad m_i > 0.$$

After renumbering we can consider the overlap to be from $\rho_k, \ldots, \rho_r$. Then

$$0 = \chi_\rho - \chi_\sigma = \sum_{i=1}^{k-1} n_i \chi_{\rho_i} + \sum_{i=k}^{r} (n_i - m_i) \chi_{\rho_i} + \sum_{i=r+1}^{s} -m_i \chi_{\rho_i}.$$

Since the χ_{ρ_i} are linearly independent we conclude $n_i = 0$ for $i = 1, \ldots, k-1$, $n_i = m_i$ for $i = k, \ldots, r$ and $-m_i = 0$ for $i = r+1, \ldots, s$. But since the n_i and m_i are positive $k = 1$, $r = s$ and $n_i = m_i$ for all $i = 1, \ldots, r$. That is, ρ and σ are equivalent.   □

The next result follows from the orthonormality of $\mathcal{X}(G)$ in a similar manner. We leave its proof to the reader as an exercise.

Corollary 12.3.3. *Let ρ be a unitary representation of G whose decomposition into irreducibles is $\rho = \sum_{i=1}^{r} n_i \rho_i$, $n_i > 0$. Then $\|\chi_\rho\|_2^2 = \sum_{i=1}^{r} n_i^2$. In particular ρ is irreducible if and only if $\|\chi_\rho\|_2^2 = 1$. Moreover, the multiplicity of an irreducible ρ_i in ρ is $\langle \chi_\rho, \chi_{\rho_i} \rangle$.*

Exercise 12.3.4. Let ρ be a unitary representation of G and χ_ρ be its character. Then of course $\chi_\rho : G \to \mathbb{C}$ is a bounded function on G. Where does it take its largest absolute value and what is this value?

We can use our irreducibility criterion to study tensor product representations. Let G and H be finite groups and ρ and σ be representations of G and H respectively. Form the representation $\rho \otimes \sigma$ of $G \times H$ on $V_\rho \otimes V_\sigma$ by defining $\rho \otimes \sigma(g, h) = \rho_g \otimes \sigma_h$.

We leave it to the reader to check that this is a representation of $G \times H$. It is actually unitary, but this does not really matter since everything is equivalent to a unitary representation anyway.

Corollary 12.3.5. *If $\rho \in \mathcal{R}(G)$ and $\sigma \in \mathcal{R}(H)$, then $\rho \otimes \sigma \in \mathcal{R}(G \times H)$. Conversely, all irreducibles of $G \times H$ arise as tensor products in this way.*

Proof. If dg and dh are normalized Haar measures on G and H respectively then $dg\,dh$ is normalized Haar measure on the finite group $G \times H$. Now $\chi_{\rho \otimes \sigma}(g, h) = \chi_\rho(g) \chi_\sigma(h)$. Hence

$$\|\chi_{\rho \otimes \sigma}\|_2^2 = \int_G \int_H \chi_\rho(g) \chi_\sigma(h) \overline{\chi_\rho(g) \chi_\sigma(h)}\, dg\, dh = \|\chi_\rho\|_2^2 \|\chi_\sigma\|_2^2 = 1 \cdot 1 = 1,$$

proving irreducibility.

Now let $\tau \in \mathcal{R}(G \times H)$ and consider, as before, all $\rho \otimes \sigma$ where $\rho \in \mathcal{R}(G)$ and $\sigma \in \mathcal{R}(H)$. Let $f \in C(G \times H)$. By the Stone Weierstrass theorem f can be uniformly approximated by variable separable functions $\sum_{i=1}^{n} g_i(x)h_i(y)$, where $g_i \in C(G)$ and $h_i \in C(H)$. But by the Peter-Weyl theorem these in turn can be uniformly approximated by $\phi(x,y) = \sum_{i=1}^{n} r_i(x)s_i(y)$, where $r_i \in R(G)$ and $s_i \in R(H)$. Hence Φ, the collection of these ϕ' are representative functions associated with certain irreducibles of $G \times H$ and they form a uniformly dense linear subspace of $C(G \times H)$. If τ is not involved, its coefficients must be perpendicular to Φ and therefore to all of $C(G \times H)$. In particular it must be orthogonal to itself, a contradiction. $\qquad\square$

As we saw earlier, linear combinations of characters of finite dimensional representations are central functions. The converse is also true. Namely,

Theorem 12.3.6. *Every central function is in the lin.sp. of $\mathcal{X}(G)$ and conversely.*

Before turning to the proof we need a pair of lemmas.

Lemma 12.3.7. *Let $\rho \in \mathcal{R}(G)$ and $r \in R(\rho)$. If r is central, then $r = \lambda\chi_\rho$, where $\lambda \in \mathbb{C}$.*

Proof. Here $r(x) = \sum_{i,j=1}^{d_\rho} c_{i,j}\rho_{i,j}(x)$. Since $r(x) = r(gxg^{-1})$ we conclude upon substituting and taking into account the linear independence of $\rho_{i,j}$ that $C = \rho(g)^t \bar{C}\rho(g)$. Alternatively, $\rho(g)\bar{C} = \bar{C}\rho(g)$ for all $g \in G$, where C is the matrix of $c_{i,j}$. By Schur's lemma $\bar{C}$ is a scalar multiple of the identity and hence so is C. $\qquad\square$

Lemma 12.3.8. *Suppose f is central and in $R(G)$, then f is in the linear span of $\mathcal{X}(G)$.*

Proof. We know $f = \sum_{i=1}^{n} r_i$, where $r_i \in R(\rho_i)$ and the ρ_i are distinct in $\mathcal{R}(G)$. Applying the assumption $f(gxg^{-1}) = f(x)$ and taking into account the linear independence of the r_i and the fact that f is a class function tells us each r_i is a class function. Hence by Lemma 12.3.7 each $r_i = \lambda_i \chi_{\rho_i}$ and $f = \sum_{i=1}^{n} \lambda_i \chi_{\rho_i}$.   □

We now prove Theorem 12.3.6.

Proof. We first define a projection operator $\# : R(G) \to R(G)^G$ via the formula $f^{\#}(x) = \int_G f(gxg^{-1})dg$. Then f is central if and only if $f = f^{\#}$. We leave these details to the reader to check. Let $f \in R(G)^G$. Since f is a representative function, we apply the $\#$ operator and get $f = f^{\#}$ is a central representative function, which by Lemma 12.3.8 is a linear combination of characters of $R(G)$.   □

Exercise 12.3.9. The reader should verify the various properties of $\#$ mentioned above as these are necessary to complete the proof of Theorem 12.3.6. For example, the operator is norm decreasing $\|f^{\#}\|_G \leq \|f\|_G$.

Corollary 12.3.10. $\mathcal{X}(G)$ *separates the conjugacy classes of* G.

Proof. Let C_x and C_y be disjoint conjugacy classes. Since these are disjoint sets we can find $f \in R(G)$ with $f|C_x = 0$ and $f|C_y = 1$. Applying $\#$ yields $f^{\#}|C_x = 0$ and $f^{\#}|C_y = 1$. Now $f^{\#}$ is a linear combination of characters. If $\chi_\rho(x) = \chi_\rho(y)$ for every $\rho \in \mathcal{R}(G)$ this would give a contradiction and hence the conclusion.   □

Thus the irreducible representations of a finite group are in bijective correspondence with the irreducible characters, and the characters are in bijective correspondence with the set of conjugacy classes. This is the significance of the *character table* of a finite group. Vertically the characters are listed and horizontally the conjugacy classes are listed. Then the table must be filled in with the value of that character on that particular conjugacy class. For example, as we saw above S_3 has exactly three characters and therefore also three conjugacy classes.

We conclude this section with the functional equation for a character of a representation in $\mathcal{R}(G)$.

Theorem 12.3.11. *Let $f = \chi_\rho$ be the character of a finite dimensional, continuous, irreducible, unitary representation ρ. Then for all x, $y \in G$,*

$$f(x)f(y) = f(1) \int_G f(gxg^{-1}y)dg.$$

Conversely, if f is a function $G \to \mathbb{C}$, not identically zero satisfying this equation, then $\frac{f}{f(1)} = \frac{\chi_\rho}{\chi_\rho(1)}$ for a unique $\rho \in \mathcal{R}(G)$.

Proof. We extend the $\#$ operator defined earlier on functions to representations. For any finite dimensional continuous unitary representation, ρ, let $\rho^{\#}(x) = \int_G \rho(gxg^{-1})dg$, giving an operator valued function on G. For $y \in G$ using invariance of dg we get

$$\rho(y)\rho^{\#}(x)\rho(y)^{-1} = \int_G \rho(y)\rho(gxg^{-1})\rho(y)^{-1}dg$$

$$= \int_G \rho((gy)x(gy)^{-1})dg = \int_G \rho(gxg^{-1})dg = \rho^{\#}(x).$$

Thus $\rho^{\#}(x)$ is an intertwining operator. If ρ is irreducible, $\rho^{\#}(x) = \lambda(x)I$ and taking traces shows

$$\lambda(x) = \frac{\operatorname{tr}(\rho^{\#}(x))}{d_\rho}.$$

On the other hand,

$$\operatorname{tr}(\rho^{\#}(x)) = \int_G \operatorname{tr}(\rho(gxg^{-1}))dg = \chi_\rho(x)$$

so that for all $x \in G$,

$$\int_G \rho(gxg^{-1})dg = \frac{\chi_\rho(x)}{d_\rho}I.$$

Hence

$$\int_G \rho(gxg^{-1})dg\rho(y) = \int_G \rho(gxg^{-1})\rho(y)dg$$

$$= \int_G \rho(gxg^{-1}y)dg = \frac{\chi_\rho(x)}{d_\rho}\rho(y).$$

Taking traces yields the functional equation

$$\int_G \chi_\rho g x g^{-1} y) dg = \frac{\chi_\rho(x)\chi_\rho(y)}{d_\rho}.$$

Conversely, let f be an arbitrary function satisfying the functional equation. From it we see $f(1) \neq 0$ for otherwise $f \equiv 0$. Let $y = 1$ in the equation. Then

$$f(x)f(1) = f(1) \int_G f(g x g^{-1}) dg = f(1) f^{\#}(x).$$

Since $f(1) \neq 0$, $f(x) = f^{\#}(x)$ so f is central. We will show that for every $\rho \in \mathcal{R}(G)$ and every $x \in G$,

$$\frac{f(x)}{f(1)} \langle f, \chi_\rho \rangle = \frac{\chi_\rho(x)}{\chi_\rho(1)} \langle f, \chi_\rho \rangle. \tag{1}$$

Having done so we complete the proof by choosing a $\rho \in \mathcal{R}(G)$ such that $\langle f, \chi_\rho \rangle \neq 0$. For then we can cancel and conclude from equation (1) that

$$\frac{f(x)}{f(1)} = \frac{\chi_\rho(x)}{\chi_\rho(1)}.$$

Since f is central such a ρ must exist by Theorem 12.3.6. The ρ satisfying (2) is unique because the characters of distinct representations are linearly independent.

It remains only to prove (1). To do so, consider

$$\int_G \int_G f(g x g^{-1} y) \overline{\chi_\rho(y)} dg dy.$$

By hypothesis this is

$$\int_G \int_G \overline{\chi_\rho(y)} \frac{f(x)f(y)}{f(1)} dg dy = \frac{f(x)}{f(1)} \int_G f(y) \overline{\chi_\rho(y)} dy = \frac{f(x)}{f(1)} \langle f, \chi_\rho \rangle.$$

On the other hand (Fubini),

$$\int_G \int_G f(g x g^{-1} y) \overline{\chi_\rho(y)} dg dy = \int_G \overline{\chi_\rho(y)} \int_G f(g x g^{-1} y) dg dy.$$

Left translating by the inverse of gxg^{-1} gives

$$\int_G \overline{\chi_\rho(gx^{-1}g^{-1}y)} \int_G f(y)dgdy.$$

Again (Fubini) this is

$$\int_G \left(\int_G f(y)\overline{\chi_\rho(gx^{-1}g^{-1}y)}dy \right)dg.$$

Using $\chi_\rho(t^{-1}) = \overline{\chi_\rho(t)}$ we get

$$\int_G \left(\int_G f(y)\chi_\rho(y^{-1}g^{-1}xg)dy \right)dg.$$

Since χ_ρ is a class function this is

$$\int_G \left(\int_G f(y)\chi_\rho(g^{-1}xgy^{-1})dy \right)dg.$$

Because both left and right translations are invariant we get $\int_G \left(\int_G f(y)\chi_\rho(gxg^{-1}y^{-1})dy \right)dg$. By the part of the theorem already proved this is just

$$\int_G \left(\int_G f(y)d_\rho^{-1}\chi_\rho(x)\overline{\chi_\rho(y)}dy \right)dg.$$

Or

$$\chi_\rho(1)^{-1}\chi_\rho(x)\langle f, \chi_\rho \rangle.$$

Thus

$$\frac{f(x)}{f(1)}\langle f, \chi_\rho \rangle = \frac{\chi_\rho(x)}{\chi_\rho(1)}\langle f, \chi_\rho \rangle.$$

$\square$

12.4   Induced Representations

We now study *induced representations* of a finite group G.

Let σ be a finite dimensional unitary representation of a subgroup H of G, on V_σ. We now define the induced representation of σ to G, written $\text{Ind}(H \uparrow G, \sigma)$. It will be of dimension $d_\sigma[G : H]$.

Consider functions $F : G \to V_\sigma$ satisfying $F(gh) = \sigma(h)^{-1}F(g)$, for $h \in H$ and $g \in G$. Such functions clearly form a complex vector space W under pointwise operations. If $\langle,\rangle_{V_\sigma}$ denotes the Hermitian inner product on V_σ we can use this to define an inner product on this space as follows. For F_1 and F_2 here, the function $g \mapsto \langle F_1(g), F_2(g)\rangle_{V_\sigma}$ is a numerical function on G, but because of the condition above it is actually a function on G/H. In particular, $g \mapsto \|F(g)\|^2_{V_\sigma}$ is a non-negative function on G/H. Now W is a Hermitian inner product space given by

$$\langle F_1, F_2\rangle = \int_{G/H} \langle F_1(g), F_2(g)\rangle_{V_\sigma} d(\mu)(g^{\cdot})$$

$$= \int_{G/H} \langle F_1(g), F_2(g)\rangle_{V_\sigma} d(\mu)(g^{\cdot})$$

$$\leq \int_{G/H} \|F_1(g)\|^2_{V_\sigma} d(\mu)(g^{\cdot}) \int_{G/H} \|F_2(g)\|^2_{V_\sigma} d(\mu)(g^{\cdot}).$$

Let G act on W by left translation $(x \cdot F)(g) = F(x^{-1}g)$, where $F \in W$ and $x, g \in G$.

Proposition 12.4.1. $\text{Ind}(H \uparrow G, \sigma)$ *is a unitary representation of G on W.*

Exercise 12.4.2. The proof of this is routine and is left to the reader. We also leave to the reader to check that the left regular representation, ρ_{reg} acting on $\mathbb{C}^n$, is an induced representation, where σ is the trivial 1 dimensional representation of $H = (1)$. (This is a good example of an induced representation to keep in mind.)

We now show W can be described as follows. Let $f : G \to V_\sigma$ and for $x \in g$, define $F_f(x) = \int_H \sigma(h)f(xh)dh$. Since the integrand is a V_σ valued function on H we can integrate it and get a V_σ valued function $F_f : G \to V_\sigma$ on G.

Lemma 12.4.3. *The F_f are in W.*

Proof.

$$F_f(gh_1) = \int_H \sigma(h)f(gH_1h)dh = \int_H \sigma(h_1^{-1}h)f(gh)dh$$

$$= \int_H \sigma(h_1^{-1})\sigma(h)f(gh)dh = \sigma(h_1^{-1})\int_H \sigma(h)f(gh)dh.$$

Thus $F_f(gh_1) = \sigma(h_1^{-1})F_f(g)$.   □

Lemma 12.4.4. *The F_f comprise all of $\mathcal{W}$.*

Proof. Let F_1 be any function satisfying the equation defining $\mathcal{W}$. We want to find an $f : G \to V_\sigma$ so that $F_1 = F_f$. Now $F_1(g) = \sigma(h)F_1(gh)$, and therefore

$$F_1(g) = \int_H F_1(g)dh = \int_H \sigma(h)F_1(gh)dh.$$

On the other hand, $F_f(g) = \int_H \sigma(h)f(gh)dh$. Therefore since σ is unitary

$$\|F_1 - F_f\|^2 = \int_{G/H}\int_H \|f(gh) - F_1(gh)\|^2_{V_\sigma}dh \, d\mu(g').$$

Taking $f = F_1$, this last integral is zero, and so $F_f = F_1$.   □

Corollary 12.4.5. $\mathrm{Ind}(H \uparrow G, \sigma)$ *is a direct sum of finite dimensional, irreducible unitary representations of G on $\mathcal{W}$.*

The following is a consequence of the fact that a function $F : G \to V_\sigma$ which satisfies the defining condition is determined by its values on coset representatives of H in G. Its proof is left to the reader.

Corollary 12.4.6. $\dim_{\mathbb{C}} \mathcal{W} = [G : H]\dim_{\mathbb{C}} V_\sigma.$

Proposition 12.4.7. *Let G be a finite group, γ a representation of G on V and $\rho \in \mathcal{R}(G)$. Then $[\gamma : \rho] = \dim_{\mathbb{C}} \mathrm{Hom}_G(V_\rho, V)$.*

Proof. Write V as the orthogonal direct sum of irreducible unitary sub-representations, $(V_i, \gamma | V_i)$, where $i \in I$ and $V = \bigoplus V_i$. Let π_i be the orthogonal projection of V onto V_i. Partition $I = I_1 \cup I_2$, where I_1 contains those representations equivalent to ρ, while I_2 contains those representations which are not equivalent to ρ. For $T \in \text{Hom}_G(V_\rho, V)$, $\pi_i \circ T \in \text{Hom}_G(V_i, V_\rho)$. So if $i \in I_2$, Schur's lemma tells us $\pi_i \circ T = 0$, while if $i \in I_1$, $\pi_i \circ T$ is a scalar multiple of the identity. Thus the former components have dimension 0 while the latter have dimension 1. Let W be the sum of the V_i for $i \in I_1$. Hence the dimension of $\text{Hom}_G(V_{rho}, V)$ is the same as that of $\text{Hom}_G(V_{rho}, W)$ which is the cardinality of I_1. $\square$

We now come to the important Frobenius Reciprocity Theorem[1] which says that each irreducible representation ρ of G is contained in the induced from σ with the same multiplicity that its restriction contains the irreducible σ of H.

Theorem 12.4.8. *Let G be a finite group, H a subgroup, σ any unitary representation of H and ρ a unitary representation of G. Then*

$$\text{Hom}_G(\rho, \text{Ind}(H \uparrow G, \sigma)) \cong \text{Hom}_H(\rho | H, \sigma). \qquad (*)$$

In particular, these have the same dimension. Hence, by Proposition 12.4.7 if ρ and σ are each irreducible, then

$$[\text{Ind}(H \uparrow G, \sigma) : \rho] = [\rho | H : \sigma].$$

Our proof of this result is functorial. In this way it does not really depend much on the particulars, for example even if G were infinite. It also works for any (not necessarily unitary, but) finite dimensional representations or if $[G : H] < \infty$ (but G was not). Nor does it depend on irreducibility!

Proof. We will prove (*) by constructing a vector space isomorphism between them. Let T be a linear G intertwining map $T : V_\rho \to \mathcal{W}$.

---

[1]F. Frobenius, (1849-1917) German mathematician made major contributions to the study of groups and their representations and characters, particularly the Frobenius Reciprocity Theorem. He also made important contributions to ordinary and partial differential equations and a number of other subjects.

Then for each $v_\rho \in V_\rho$ we know $T(v_\rho) \in \mathcal{W}$ and so $T(v_\rho)(1) \in V_\sigma$. This gives us a linear map $T^\star$ from V_ρ to V_σ. So $T^\star \in \text{Hom}_{\mathbb{C}}(V_\rho, V_\sigma)$.

Now let's consider how the action of H fits into this picture. Let $h \in H$. Then $T^\star(\rho_h(v_\rho)) = T(\rho_h(v_\rho))(1)$ and since $F(h^{-1}) = \sigma(h)F(1)$ we get

$$L_h T(v_\rho)(1) = T(v_\rho)(h^{-1}) = \sigma(h)T(v_\rho)(1) = \sigma(h)T^\star(v_\rho).$$

This says $T^\star \rho_h = \sigma_h T^\star$, for all $h \in H$. Thus $T^\star \in \text{Hom}_H(\rho|H, \sigma)$. We now construct the inverse of this map.

Let $S \in \text{Hom}_H(\rho|H, \sigma)$ and define $S_\star : V_\rho \to \mathcal{W}$ by $S_\star(v_\rho)(g) = S(\rho_g^{-1})(v_\rho) \in V_\sigma$. Since $S_\star(v_\rho)$ is a mapping from G to V_σ it has a chance of being in $\mathcal{W}$. $S_\star$ is linear.

$S_\star(v_\rho)$ satisfies our equation.

$$\begin{aligned} S_\star(v_\rho)(gh) &= S(\rho(gh))^{-1}(v_\rho) = S(\rho(h)^{-1}\rho(g)^{-1})(v_\rho) \\ &= \sigma(h)^{-1}S(\rho(g)^{-1}(v_\rho)) = \sigma(h)^{-1}S_\star(v_\rho)(g). \end{aligned}$$

Hence $S_\star(v_\rho) \in \mathcal{W}$ and we have a linear map $S_\star : V_\rho \to \mathcal{W}$.

For $g \in G$, one checks easily that $\text{Ind}_g S_\star = S_\star \rho_g$. Hence $S_\star \in \text{Hom}(\mathcal{W}, V_\rho)$. Thus $S \mapsto S_\star$ goes from $\text{Hom}_H \to \text{Hom}_G$ and is linear. It remains only to see that these maps invert one another.

Now

$$\begin{aligned} T^\star(v_\rho)_\star(g) &= T^\star(\rho(g)^{-1})(v_\rho) = T\rho(g)^{-1}(v_\rho)(1) \\ &= L_{g^{-1}}T(v_\rho)(1) = T(v_\rho)(g). \end{aligned}$$

Since this holds for all $g \in G$ and $v_\rho \in V_\rho$ we conclude $T^{\star\star} = T$. Also, $S_\star(v_\rho)(1) = S(v_\rho)$. Hence, $(S_\star)^\star(v_\rho) = S(v_\rho)$, and so $(S_\star)^\star = S$. $\square$

12.4.1   Some Consequences of Frobenius Reciprocity

We now study the relationship between the representations of G and those of a proper subgroup H. However, before doing so we shall need the following propositions.

Proposition 12.4.9. *Let G be a finite group and ρ_0 be a faithful finite dimensional unitary representation (as guaranteed by a corollary to the Peter Weyl theorem). Then each irreducible representation ρ of G is an irreducible component of $\otimes^n \rho_0 \otimes^m \bar{\rho}_0$ for some non-negative integers n and m.*

Proof. Consider the representative functions Φ associated with all irreducible subrepresentations of $\otimes^n \rho_0 \otimes^m \bar{\rho}_0$ as n and m vary. Since ρ_0 is faithful, Φ separates the points of G. It is clearly stable under conjugation, contains the constants and is a subalgebra of the algebra of all functions on G. By the Stone-Weierstrass theorem Φ is dense in $C(G)$, the space of all complex functions on G. Let $\sigma \in \mathcal{R}(G)$. If σ is not equivalent to some irreducible component of $\otimes^n \rho_0 \otimes^m \bar{\rho}_0$ for some n and m, then all the coefficients in $R(\sigma)$ are perpendicular to Φ. On the other hand, given σ_{ij} we can choose $f_\nu \to \sigma_{ij}$ on G. Therefore $\langle f_\nu, \sigma_{ij} \rangle = 0 \to \langle \sigma_{ij}, \sigma_{ij} \rangle \neq 0$, a contradiction. $\square$

Proposition 12.4.10. *Let G be a finite group and H a proper subgroup. Then there exists a $\rho \in \mathcal{R}(G) - \{1\}$ whose restriction to H contains the 1-dimensional trivial representation. That is, there exists a $v_0 \in V_\rho$ with $\rho_h(v_0) = v_0$ for all $h \in H$.*

Proof. If $\rho \in \mathcal{R}(G) - \{1\}$, the orthogonality relations show that $\int_G \rho_{ij}(x)\overline{1(x)}dx = 0$ for all $i, j = 1 \ldots d_\rho$. Hence $\int_G r(x)dx = 0$ for all $r \in R(\rho)$ and all such ρ. For each such ρ, $\rho|_H$ is a direct sum of irreducibles, $\sigma^1, \ldots, \sigma^m$ of H. If the statement of the proposition were false, none of the σ^i would be 1_H. If $r \in R(\rho)$, then $r|_H$ is a linear combination of the coefficients of the σ^i. By the orthogonality relations on H, $\int_H \sigma^i_{kl}(h)\overline{1(h)}dh = 0$ for all i. Hence $\int_H r(h)dh = 0$. Thus

$$\int_G r(x)dx = \int_H r(h)dh \text{ for all } r \in R(\rho), \text{and } \rho \in \mathcal{R}(G) - \{1\}. \quad (*)$$

On the other hand, consider $\rho = 1_G \in \mathcal{R}(G)$. Here $\rho|_H = 1$ and the representative functions associated with this are $r(x) = \lambda \cdot 1 = \lambda$ and so $\int_G r(x)dx = \lambda = \int_H r(h)dh$. Hence $(*)$ holds for all $\rho \in \mathcal{R}(G)$. Now let f be a complex function on G and $\epsilon > 0$. Choose $r \in R(G)$ so that $\|f - r\|_G < \epsilon$. Then

$$|\int_H f(h)dh - \int_G f(g)dg| \leq |\int_H f(h)dh - \int_H r(h)dh|$$
$$+ |\int_H r(h)dh - \int_G r(g)dg|$$
$$+ |\int_G r(g)dg - \int_G f(g)dg|$$
$$\leq 2\epsilon$$

and since ϵ is arbitrary, $\int_G f(g)dg = \int_H f(h)dh$ for all such f. Now $H \neq G$ so there must be another coset, $x_0 H$. Since $x_0 H$ is disjoint from H there is a non-negative real valued function, f, which is $\equiv 1$ on $x_0 H$ and $\equiv 0$ on H. Thus $\int_G f(g)dg > 0$ and $\int_H f(h)dh = 0$, a contradiction.   □

Because of Frobenius reciprocity and the fact that $[\rho|_H : 1] \geq 1$ this proposition has the following corollary.

Corollary 12.4.11. *Let G be a finite group and H a proper subgroup. Then there exists $\rho \in \mathcal{R}(G) - \{1\}$ for which $[\mathrm{Ind}(H \uparrow G, 1) : \rho] \geq 1$.*

We now prove the following theorem,

Theorem 12.4.12. *Let H be a subgroup of the finite group G. For each $\sigma \in \mathcal{R}(H)$ there is some $\rho \in \mathcal{R}(G)$ with $[\rho|_H : \sigma] \geq 1$.*

Proof of Theorem 12.4.12. Let ρ_0 be faithful representation of G (and also of H). As we saw in Proposition 12.4.9, σ is a subrepresentation of $\rho_0^{(n)} \otimes \rho_0^{-(m)}|_H$ for some choice of n and m. Therefore, σ is a subrepresentation of some irreducible component of ρ since ρ is an irreducible component of some $\rho_0^{(n)} \otimes \rho_0^{-(m)}$. Hence, some irreducible component ρ of $\rho_0^{(n)} \otimes \rho_0^{-(m)}$ must restrict to σ.

By Frobenius reciprocity we get the following corollary.

Corollary 12.4.13. *Let H be a subgroup of the finite group G. For each $\sigma \in \mathcal{R}(H)$ there is some $\rho \in \mathcal{R}(G)$ with $[\mathrm{Ind}(H \uparrow G, \sigma) : \rho] \geq 1$.*

We conclude this chapter with the following result connected with equivariant imbeddings.

Theorem 12.4.14. *Let H be a subgroup of the finite group G. Then there exists a finite dimensional unitary representation ρ of G on V_ρ and a non-zero vector $v_0 \in V_\rho$ so that $H = \mathrm{Stab}_G(v_0)$.*

Proof of Theorem 12.4.14. We may assume $H < G$ since if $H = G$ we may take ρ to be the trivial 1-dimensional representation and v_0 any non-zero vector. We will now prove:

If $g_0 \in G - H$, then there exists a representation ρ of G on V and $v_0 \neq 0 \in V_\rho$ such that $\rho_{g_0}(v_0) \neq 0$ and $\rho_h(v_0) = 0$ for all $h \in H$. $(**)$

Suppose we can do this. Then $G \supseteq \mathrm{Stab}_G(v_0) \supseteq H$ and g_0 not in $\mathrm{Stab}_G(v_0)$. Replacing G by $\mathrm{Stab}_G(v_0)$ we can apply $(**)$ again to this subgroup. In this way we get a descending chain of subgroups terminating in H. This would prove Theorem 12.4.14.

Proof of $(**)$. Since H and Hg_0^{-1} are disjoint sets we can find a function f on G for which $f|_H = \alpha$ and $f|_{Hg_0^{-1}} = \beta$, where, say, $\alpha < \beta$ (the same argument works if $\beta < \alpha$). Approximate f by $r \in R(G)$ to within $\epsilon = \frac{\beta - \alpha}{2}$. Let $F(g) = \int_H r(hg)dh$. Since $r \in R(G)$, also $F \in R(G)$. For $h_1 \in H$,

$$F(h_1) = \int_H r(hh_1)dh = \int_H r(h)dh \leq \epsilon + \alpha.$$

So $F|_H \leq \epsilon + \alpha$. On the other hand,

$$F(h_1 g_0^{-1}) = \int_H r(hh_1 g_0^{-1})dh = \int_H r(hg_0^{-1})dh \geq \beta - \epsilon.$$

So $F|_{Hg_0^{-1}} \geq \beta - \epsilon$. In particular, $F(1) \leq \epsilon + \alpha$ and $F(g_0^{-1}) \geq \beta - \epsilon$ so $F(1) \neq F(g_0^{-1})$. Now apply ρ_{reg}, the left regular representation of G. Because $\rho_{\mathrm{reg}}(g_0)F(1) = F(g_0^{-1})$, we see $\rho_{\mathrm{reg}}(g_0)F \neq F$. But,

$$\rho_{\mathrm{reg}}(h_1)(F)(g) = F(h_1^{-1}g) = \int_H r(hh_1^{-1}g)dh = \int_H r(hg)dh = F(g).$$

Thus $\rho_{reg}(h_1)(F) = F$ for all $h_1 \in H$. Since $F \in R(G)$ it lies in a (finite dimensional) ρ_{reg}-invariant subspace V_ρ of the functions on G. So there is a finite dimensional unitary representation ρ of G and a non-zero vector F with $\rho_h(F) = F$ for all $h \in H$ and $\rho_{g_0}(F) \neq F$, proving (**).

Exercise 12.4.15. Let G be a finite group and H be a subgroup.

(1) Each $\sigma \in \mathcal{R}(H)$ is an irreducible component of the restriction $\rho|_H$ of some $\rho \in \mathcal{R}(G)$.

(2) There is a finite dimensional representation ρ of G whose restriction to H is faithful.

(3) The restriction map $R(G) \to R(H)$ is surjective.

12.5   Estimates and Divisibility Properties

In this section we study the possibilities for the degree of an irreducible representation of a finite group G. $\mathcal{Z}(G)$ denotes its center, $|G|$ its order and $[G : \mathcal{Z}(G)]$, the index of $\mathcal{Z}(G)$. For a finite dimensional representation ρ, d_ρ is its degree and χ_ρ its character.

Proposition 11.8.7 gives us information concerning the values of a character on a finite group.

Proposition 12.5.1. *Let ρ be any finite dimensional representation of a finite group G. Then $\chi_\rho(G) \subseteq \mathbb{A}_0$.*

Proof. Let $n = |G|$. Then $g^n = 1$ for all $g \in G$. Hence also $\rho(g)^n = 1$. Since each $\rho(g)$ is unitary and therefore diagonalizable. The eigenvalues of each $\rho(g)$ are n-th roots of unity and so are in $\mathbb{A}_0$. Hence their sum $\chi_\rho(g)$ is also in $\mathbb{A}_0$. $\qquad\square$

Theorem 12.5.2. *Let G be a finite group and ρ be an irreducible representation. Then d_ρ divides $[G : \mathcal{Z}(G)]$.*

Some obvious consequences of Theorem 12.5.2 are: $d_\rho \leq [G : \mathcal{Z}(G)]$, d_ρ divides $|G|$, $d_\rho \leq |G|$. Even this last statement is not completely trivial since as we saw earlier, $\sum_{\rho \in \tilde{G}} d_\rho^2 = |G|$. From which it follows that $d_\rho \leq \sqrt{|G|}$ and therefore $d_\rho \leq |G|$.

We will prove this theorem by a sequence of results which are of interest in their own right (particularly, Proposition 12.5.1). Here $\mathbb{A}$ denotes the field of algebraic numbers. As above, $\mathcal{R}(G)$ denotes the set of equivalence classes of finite dimensional representations ρ of G over $\mathbb{C}$.

Proof. We first show d_ρ divides $|G|$. For $x, y \in G$, let $m(x,y) = \chi_\rho(xy^{-1})$. Then $M = (m(x,y))$ is an $n \times n$ matrix all of whose entries all lie in $\mathbb{A}$ by Proposition 12.5.1. We consider the linear operator $\chi_\rho * \cdot$ on $\mathbb{C}^n (= \mathcal{F}(G))$. That is, $f \mapsto \chi_\rho * f$, (convolution), where $\chi_\rho * f(x) = \sum_{y \in G} \chi_\rho(xy^{-1})f(y)$. Then

$$\chi_\rho * \chi_\rho(x) = \sum_{y \in G} \chi_\rho(xy^{-1})\chi_\rho(y) = \frac{|G|}{d_\rho}\chi_\rho(x).$$

Since $\chi_\rho \neq 0$ (because $\chi_\rho(1) = d_\rho$), the linear operator of convolution by χ_ρ has an eigenfunction with eigenvalue $\frac{|G|}{d_\rho}$. As above, $\frac{|G|}{d_\rho} \in \mathbb{A}$. It is clearly in $\mathbb{Q}$ so it is in $\mathbb{Z}$. Hence d_ρ divides $|G|$. $\square$

We now sharpen this to show d_ρ actually divides $[G : \mathcal{Z}(G)]$.

Lemma 12.5.3. *Let a, b, c be positive integers with $c|b$. If $a^k | \frac{b^k}{c^{k-1}}$ for all positive integers k, then $a| \frac{b}{c}$.*

Proof. Since $c|b$ also $c^k|b^k$ for all k. In particular, $c^{k-1}|b^k$ for all k. Let p be a prime p^x, p^y, p^z be the largest powers of p dividing a, b, c respectively (including the possibility that some of these exponents are 0). By hypothesis for each k, $b^k = j_k a^k c^{k-1}$. Hence by uniqueness of prime factorization, $ky \geq kx + (k-1)z$ for all k. Then $kx \leq ky - (k-1)z$ and choosing $k = z+1$ we get $x \leq y - z + \frac{z}{z+1}$. Since x, y, z are integers this means that actually $x \leq y - z$. Therefore, $p^x|a$, $p^y|b$ and $p^z|c$. Since this holds for each prime p it follows that $a|\frac{b}{c}$. $\square$

To complete the proof, for each positive integer k, let G^k be the direct product of G, k times and $\mathcal{Z}(G)^k$ the direct product of $\mathcal{Z}(G)$, k

times. Then $\mathcal{Z}(G)^k$ is the center of G^k. Within $\mathcal{Z}(G)^k$ consider

$$H_k = \{(z_1, \ldots z_k) : \prod_{i=1}^{k} z_i = 1\}.$$

H_k is a central and therefore normal subgroup of G^k. Its order is $|\mathcal{Z}(G)|^{k-1}$.

Now take tensor products. $\otimes^k \rho$ acts on $\otimes^k V_\rho$ by

$$\otimes^k \rho(g_1, \ldots g_k) = \rho(g_1) \otimes \ldots \otimes \rho(g_k).$$

This is an irreducible representation of G^k of degree d_ρ^k. Since ρ is irreducible, by Schur's Lemma 6.9.1 of Volume I, $\rho(z) = \lambda(z)I$ for all $z \in \mathcal{Z}(G)$, where λ is a multiplicative character of $\mathcal{Z}(G)$. For $(z_1, \ldots z_k) \in H_k$, a direct calculation shows $\otimes^k \rho(z_1, \ldots z_k) = I_{\otimes^k V_\rho}$. Thus $\otimes^k \rho$ is trivial on H_k and so gives rise to an irreducible representation of the finite group G^k/H_k. Hence by what we already know, $d(\otimes^k \rho) = d_\rho^k$ divides $|G|^k/|\mathcal{Z}(G)|^k$, for all k. Since $|\mathcal{Z}(G)|$ divides $|G|$ it follows from Lemma 12.5.3 that d_ρ divides $|G|/|\mathcal{Z}(G)| = [G : \mathcal{Z}(G)]$.

12.6   Burnside's $p^a q^b$ Theorem

Here we prove an important (and classical) result of Burnside[2], namely, that a finite group of order $p^a q^b$, where p and q are primes and a and b are positive integers, is solvable (substantially extending the fact that a group of order p^a is nilpotent as well as generalizing some earlier results we got on groups of order pq). This result plays an important role in the famous theorem of Feit and Thompson [40] that all finite groups of odd

---

[2]William Burnside, (1852-1927) an English mathematician worked on finite groups and their representations. He was a competitor of Frobenius. In addition to the p, q theorem he is remembered for *Burnside's problem* (1902), one of the oldest and most influential questions in group theory and is a direct generalization of the Fundamental Theorem of Abelian Groups. It asks whether a finitely generated group in which every element has finite order must be finite. Although false in its most general form, less general variants of it have been proved, or remain open (e.g. of bounded order). Burnside was a fellow of the Royal Society.

order are solvable. The most accessible proof[3] of Burnside's theorem is via representation theory which we give below.

We begin with a Lemma.

Lemma 12.6.1. *Let ρ be an irreducible representation of a finite group G on a finite dimensional vector space V, and let C be a conjugacy class of G with*

$$gcd(|C|, d_\rho|) = 1.$$

Then, for any $g \in C$, $\chi_\rho(g) = 0$ or $\rho(g)$ acts as a scalar linear operator on V.

Proof. Since $\gcd(|C|, d_\rho|) = 1$ there exist a and b such that

$$a|C| + bd_\rho = 1.$$

Then,

$$a\frac{\chi_\rho(g)|C|}{d_\rho} + b\chi_\rho(g) = \frac{\chi_\rho(g)}{d_\rho}.$$

Now, in the proof of the Theorem 12.5.2 we have seen that $\frac{\chi_\rho(g)|C|}{d_\rho} = \lambda_C$ is an algebraic integer, also $\varepsilon = \frac{\chi_\rho(g)}{d_\rho}$ is an algebraic integer. Now, let n be the order of g in G. Then,

$$\chi_\rho(g) = \zeta_1 + \ldots + \zeta_{d_\rho}$$

is the sum of d_ρ, n-th roots of unity. Here, there are two possibilities: in the first all ζ_i are equal and so since $\rho(g)$ is diagonalizable it must be a scalar. In the second, they are not equal, and so

$$|\varepsilon| = \frac{1}{d_\rho}|\zeta_1 + \ldots + \zeta_{d_\rho}| < 1.$$

---

[3]However, for many years, it defied the group theorists effort to find a purely group theoretic proof. It was only in 1970 that, following up on ideas of J. G. Thompson, D. Goldschmidt [51] gave the first group theoretic proof for the case when p, q are odd primes. Goldschmidt's proof used a rather deep result in group theory, called Glauberman's $Z(J)$-theorem. A couple of years later, Matsuyama completed Goldschmidt's work by supplying a group theoretic proof in the remaining case $p = 2$. Slightly ahead of Matsuyama, Bender gave a proof for all p, q, using (among other things) a variant of the notion of the Thompson subgroup of a p-group.

But if L is the cyclotomic field spanned by n-th roots of unity, any element $\sigma \in Gal(L/\mathbb{Q})$ preserves the set of these roots of unity. Thus

$$|\sigma(\varepsilon)| = \frac{1}{d_\rho}|\sigma(\zeta_1) + \ldots + \sigma(\zeta_{d_\rho})| < 1,$$

which implies that the norm

$$1 > N(\varepsilon) = \prod_{\sigma \in Gal(L/\mathbb{Q})} \sigma(\epsilon) \in \mathbb{Q}.$$

But each $\sigma(\epsilon)$ is an algebraic integer. Therefore $N(\varepsilon)$ is also an algebraic integer. Thus $N(\varepsilon) \in \mathbb{Z}$, and since it has absolute value < 1, it must be 0. So $\chi_\rho((g)) = 0$. $\qquad\square$

Theorem 12.6.2. *Let G be a finite group, C a conjugacy class in G of order p^a, where p is a prime and $a > 0$. Then, G is not a simple group.*

Proof. We know the regular representation has a decomposition into irreducible representations as

$$\mathbb{C}[G] = \bigoplus_{i=1}^{r} d_i V_i,$$

where each irreducible subspace appears d_i times. We also know $\chi_{reg}(g) = |G|$ if $g = 1$, and $= 0$ otherwise. Now, let $g \in C$. Since each character is additive,

$$1 + \sum_{\chi \neq 1} d_\chi(g)\chi(g),$$

where χ is any non-trivial irreducible character. Consider the irreducible representations such that $p \nmid d_\chi$. From Lemma 12.6.1 we see that $\chi(g)$ is a scalar or 0. But at least one such irreducible must have $\chi(g) \neq 0$. Otherwise, one could write $\frac{1}{p}$ as an integral combination of characters, which means that $\frac{1}{p}$ is an algebraic integer, which is impossible. Therefore, if g acts on V_χ by a scalar matrix, then G has a proper normal subgroup, which is the pre-image of the normal subgroup $\mathbb{C}^* \, \mathrm{Id} \subseteq \mathrm{GL}(V_\chi)$. This pre-image can not be equal to G since V_χ is irreducible. Hence, G is not simple. $\qquad\square$

Now we are ready to prove Burnside's Theorem.

Theorem 12.6.3. *(Burnside's p, q-Theorem) Let G be a group of order $p^a q^b$, where p and q are primes and a and b are positive integers. Then G is solvable.*

Proof. We can assume p and q are distinct since a p group is nilpotent. Using induction on $|G|$, and the Sylow Theorems, it suffices to show that G is not a simple group. Let H be a Sylow q-group and g be a non-trivial element of the center of H. If g is central in G, G is clearly non simple. Otherwise, $Z_G(g)$ the centralizer of g in G is a proper subgroup of G containing H. Then the conjugacy class of g has cardinality $[G : Z_G(g)] = p^k$ for some $k \geq 1$, so G is again non simple by Theorem 12.6.2.   □

In the last three sections we deal with representation over domains other than $\mathbb{C}$.

12.7   Clifford's Theorem

Here we analyze an important question in representation theory, namely what occurs when one restricts a finite dimensional irreducible representation to a normal subgroup. This question was addressed by A.H. Clifford in [26]. The result holds over any field k and does not require N to have finite index in G.

Suppose G is a group, N a normal subgroup and $\rho : G \to \mathrm{GL}(V)$ is a finite dimensional representation of G on V, a vector space over a field k. Let $\sigma = \rho|N$ and G operate on σ via $\sigma^g(n) = \sigma(gng^{-1})$. Since N is normal in G and σ is a representation of N on V, for each $g \in G$, σ^g is also a representation of N on V called the g-conjugate of σ. It could be that σ is actually irreducible. But if not, there must be a non-trivial σ-invariant subspace U of V which may or may not be irreducible. Continuing in this way we eventually arrive at a non-trivial σ-invariant irreducible subspace W of V.

Theorem 12.7.1. *Let G be a group, N a normal subgroup and $\rho : G \to \mathrm{GL}(V)$ a finite dimensional irreducible representation of G on V and as* *above let W be a non-zero $\rho|N = \sigma$ invariant irreducible subspace of V. Then $(\rho|N, V)$ decomposes as,*

$$\rho|N = \nu(\sigma \oplus \sigma^{g_1} \oplus \ldots \oplus \sigma^{g_n})$$

and correspondingly,

$$V = \nu W \oplus \nu \rho^{g_1}(W) \oplus \ldots \oplus \nu \rho^{g_n}(W).$$

That is, it breaks up into a finite number of inequivalent conjugate representations all with the same degree and the same multiplicity (ν).

The decomposition of $(\rho|N, V)$ into the inequivalent components $(\nu \sigma^{g_i}, \nu \rho^{g_i}(W))$ constitutes what is called *a system of imprimitivity*

Proof. For $g \in G$ and $n \in N$, since $gng^{-1} \in N$, on all of V we have,

$$\sigma^g(n) = \sigma(gng^{-1}) = \rho(gng^{-1}) = \rho(g)\rho(n)\rho(g)^{-1} = \rho(g)\sigma(n)\rho(g)^{-1}.$$

In particular,

$$\sigma^g(n)(\rho(g)(W)) = \rho(g)\sigma(n)\rho(g)^{-1}(\rho(g)(W))$$
$$= \rho(g)\sigma(n)(W) = \rho(g)(W),$$

because W is σ invariant. Thus, σ^g leaves $\rho(g)(W)$ invariant and when σ acts on W, σ^g acts on $\rho(g)(W)$ a subspace of the same dimension as W. In this way for each fixed $g \in G$ we get a family of representations $\{\sigma^g : g \in G\}$ of N, all of the same degree and irreducible since σ is.

If $W \neq V$, i.e. if $\rho|N$ is not irreducible, there must be a $g \in G$ with $\rho(g)(W) \neq W$, since otherwise W would be $\rho(G)$-invariant and this would contradict the assumption that ρ itself is irreducible. But then, $\rho(g)(W) \cap W$ would be a proper N invariant subspace of $\rho(g)W$, a contradiction since this is also irreducible. We conclude $\rho(g)(W) \cap W = (0)$. Then the sum $W + \rho(g)(W)$ is direct and σ-invariant. If this direct sum of subspaces is not all of V we continue and get g' so that $(W + \rho(g)(W)) + \rho(g')(W)$ is a direct sum of invariant irreducible subspaces of

V, etc. and because V is finite dimensional we must eventually exhaust V. Thus,

$$V = \sum_{i=0}^{n} \rho(g_i)(W),$$

where $g_0 = 1$, the sum is direct and each $\rho(g_i)(W)$ is a σ-invariant irreducible subspace of V. (This already shows $\rho|N$ is completely reducible.) Moreover as we have seen, each σ^{g_i} is a conjugate of σ. Since g was arbitrary, all conjugates (up to equivalence of representations) must occur because the only restriction above on choosing g was that $\rho(g)(W) \neq W$, while if $\rho(g)(W) = W$, the two conjugates are equivalent. Thus these spaces are just permuted among themselves by G. Hence $\nu = [G : \mathrm{Stab}_G(\sigma)]$. Because G permutes these direct summands, $\nu\rho(g_i)(W)$, one says $\nu\{\rho(g_i)(W) : i = 0, \ldots, n\}$ constitutes a *system of imprimitivity* for (ρ, V). Of course, if $\rho|N$ is irreducible, all claims are also valid.   □

Corollary 12.7.2. *Let G be a group, N a normal subgroup and $\rho : G \rightarrow \mathrm{GL}(V)$ any finite dimensional representation of G on V. If ρ is completely reducible, then so is $\rho|N$.*

Proof. When ρ is irreducible it follows from Clifford's Theorem in particular that $\rho|N$ is completely reducible. Applying this to each irreducible component, the same is true if ρ is completely reducible.   □

12.8   Some Applications of Real Representation Theory

We first apply representation theory to the quoternions $\mathbb{H}$ and from this we will get useful information about the relationship of $\mathrm{SO}(3, \mathbb{R})$ and $\mathrm{SO}(4, \mathbb{R})$ to the 3 sphere S^3. As we know (see Chapter 5 of Volume I) the quaternions form a division algebra; that is $\mathbb{H}^{\times}$ is a group. Accordingly, we define the representation ρ of $\mathbb{H}^{\times}$ on $\mathbb{H}$ by $\rho_q(x) = qxq^{-1}$, where $q \in \mathbb{H}^{\times}$ and $x \in \mathbb{H}$. On checks that ρ is indeed a representation. Writing $\mathbb{H} = \mathcal{Z}(H) \oplus \Im(\mathbb{H})$ the direct sum of the center and the purely imaginary quoternions we see that ρ is $\mathbb{R}$-linear. That is, for $x, y \in \mathbb{H}$ and $t \in \mathcal{Z}(H)$ one has, $\rho_q(x + y) = \rho_q(x) + \rho_q(y)$ and $\rho_q(tx) = t\rho_q(x)$. Notice also that ρ depends continuously on both q and x. Now consider

$$S^3 = \{q \in \mathbb{H}^\times \; : \; N(q) = 1\}.$$

This is clearly a subgroup homeomorphic to the 3 sphere (and one of the very few times a sphere carries a group structure). Since $N(qxq^{-1}) = N(x)$ for $q \in S^3$ and $x \in \mathbb{H}$, S^3 operates orthogonally $\mathbb{H}$. It leaves $\mathcal{Z}(H)$ pointwise fixed and, in particular, invariant. Therefore, it also stabilizes the orthogonal complement $\Im(\mathbb{H}) = \mathbb{R}^3$. Thus, on restriction we have a continuous homomorphism, which we again call $\rho : S^3 \to O(3, \mathbb{R})$. Since S^3 is connected and ρ is continuous so is its image. Therefore,

$$\rho : S^3 \to \mathrm{SO}(3, \mathbb{R}).$$

Let us compute $\mathrm{Ker}(\rho)$. Suppose $\rho_q = \mathrm{Id}$ on $\Im(\mathbb{H})$, where $q \in S^3$. Then since it acts trivially on $\mathcal{Z}(H)$, $\rho_q = \mathrm{Id}$ on all of $\mathbb{H}$. Therefore, $q \in \mathcal{Z}(H)$, and since it has unit length, $q = \pm 1$ and conversely. Thus, $\mathrm{Ker}(\rho) = \pm 1$. Now, because these are Lie groups and this kernel is finite (and so discrete), the image $\rho(S^3)$ has the same dimension as S^3 itself, namely 3. But $\mathrm{SO}(3, \mathbb{R})$ also has dimension 3. It follows that, since $\mathrm{SO}(3, \mathbb{R})$ is connected, these two Lie groups must coincide, that is ρ is *surjective*. Because S^3 is not just connected, but also simply connected, we have constructed the universal covering space of $\mathrm{SO}(3, \mathbb{R})$ and know that its fundamental group is ± 1, the group of order 2.

Let us identify S^3. Each element of S^3 can be identified as a pair of complex numbers z, w subject to $|z|^2 + |w|^2 = 1$ and therefore can be associated with

$$g = \begin{pmatrix} z & w \\ -\bar{w} & \bar{z} \end{pmatrix}$$

with $\det(g) = 1$.

This map is easily seen to be continuous and by compactness, a homeomorphism. We invite the reader to also show it is a homomorphism. Hence, $S^3 = \mathrm{SU}(2, \mathbb{C})$. Since S^3 is simply connected, we have the universal covering $\mathrm{SU}(2, \mathbb{C}) \to \mathrm{SO}(3, \mathbb{R})$ of $\mathrm{SO}(3, \mathbb{R})$.

Now we consider the representation σ in which $\mathbb{H}^\times \times \mathbb{H}^\times$ acts on $\mathbb{H}$ by $\sigma_{q,r}(x) = qxr^{-1}$. This is also a continuous $\mathbb{R}$-linear representation and, as before, we restrict to $S^3 \times S^3$. Then, for all q, r, $N(qxr^{-1}) = N(x)$. Thus, $S^3 \times S^3$ acts on $\mathbb{R}^4$ orthogonally so the map $\sigma : S^3 \times S^3 \to O(4, \mathbb{R})$. Now, just as before, the product $S^3 \times S^3$ is connected. Hence

$$\sigma : S^3 \times S^3 \to SO(4, \mathbb{R}).$$

Next, we check $\text{Ker}(\sigma)$. Suppose $qxr^{-1} = x$ for all $x \in \mathbb{H}$. Letting $x = 1$ we see that $q = r$. Therefore, $qxq^{-1} = x$ for all x. It follows that $q \in \mathcal{Z}(\mathbb{H})$ and since it has norm 1, $q = \pm 1$. Thus, the kernel consists of $(1, 1)$ and $(-1, -1)$. In particular, it is finite and therefore discrete. From this it follows that the image $\sigma(S^3 \times S^3)$ has dimension 6. Since $SO(4, \mathbb{R})$ is connected and also has dimension 6, we see σ is surjective. Now, because $S^3 \times S^3$ is simply connected, it follows that the fundamental group of $SO(4, \mathbb{R})$ is the 2 element group and the universal covering group of $SO(4, \mathbb{R})$ is $SU(2, \mathbb{C}) \times SU(2, \mathbb{C})$. Hence, $SO(4, \mathbb{R})/\{\pm I\}$ is isomorphic to $SO(3, \mathbb{R}) \times SO(3, \mathbb{R})$, a situation very particular to dimension 4.

One can also apply these ideas in the case of non-compact groups. Consider $SO_o(1, 2)$ and $SO_o(1, 3)$, the connected components of the group of isometries of hyperbolic 2 and 3 space, H^2 and H^3. That is, we consider the forms $q_{12}(X) = x^2 - y^2 - z^2$ on $\mathbb{R}^3$ and $q_{13}(X) = x^2 - y^2 - z^2 - t^2$ on $\mathbb{R}^4$ and the corresponding connected groups of isometries.

Let $SL(2, \mathbb{R})$ act on the space of 2×2 real symmetric matrices $\mathcal{S}$ by $\rho_g(S) = gSg^t$. Then $\mathcal{S}$ is a real vector space of dimension 3 and ρ is a continuous real linear representation of $SL(2, \mathbb{R})$. Since $\det g = 1$, $\det(\rho_g(S)) = \det(gSg^t) = \det S$. Now if

$$S = \begin{pmatrix} a & z \\ z & b \end{pmatrix},$$

then $\det S = ab - z^2$. Consider the change of variables $x = \frac{a+b}{2}$ and $y = \frac{a-b}{2}$. This is an $\mathbb{R}$-*linear* change of variables and $a = x+y$ and $b = x-y$. Therefore, $\det S = x^2 - y^2 - z^2$. Thus ρ preserves a $(1, 2)$-form on $\mathbb{R}^3$.

Since $\mathrm{SL}(2,\mathbb{R})$ is connected so is its image $\rho(G)$ in $\mathrm{GL}(3,\mathbb{R})$. A direct calculation shows $\mathrm{Ker}\,\rho = \pm I$. Since this is finite and therefore discrete, $\rho(G)$ has dimension 3 just as does $\mathrm{SL}(2,\mathbb{R})$. Since $\rho(G) \subseteq \mathrm{SO}_o(1,2)$ and this connected group also has dimension 3, ρ is surjective and therefore a covering. It induces an isomorphism between $\mathrm{SO}_o(1,2) = \mathrm{SO}(1,2)$ and $\mathrm{SL}(2,\mathbb{R})/\{\pm I\} = \mathrm{PSL}(2,\mathbb{R})$.

Similarly, let $\mathrm{SL}(2,\mathbb{C})$ act on the space of 2×2 complex Hermitian matrices $\mathcal{H}$ by $\rho_g(H) = gHg^*$. Then $\mathcal{H}$ is a real vector space of dimension 4 and ρ is also a continuous real linear representation of $\mathrm{SL}(2,\mathbb{C})$. Since $\det g = 1$, $\det(\rho_g(H)) = \det(gHg^*) = \det H$. Now if

$$H = \begin{pmatrix} a & z \\ \bar{z} & b \end{pmatrix},$$

then $\det H = ab - |z|^2$. Consider the change of variables $x = \frac{a+b}{2}$ and $y = \frac{a-b}{2}$. This is an $\mathbb{R}$ linear change of variables, $a = x + y$ and $b = x - y$. Therefore, $\det H = x^2 - y^2 - |z|^2 = x^2 - y^2 - u^2 - v^2$. Thus ρ preserves a $(1,3)$-form on $\mathbb{R}^4$. Since $\mathrm{SL}(2,\mathbb{C})$ is connected so is its image $\rho(G)$ in $\mathrm{GL}(4,\mathbb{R})$. A direct calculation shows $\mathrm{Ker}(\rho) = \pm I$. Since this is finite, and therefore discrete, $\rho(G)$ has dimension 6 just as $\mathrm{SL}(2,\mathbb{C})$ does. But $\rho(G) \subseteq \mathrm{SO}_o(1,3)$. Because this connected group also has dimension 6, ρ is surjective and therefore a covering. It induces an isomorphism between $\mathrm{SO}_o(1,3)$ and $\mathrm{SL}(2,\mathbb{C})/\{\pm I\} = \mathrm{PSL}(2,\mathbb{C})$. Since $\mathrm{SL}(2,\mathbb{C})$ is simply connected (its maximal compact is $\mathrm{SU}(2,\mathbb{C})$ which is simply connected), it is the universal cover of $\mathrm{SO}_o(1,3)$.

12.9   Finitely Generated Linear Groups

The story begins with Burnside's conjecture that a finitely generated torsion group G whose torsion elements are of *bounded order* is necessarily finite. This is of course true even without assuming bounded order when G is Abelian because of the Fundamental Theorem of Abelian Groups (see 6.4.5 of Volume I). More generally, this is also true when each of the conjugacy classes of G is finite. To see this we need the following lemma.

Lemma 12.9.1. *Let G be a finitely generated group with the index of the center $[G : \mathcal{Z}(G)] = n$ finite. Then the center, $\mathcal{Z}(G)$, is also finitely generated.*

Proof. If not, then there exists an infinite strictly ascending chain of finitely generated subgroups Z_i of $\mathcal{Z}(G)$. For each i let G_i be the group generated by Z_i together with the $n-1$ coset representatives of $G-\mathcal{Z}(G)$. Then, the G_i is an infinite ascending chain of distinct finitely generated subgroups of G. Hence G can not be finitely generated. □

Proposition 12.9.2. *Let G be a finitely generated torsion group. If all the conjugacy classes of G are finite, then G is finite.*

Proof. Let (g_i) be a finite generating set for G. Then the centralizers, $Z_G(g_i)$, each have finite index in G. Therefore $\cap_i Z_G(g_i) = \mathcal{Z}(G)$ also has finite index in G. By Lemma 12.9.1 $\mathcal{Z}(G)$ is itself finitely generated and of course a torsion group. By the Fundamental Theorem of Abelian Groups $\mathcal{Z}(G)$ is finite and hence so is G. □

We mention the general Burnside conjecture. Namely, is an arbitrary finitely generated torsion group G necessarily finite? When G is a linear group, that is, a subgroup of $\mathrm{GL}(n, \mathbb{C})$ this was proved by Schur in the 1930s, see 12.9.4 below. Novikov gave a counterexample to the general statement in the 1950s. This lead to the question of what happens when G has bounded exponent (in other words G is of bounded order). This has been completely settled in the affirmative by Efim Zelmanov for which he was awarded the Fields Medal in 1994.

Schur's result leads us directly to the study of finitely generated linear groups and *Selberg's Lemma* [116] as follows.

Theorem 12.9.3. *Let G be a finitely generated subgroup of $\mathrm{GL}(n, k)$ where k is a field of characteristic zero. Then G contains a torsion free subgroup H of finite index.*

In particular, this implies the result of Schur. Using techniques rather similar to those of Selberg (see pp. 252-254 of [30]), Schur proved:

Corollary 12.9.4. *A finitely generated torsion subgroup of* GL(n, k) *is finite.*

The applications of 12.9.3 are wide ranging. If G is any connected Lie group and Γ is a lattice in G, then Γ is finitely generated (see [1] p. 373). Since G/Γ carries a manifold structure if Γ is torsion free, Theorem 12.9.3 tells us that Γ is commensurable to a lattice Γ' for which G/Γ' is a manifold. Since, as we mentioned, SL$(n, \mathbb{Z})$ is a lattice in SL$(n, \mathbb{R})$, it is a finitely generated linear group. In the 19th century, using congruence subgroups, Minkowski proved Selberg's lemma when $n \geq 3$, for SL$(n, \mathbb{Z})$ and GL$(n, \mathbb{Z})$, which contains SL$(n, \mathbb{Z})$ as a subgroup of index 2.

Before embarking on the proof of Selberg's Lemma we will have need of the concept of a *residually finite group* which we encountered in 1.14.47 of Volume I. This is one in which there is a separating family of homomorphisms to finite groups. Alternatively, the intersection of normal subgroups of finite index is trivial.

Proof. Let G be generated by $g_1, \ldots g_r$ and A be the set of all entries in the field k of the matrices in GL(n, k) coming from G. Then A is a finitely generated commutative subring of k, with identity. Since k is a field, A is without zero divisors. In this way we may consider

$$A = \mathbb{Z}[x_1, \ldots, x_s]$$

where we have polynomials with integer coefficients in the s variables and some of these also have relations. Consider the ideal $M_p = (p, x_1, \ldots, x_s)$, where p is a prime of $\mathbb{Z}$. M is easily seen to be a maximal ideal in A. Moreover, A/M_p is clearly finite. Furthermore, from the uniqueness of prime factorization in $\mathbb{Z}$, $\cap M_p = (0)$ so that for example the Jacobson radical, $J(A) = (0)$. The natural ring homomorphism $\pi_{M_p} : A \to A/M_p$ has the property that,

$$\bigcap_{M_p} \mathrm{Ker}(\pi_{M_p}) = (0).$$

From this it follows that as p varies, the congruence homomorphisms,

$$\pi_{M_p} : \mathrm{GL}(n, A) \to \mathrm{GL}(n, A/M_p),$$

form a separating family and since they each take values in a finite group, $GL(n, A)$ is residually finite.

By the second isomorphism theorem for each M_p, $G \cap \text{Ker}(\pi_{M_p})$ has finite index in G. Therefore it remains to prove that for one of these $G \cap \text{Ker}(\pi_{M_p})$ is torsion free. As A is isomorphic, as a ring, to $\mathbb{Z}[x_1, \ldots z_s]$, $M_p \cap \mathbb{Z}$ is the ideal (p) for a prime number p. Choose a p that does not divide n. For this p, $G \cap \text{Ker}(\pi_{M_p})$ is torsion free. Suppose $g \in G \cap \text{Ker}(\pi_{M_p})$ has finite order ν. Then $g^\nu = 1$ and its minimal polynomial is $x^\nu - 1$. Since the roots of this are the νth-roots of unity, the minimal polynomial has distinct roots and because the characteristic of A is zero, g is diagonalizable. Moreover, the sum of these roots is zero so $\text{tr}(g) = 0$. Now $\pi_{M_p} g$ is congruent to $I \mod (M_p)$ and therefore its trace is $n \in \mathbb{Z}$. Hence 0 is congruent to $n \mod p$, a contradiction. $\qquad\square$

Along the way we have also proved an important theorem of Malcev in 1940. Since $\pi_{M_p} : GL(n, A) \to GL(n, A/M_p)$ form a separating family with values in finite groups, upon restriction to G we get:

Corollary 12.9.5. *(Malcev) Any finitely generated linear group (in characteristic 0) is residually finite.*

We also state the following results of Jordan and of Schur. These can be found in Curtis and Reiner p. 258 of [30].

Theorem 12.9.6. *Let $\mathbb{Z}^+$ denote the positive integers. There is a function $f : \mathbb{Z}^+ \to \mathbb{Z}^+$ so that for every finite subgroup G of $GL(n, \mathbb{C})$ there is some normal Abelian subgroup A of G so that $[G, A] \leq f(n)$.*

Schur both generalized and strengthened this as follows.

Theorem 12.9.7. *Let G be a torsion subgroup of $GL(n, \mathbb{C})$. Then G is conjugate to a subgroup of $U(n, \mathbb{C})$ and moreover there is some normal Abelian subgroup A of G so that,*

$$[G, A] \leq (\sqrt{8n} + 1)^{2n^2} - (\sqrt{8n} - 1)^{2n^2}.$$

Finally, we mention the Tits alternative. See [121].

Theorem 12.9.8. *Let G be any subgroup of* $\mathrm{GL}(n, k)$, *where k is any field. Then, either G has a normal solvable subgroup of finite index, or contains a non-Abelian free group.*

12.9.1   Exercises

1. Prove Theorem 12.9.7.

2. Show that a finitely generated Abelian group has a faithful finite dimensional representation, ρ. Can ρ be taken to be unitary?

3. Continuing item 2. Is the same true if G has all its conjugacy classes finite?

4. Let G be an arbitrary group with $G/\mathcal{Z}(G)$ finite. Is $G/[G, G]$ necessarily finite?

5. Let G be a subgroup of $\mathrm{GL}(n, \mathbb{C})$ with only a finite number of conjugacy classes. Prove G is finite.

6. Prove $\mathrm{GL}(n, A)$ and in particular G has no unipotent elements.

7. Show $I_n + E_{i,j}$, where $E_{i,j}$ are the matrix units for the $n \times n$ matrices, are a generating set for $\mathrm{SL}(n, \mathbb{Z})$.

8. Is the discrete group $\mathrm{GL}(n, \mathbb{Z})$ a lattice in $\mathrm{GL}(n, \mathbb{R})$? Why, or why not?

12.10   Pythagorean Triples (continued)

Here we continue considering Pythagorean triples, but where group theory enters the picture. For the initial discussion of Pythagorean triples see Chapter 0 of Volume I.

Definition 12.10.1. Here $\mathrm{SL}^{\pm}(2, \mathbb{Z})$ denotes the 2×2 matrices with entries in $\mathbb{Z}$ and determinant ± 1, while $\mathrm{SL}(2, \mathbb{Z})$ indicates those whose determinant is merely 1.

Hence, if

$$\begin{pmatrix} s_{11} & s_{12} \\ s_{21} & s_{22} \end{pmatrix} = s \in \mathrm{SL}^{\pm}(2, \mathbb{Z}),$$

this group acts linearly on $\mathbb{R}^2$ taking $(a, b)^t$, where $a, b \in \mathbb{Z}$ and $a > b$, to $(a', b')^t$, where $a' = s_{11}a + s_{12}b$ and $b' = s_{21}a + s_{22}b$.

Now suppose s_{11} and s_{22} are both odd and s_{12} and s_{21} are both even, then clearly a' is even and b' is odd. Moreover, if $\gcd(a, b) = 1$, then also $\gcd(a', b') = 1$. This is because $s^{-1}(a', b')^t = (a, b)^t$, and since s^{-1} is also in $\mathrm{SL}^{\pm}(2, \mathbb{Z})$ because it is a group, the ideal in $\mathbb{Z}$ generated by a', b' contains the ideal generated by a, b, which is all of $\mathbb{Z}$. Hence, also $\gcd(a', b') = 1$.

Lemma 12.10.2. *Given (a, b) with $\gcd(a, b) = 1$ we can find $s \in \mathrm{SL}(2, \mathbb{Z})$ mapping (a, b) to $(0, 1)$.*

Proof. By assumption there are integers α^* and β^* so that $\alpha^* a + \beta^* b = 1$. Let $\alpha = \alpha^* + kb$ and $\beta = \beta^* - ka$, where k is an integer. Then for all k,

$$\alpha a + \beta b = \alpha^* a + kab + \beta^* b - kab = \alpha^* a + \beta^* b = 1.$$

Choose k so that $\alpha = s_{11}$ is even by taking $k = 0$ if α^* is even and $k = 1$ if α^* is odd. Then $\beta = s_{12}$ will always be odd. Finally, take $s_{21} = b$ and $s_{22} = -a$. Then s_{21} is odd and s_{22} is even and of course $\det s = 1$.   □

Let $\mathcal{A}$ denote the points (vectors in $\mathbb{Z}^2 \subseteq \mathbb{R}^2$) of the form $(a, b)^t$, and $\Phi : \mathcal{A} \to \mathcal{P}$ stand for the map given by the equation above. Then, Φ is bijective. We now compose to see what the effect of $\mathrm{SL}^{\pm}(2, \mathbb{Z})$ is on $(x, y, z) = \Phi(a, b)$. A direct calculation shows,

- $x' = (a')^2 + (b')^2 = (s_{11}^2 - s_{21}^2)a^2 + 2(s_{11}s_{12} - s_{21}s_{22})ab + (s_{12}^2 - s_{22}^2)b^2$.

- $y' = 2a'b' = 2(s_{11}s_{21}a^2 + (s_{11}s_{22} + s_{12}s_{21})ab + s_{12}s_{22}b^2)$.

- $z' = (a')^2 + (b')^2 = (s_{11}^2 + s_{21}^2)a^2 + 2(s_{11}s_{12} + s_{21}s_{22})ab + (s_{12}^2 + s_{22}^2)b^2$.

And so

- $y' = (s_{11}s_{22} + s_{12}s_{21})x + (s_{11}s_{21} - s_{12}s_{22})y + (s_{11}s_{21} + s_{12}s_{22})z.$
- $x' = (s_{11}^2 - s_{21}^2)a^2 + 2(s_{11}s_{12} - s_{21}s_{22})ab + (s_{12}^2 - s_{22}^2).$

Thus we have a commutative diagram,

$$
\begin{array}{ccc}
\mathcal{A} & \xrightarrow{\;\Phi\;} & \mathcal{P} \\[2pt]
{\scriptstyle s}\Big\downarrow & & \Big\downarrow{\scriptstyle s} \\[2pt]
\mathcal{A} & \xrightarrow[\;\Phi\;]{} & \mathcal{P}
\end{array}
$$

where $s \in \mathrm{SL}^{\pm}(2, \mathbb{Z})$.

Now the basic equation for a Pythagorean triple is $x^2 + y^2 - z^2 = 0$, which is a non-degenerate, but indefinite quadratic form in 3 variables. This suggests we consider the group, $G = O(2, 1)$ (the isometry group of the hyperbolic plane) and its $\mathbb{Z}$ points $G_{\mathbb{Z}}$. Because this form is non-degenerate and G is the *orthogonal* group of the form, $\det g = \pm 1$ for all $g \in G$ (prove) and hence $G_{\mathbb{Z}}$ is just the points of G with integer coordinates. This linear group evidently takes Pythagorean triples *onto* Pythagorean triples. Let H be the subgroup of $G_{\mathbb{Z}}$ which preserves $\mathcal{P}$.

We must check that the transformations of $\mathcal{P}$ in the above diagram are just those of H, thus giving a commutative diagram where the mappings on the left come from $\mathrm{SL}^{\pm}(2, \mathbb{Z})$, while those on the right come from H.

In order to do this we examine the representation

$$\rho : \mathrm{SL}_2^{\pm}(\mathbb{R}) \to O(2, 1)$$

given explicitly as follows

$$\begin{pmatrix} s_{11} & s_{12} \\ s_{21} & s_{22} \end{pmatrix} = s \mapsto \rho(s),$$

where

$$\rho(s)$$
$$= \begin{pmatrix} \frac{1}{2}(s_{11}^2 - s_{12}^2 - s_{21}^2 + s_{22}^2) & s_{11}s_{21} - s_{12}s_{22} & \frac{1}{2}(s_{11}^2 - s_{12}^2 + s_{21}^2 - s_{22}^2) \\ s_{11}s_{12} - s_{21}s_{22} & s_{12}s_{21} + s_{11}s_{22} & s_{11}s_{12} + s_{21}s_{22} \\ \frac{1}{2}(s_{11}^2 + s_{12}^2 - s_{21}^2 - s_{22}^2) & s_{11}s_{21} + s_{12}s_{22} & \frac{1}{2}(s_{11}^2 + s_{12}^2 + s_{21}^2 + s_{22}^2) \end{pmatrix}.$$

The representation ρ is a two-sheeted representation taking $\mathrm{SL}_2^{\pm}(\mathbb{R})$ onto $\mathrm{O}(2,1)$ which takes $\mathrm{SL}_2^{\pm}(\mathbb{Z})$ into $G_{\mathbb{Z}}$. We now show $\mathrm{SL}_2(\mathbb{Z})$ actually maps onto $G_{\mathbb{Z}}$.

This proves the following result.

Theorem 12.10.3. *H acts transitively on $\mathcal{P}$.*

Since H is a quotient group of $\mathrm{SL}^{\pm}(2,\mathbb{Z})$ this group also acts transitively on $\mathcal{P}$.

Proof. Because of the equivariance and Lemma 12.10.2 every element of $\mathcal{P}$ can be carried to either $(0,1,1)$ or $(0,-1,1)$ by an element of H. Moreover, these two triples can be interchanged by $a' \mapsto b$, $b' \mapsto -a$ since this induces the map $x' \mapsto x$, $y' \mapsto -y$, $z' \mapsto z$ and so H acts transitively on $\mathcal{P}$. $\square$

Exercise 12.10.4. Prove the *orthogonality relations* for the group $\mathrm{O}(2,1)$.

That is, for $s \in \mathrm{O}(2,1)$ and $i = 1,2,3$

$$\rho_{i,1}^2(s) + \rho_{i,2}^2(s) - \rho_{i,3}^2(s) = 1,$$

and when $i \neq j$,

$$\rho_{i,1}(s)\rho_{j,1}(s) + \rho_{i,2}(s)\rho_{j,2}(s) - \rho_{i,3}(s)\rho_{j,3}(s) = 0.$$

Exercise 12.10.5. For any $(x,y,z) \in \mathcal{P}$, show that $\mathrm{Stab}_H(x,y,z)$ is isomorphic with $\mathbb{Z}_2 \times \mathbb{Z}$. Suggestion: It is sufficient to do it for $(3,4,5)$.

12.10.1   Exercises

Exercise 12.10.6. Let G be a finite group and k be a field of characteristic 0. Is it still true that $|G|$ is the sum of the squares of the inequivalent finite dimensional irreducible representations?

Exercise 12.10.7. Prove that:

1. If ζ is a complex root of unity, then $\zeta^{-1} = \bar{\zeta}$.

2. If G is a finite group and M is a finite dimensional $\mathbb{C}G$-module with character χ_M, then $\chi_M(g^{-1}) = \overline{\chi_M(g)}$. (Hint: Every eigenvalue of $\rho(g)$ is a root of unity.)

3. If M is a finite dimensional $\mathbb{C}G$-module, its dual M^* is the dual vector space with action of $g \in G$ defined by $(gf)(v) = f(g^{-1}v)$ for $f \in M^*$, $v \in M$. Prove that M^* is irreducible if and only if M is irreducible.

4. Show that the character χ_{M^*} of M^* is related to the character χ_M of M by $\chi_{M^*}(g) = \overline{\chi_M(g)}$. Deduce that $M \cong M^*$ if and only if $\chi_M(g)$ is a real number for every $g \in G$.

5. Show that $g \in G$ is conjugate to g^{-1} if and only if $\chi(g)$ is a real number for every character χ.

Exercise 12.10.8. Let n be a positive integer and let ζ, ζ_1 be primitive n-th roots of unity.

1. Show that $(1 - \zeta)/(1 - \zeta_1)$ is an algebraic integer.

2. If $n > 6$ is divisible by at least two primes, show that $1 - \zeta$ is a unit in the ring $\mathbb{Z}[\zeta]$.

The following two exercises concerning representations of Lie algebras (or Lie groups).

Exercise 12.10.9. Let $V \neq 0$ be a finite-dimensional vector space over the complex numbers $\mathbb{C}$ and X, Y, and Z be linear operators on V which satisfy

$$XY - YX = Z \quad XZ = ZX \quad YZ = ZY.$$

If V has no proper subspace invariant under all three operators, prove that $\dim_{\mathbb{C}} V = 1$. In other words, a finite dimensional complex representation of a Heisenberg Lie algebra must be 1-dimensional.

Exercise 12.10.10. Let F be an algebraically closed field of prime characteristic p and let V be an F-vector space of dimension precisely *p.* Suppose A and B are linear operators on V such that $AB - BA = B$. If B is non-singular, prove that V has a basis $\{v_1, v_2, \ldots, v_p\}$ of eigenvectors of A such that $Bv_i = v_{i+1}$ for $1 \leq i \leq p - 1$ and $Bv_p = \lambda v_1$ for some $0 \neq \lambda \in F$. This is a result about finite dimensional representations of the $Ax + B$ Lie algebra.

Exercise 12.10.11. Let G be a finite group which has the property that for any element $g \in G$ of order n, and an integer r prime to n, the elements g and g^r lie in the same conjugacy class. Then, show that the character of every representation of G takes values in $\mathbb{Q}$ (in fact even the integers $\mathbb{Z}$). (Hint: Use Galois theory.)

Chapter 13

Representations of Associative Algebras

In this final chapter we study the salient facts concerning *finite dimensional* mostly non-commutative, associative algebras, $\mathcal{A}$, with identity, over a field k. These play an important role in non-commutative ring theory, in representation theory, in non commutative geometry, as well as in quantum field theory. Our study will be by means of representation theory. The definition of a representation of an associative algebra being quite analogous to that of a group. Namely, an algebra homomorphism with values in the algebra of matrices, all over a field k. One big advantage to restricting ourselves to finite dimensional associative algebras is that we avoid the necessity of having to deal with the chain conditions.

Definition 13.0.1. A *representation of a finite dimensional associative algebra* $\mathcal{A}$ over k is an algebra homomorphism $\rho : \mathcal{A} \to \mathrm{End}_k(V)$. We say $\mathcal{A}$ or ρ acts on V, or V is the *representation space* of ρ. The $\dim_k V$ is called $\deg(\rho)$. A subspace W of V is called ρ *invariant* if W is $\rho(\mathcal{A})$-stable and ρ is called *irreducible* if it has no non-trivial invariant subspaces. ρ is *completely reducible* if every invariant subspace of V has a complementary invariant subspace; alternatively, if V is a direct sum of a finite number of irreducible invariant subspaces. Ideals in $\mathcal{A}$ will always mean two sided k-ideals.

One sees easily that $\mathrm{Ker}(\rho)$ is a two-sided ideal in $\mathcal{A}$, that $\rho(\mathcal{A})$ is a subalgebra of $\mathrm{End}_k(V)$ and that the first isomorphism theorem holds, namely that,

$$\mathcal{A}/\mathrm{Ker}(\rho) = \rho(\mathcal{A}).$$

The novel feature here is that since we know from Volume I Chapter 5 that $\mathrm{End}_k(V)$ is a simple k-algebra, any representation of it is either identically zero or is injective.

Example 13.0.2. For a finite dimensional associative algebra $\mathcal{A}$, its *left regular representation* λ is defined as follows. For $x \in \mathcal{A}$ we let x act on each $a \in \mathcal{A}$ by $\lambda(x)(a) = xa$. Then for each $x \in \mathcal{A}$, $\lambda(x) \in \mathrm{End}_k(\mathcal{A})$ and

$$\lambda : \mathcal{A} \to \mathrm{End}_k(\mathcal{A}).$$

One sees easily that λ is a k-algebra homomorphism. The multiplicative part $\lambda(ab) = \lambda(a)\lambda(b)$ being due to the associative law in $\mathcal{A}$. Evidently, $\mathrm{Ker}\,\lambda = \{x \in \mathcal{A} : x\mathcal{A} = (0)\}$ and since $\mathcal{A}$ has an identity λ is faithful. There are also the right regular representation gotten by right multiplication by the inverse which has similar properties and, because of *associativity*, the two-sided regular representation given by $a \mapsto xay^{-1}$ for $a \in \mathcal{A}$. Restricting to the diagonal, i.e. taking $x = y$ gives the conjugation representation. As we saw at the end of Chapter 12, when $k = \mathbb{R}$, analyzing the conjugation representation of the Quaternions enabled us to prove $\mathrm{SO}(4,\mathbb{R})/\pm I$ is isomorphic to $\mathrm{SO}(3,\mathbb{R}) \times \mathrm{SO}(3,\mathbb{R})$.

We ask the reader to check that the two-sided regular representation is a representation.

Example 13.0.3. If $\mathcal{A}$ is a division algebra over k, then as we know λ is faithful. Moreover, it is also irreducible since if W is a non-zero invariant subspace of $\mathcal{A}$, and $w_0 \neq 0$ is in W, then $w_0^{-1} \cdot w_0 = 1 \in W$. Hence $a \cdot 1 = a \in W$ for all a. So $W = \mathcal{A}$.

13.1   The Double Commutant Theorem

Let V be a finite dimensional vector space and k be a field. For a subset $\mathcal{S}$ of $\text{End}_k(V)$ denote its centralizer $\{X \in \text{End}_k(V) : XS = SX, S \in \mathcal{S}\}$ by $\mathfrak{z}(\mathcal{S})$. Clearly, $\mathfrak{z}(\mathcal{S})$ is a subalgebra of $\text{End}_k(V)$. Now, let T be a linear operator on V. Then $\mathfrak{z}(T)$ contains $k[T]$, the set of all polynomials in T. It is also clear that $\mathfrak{z}(\mathfrak{z}(T))$ contains T and since it is a subalgebra it also contains $k[T]$. It is our purpose to prove that these sets are actually equal. This is called the *double commutant theorem*.

Theorem 13.1.1. *For any $T \in \text{End}_k(V)$, $\mathfrak{z}(\mathfrak{z}(T)) = k[T]$.*

Corollary 13.1.2. *Let $k = \mathbb{R}$ or $\mathbb{C}$ and $f(x) = \sum_{n=0}^{\infty} a_n x^n$ be an everywhere convergent power series, then $f(T) \in k[T]$.*

This is because if U commutes with T, then it also commutes with $f(T)$.

A second consequence of the theorem is that we recover Proposition 5.9.7 of Volume I.

Corollary 13.1.3. $\mathcal{Z}(\text{End}_k(V)) = \{\lambda I : \lambda \in k\}$.

This follows immediately by taking $T = I$.

To prove the theorem we consider a finitely generated module M over a Euclidean ring D. M is the direct sum of cyclic submodules $M_{z_1} \oplus \ldots \oplus M_{z_s}$. Let $D' = \text{Hom}_D(M, M)$ and $\mathcal{Z}(D')$ be its center.

Lemma 13.1.4. *D' is a ring and contains the maps ϕ_a, where ϕ_a, takes $x \mapsto ax$, for $x \in M$ and $a \in D$. Moreover, each ϕ_a is in $\mathcal{Z}(D')$.*

Proof. If $\phi \in D'$ this means that ϕ is an additive map $M \to M$ and $\phi(bx) = b\phi(x)$ for all $b \in D$ and $x \in M$. If ψ is another such map, then so is $\phi\psi$. Also, $\phi\phi_a(x) = \phi(ax) = a\phi(x) = \phi_a(\phi(x))$ for all x. Hence $\phi(\phi_a) = \phi_a\phi$. $\square$

Our theorem depends on the following proposition.

Proposition 13.1.5. $\mathcal{Z}(D') = \{\phi_a : a \in D\}$.

Proof. For each $i = 1, \ldots, s$ let $\gamma_{i_s}(z_s) = z_i$ and $\gamma_{i_s}(z_j) = 0$ for $j \neq s$. Then for $m \in M$, m can be written uniquely as $\sum a_j(m)z_j$, where $a_j(m) \in D$. Define $\gamma_{i_s}(m) = \sum_j a_j(m)\gamma_{i_s}(z_j) = a_s(m)z_i$. Then each γ_{i_s} is an endomorphism of M.

If $\phi \in \mathcal{Z}(D')$, then

$$\phi(z_s) = \phi(\gamma_{ss}(z_s)) = \gamma_{ss}(\phi(z_s)) = \gamma_{ss}\left(\sum a_j z_j\right)$$

for suitable $a_j \in D$. The latter is $\sum a_i \gamma_{ss}(z_i) = a_s z_s$. Also, $\phi(z_i) = \phi(\gamma_{i_s}(z_s)) = \gamma_{i_s}(\phi(z_s)) = \gamma_{i_s}(\sum a_j z_j) = \sum a_j \gamma_{i_s}(z_j) = a_s z_i$. Since the z_i generate M this means that ϕ is multiplication by a_s.   □

The double commutant theorem now follows from the Proposition by taking $M = V$ and $D = k[T]$. For then

$$D' = \{S \in \mathrm{End}_k(V) : ST = TS\} = \mathfrak{z}(\mathfrak{T}),$$

while

$$\mathcal{Z}(D') = \{S \in \mathrm{End}_k(V) : ST = TS, SU = US\},$$

for all U such that $UT = TU$. Since T is itself such a U, the first condition is redundant. This means $\mathcal{Z}(D') = \mathfrak{z}(\mathfrak{z}(\mathfrak{T}))$. By the Proposition, $\mathfrak{z}(\mathfrak{z}(\mathfrak{T}))$ consists only of ϕ_a, where $a \in D$. That is, each $S \in \mathfrak{z}(\mathfrak{z}(\mathfrak{T})) = \phi_a$, where $a = p(T)$. Thus for $v \in V$, $S(v) = p(T)(v)$ and so $S = p(T)$.

13.2   Burnside's Theorem

A basic result in the representation theory of associative algebras is Burnside's theorem which states that if ρ is an *irreducible* representation of an associative algebra $\mathcal{A}$ on V over an *algebraically closed* field K of characteristic zero and V is a finite dimensional vector space over K, then ρ is surjective, i.e. $\rho(\mathcal{A}) = \mathrm{End}_K(V)$.

Of course, we must exclude the cases where $\dim_K V = 1$ and $\mathcal{A} = (0)$ by assuming $\deg(\rho) \geq 2$. Under the same assumptions on V and K this immediately reduces to the following:

Corollary 13.2.1. *When K is algebraically closed of characteristic zero and $\dim(V) \geq 2$, a non-trivial subalgebra $\mathcal{A}$ of $\mathrm{End}_K(V)$ acting irreducibly on V must be all of $\mathrm{End}_K(V)$.*

Hence another way to state the conclusion of Burnside's theorem is $\dim \mathcal{A} = (\dim V)^2$. From this we see that when K is not algebraically closed the result is false. For consider $M_2(\mathbb{R})$ and in it the group G of order 3 generated by the rotation of 120°. Then G, and therefore also the group algebra $\mathbb{R}(G)$, acts irreducibly on $\mathbb{R}^2$. But the group algebra has dimension 3 since this is the order of G.

A more elaborate counter-example when the field is $\mathbb{R}$ is the following.

Let $\mathcal{A}$ be $\mathbb{H}$, the quaternion algebra (see Chapter 5 of Volume I). Then $\mathcal{A}$ is a real algebra of dimension 4. Letting it act on itself by left translation since $\mathbb{H}$ is a division algebra we get a faithful representation of $\mathcal{A}$ as a subalgebra of $M_4(\mathbb{R}) = \mathrm{End}_\mathcal{A}$ of dimension 4. Moreover, because $\mathbb{H}$ is a division algebra, it acts irreducibly on itself. But for Burnside's theorem to hold over $\mathbb{R}$ the degree would have to be 16.

To prove Burnside's theorem we need the following proposition and the lemma that follows.

Proposition 13.2.2. *Let $\mathcal{W}$ be an irreducible space under the left regular representation of $\mathcal{A}$. Then the K-linear span of $\{AB : A \in \mathcal{A}, B \in \mathcal{W}\}$ is all of $\mathcal{W}$.*

Proof. Let $\mathcal{U}$ be the subspace of $\mathcal{W}$ spanned by $\{AB : A \in \mathcal{A}, B \in \mathcal{W}\}$. Let $A' \in \mathcal{A}$. Then $A'(AB) = (A'A)B$ and since $A'A \in \mathcal{A}$, by definition each $A' \in \mathcal{A}$ leaves $\mathcal{U}$ stable. Since $\mathcal{U}$ is a non-zero $\mathcal{A}$ stable subspace of $\mathcal{W}$ and the latter is irreducible, $\mathcal{U} = \mathcal{W}$. $\square$

Lemma 13.2.3. *For $X, Y \in M(n, k) = \mathrm{End}_k(V)$, the bilinear map, $(X, Y) \mapsto \mathrm{tr}(XY)$, is non-degenerate. Here the field need not be algebraically closed.*

Proof. Since $\mathrm{tr}(XY)$ is a symmetric function it suffices to prove that if $\mathrm{tr}(XY) = 0$ for every Y, then $X = 0$. So suppose $\mathrm{tr}(XY) = 0$ for every Y. Let V be an n-dimensional k space on which these maps operate.

If $X \neq 0$, there would be a $v_1 \neq 0 \in V$, with $X(v_1) \neq 0$. Extend $X(v_1)$ to a basis $\{X(v_1), v_2, ..., v_n\}$ of V (in any way) and define a linear function $Y : V \to V$ by $Y(X(v_1)) = v_1$ and $Y(v_i) = 0$ for $i \geq 2$. Then $(XY)(X(v_1)) = X(v_1)$ and $(XY)(v_i) = 0$ for all $i \geq 2$. Thus with respect to this basis the matrix (XY) has all rows zero except for the first which has a 1 in the upper left corner and zeros elsewhere. For this Y, $\text{tr}(XY) = 1$, a contradiction. $\qquad\square$

We are now in a position to prove Burnside's theorem.

Proof. Let $\mathcal{B}$ be the orthocomplement of $\mathcal{A}$ in $\text{End}_K(V)$ with respect to the form $(A, B) \mapsto \text{tr}(AB)$ on $\text{End}_K(V)$. That is, $\mathcal{B}$ is the subspace defined by

$$\mathcal{C} = \{B \in \text{End}_K(V) : \text{tr}(AB) = 0 \text{ for all } A \in \mathcal{A}\}.$$

If we show $\mathcal{B} = (0)$, then since this form is non-degenerate, $\mathcal{A} = \text{End}_K(V)$.

Now $\mathcal{B}$ is $\mathcal{A}$-invariant under left translation. That is, if B is perpendicular to $\mathcal{A}$ and $A \in \mathcal{A}$, then AB is also perpendicular to $\mathcal{A}$. This is because if $A' \in \mathcal{A}$, then $\text{tr}(A(A'B)) = \text{tr}((AA')B) = 0$ for all $A \in \mathcal{A}$ since $AA' \in \mathcal{A}$. Let λ be the left regular representation of $\mathcal{A}$ on the vector space $\mathcal{C}$ given by $\lambda_A(B) = AB$. Then, $\lambda : \mathcal{A} \to \text{End}_k(\mathcal{B})$. We leave to the reader the easy verification that λ is a (finite dimensional) associative algebra representation of $\mathcal{A}$ on $\mathcal{B}$.

Suppose $\mathcal{C} \neq (0)$. By finite dimensionality, $\mathcal{B}$ must have a non-trivial $\mathcal{A}$-invariant *irreducible* subspace $\mathcal{W}$ and we consider the left regular representation λ of $\mathcal{A}$, but now on $\mathcal{W}$. Here

$$\mathcal{W} \subseteq \mathcal{B} \subseteq \mathcal{A} \subseteq \text{End}_K(V).$$

For each $v \in V$ we define a map $T_v : \mathcal{W} \to V$ by $T_v(B) = B(v)$. Evidently, T_v is a linear map. Moreover, each T_v is an intertwining operator for the left regular representation λ. There is the given action of $\mathcal{A}$ on V on the right. For if $A \in \mathcal{A}$ and $B \in \mathcal{W}$, since $B \in \mathcal{B}$, $T_v(AB) = AB(v) = A(B(v)) = A(T_v(B))$. Thus for each $B \in \mathcal{W}$, $T_v(\lambda_A(B)) = A(T_v(B))$. Thus we have a commutative diagram

$$\begin{array}{ccc} \mathcal{W} & \xrightarrow{T_v} & V \\[2pt] \downarrow{\scriptstyle\lambda_A} & & \downarrow{\scriptstyle A} \\[2pt] \mathcal{W} & \xrightarrow{T_v} & V \end{array}$$

making each T_v an intertwining operator.

Since in its natural action $\mathcal{A}$ acts irreducibly on V and by our assumption $\lambda(\mathcal{A})$ acts irreducibly on $\mathcal{W}$ and we are over an algebraically closed field, Schur's Lemma 6.9.1 of Volume I tells us each T_v is a scalar multiple of the identity. That is, $T_v = c_v I$, where $c_v \in K$. Now if all $T_v = 0$, that is $B(v) = 0$ for all $v \in V$ and $B \in \mathcal{W}$, this means $\mathcal{W} = 0$, a contradiction. Thus for some $v_1 \in V$, $c(v_1) \neq 0$ and so T_{v_1} is invertible. In particular, T_{v_1} is a vector space isomorphism from $\mathcal{W}$ onto V. Now evidently $T_{v_1}^{-1}T_v$ is also an intertwining operator for each $v \in V$, so that for all $v \in V$, $T_{v_1}^{-1}T_v = d(v)I$. In particular, taking $v = v_1$, we see $d(v_1) = 1$.

Extend v_1 to a basis $v_1, \ldots v_n$ of V and let $v_1^*, \ldots v_n^*$ be the corresponding dual basis of V^*. Then $AB(v_j) = \sum_{i=1}^n c_{i,j}(A,B)v_i$, for $c_{i,j}(A,B) \in K$. Hence, $v_j^*(AB(v_j)) = c_{j,j}(A,B)$ and so

$$0 = \operatorname{tr}(AB) = \sum_{j=1}^n v_j^*(AB(v_j)) = \sum_{j=1}^n v_j^*(AT_{v_j}B).$$

Since the T_v are intertwining operators and also $T_{v_j} = d_{v_j}T_{v_1}$, this last term is

$$\sum_{j=1}^n v_j^*(T_{v_j}AB) = \sum_{j=1}^n v_j^* d_{v_j} T_{v_1} AB = \sum_{j=1}^n d_{v_j} v_j^*(T_{v_1}AB).$$

Thus, for all $A \in \mathcal{A}$ and $B \in \mathcal{W}$,

$$\sum_{j=1}^n d_{v_j} v_j^*(T_{v_1}AB) = 0. \qquad (\star)$$

However, because $\mathcal{W}$ is an irreducible space under the left regular representation of $\mathcal{A}$ on $\mathcal{W}$, Proposition 13.2.2 tells us the K-linear span of $\{AB : A \in \mathcal{A},\ B \in \mathcal{W}\}$ is all of $\mathcal{W}$. Because T_{v_1} is a bijective linear map $\mathcal{W} \to V$, $\{T_{v_1}AB : A \in \mathcal{A}, B \in \mathcal{W}\}$ is all of V. Hence, by equation $(\star)$, $\sum_{j=1}^{n} d(v_j)v_j^* = 0$ and since the v_j^* are linearly independent the $d(v_j)$ must all be zero. This contradicts the fact that $d(v_1) = 1$. $\quad\square$

Burnside's theorem has the following group theoretic corollary, called the *group theoretic* version of Engel's theorem. Notice that there are no other assumptions on G such as *connectedness* when $K = \mathbb{C}$. Also notice that, in the end Corollary 13.2.5, K need not be algebraically closed.

Corollary 13.2.4. *Let G be a subgroup of $\mathrm{GL}(V)$, where V is a finite dimensional vector space over an algebraically closed field K of characteristic zero consisting of* unipotent *elements. Then there is a non-zero $v \in V$ such that $gv = v$ for all $g \in G$.*

Proof. We consider the $K(G)$-module V. Since we are seeking a $v \neq 0 \in V$ fixed by all of G, it is sufficient to find v in a submodule. Let W be a non-trivial, irreducible, G invariant subspace. Of course the restriction of a unipotent operator to this subspace remains unipotent. By Burnside's theorem $K(G) = \mathrm{End}_K(W)$; that is the K-linear span of G is all of $\mathrm{End}_K(W)$. On the other hand, each $g \in G$ is of the form $g = I + N$, where N is nilpotent (here everything is restricted to W). Hence $\mathrm{tr}_W(g) = \mathrm{tr}_W(I) + \mathrm{tr}_W(N)$. Since N is a nilpotent operator on W its trace is zero and since $\mathrm{tr}_W(I) = \dim W$, we see $\mathrm{tr}_W(g) = \dim W$ for all $g \in G$. Now let g be fixed and $h \in G$. Then $\mathrm{tr}_W(Nh) = \mathrm{tr}_W((g-I)h) = \mathrm{tr}_W(gh) - \mathrm{tr}_W(h)$ and since both gh and h are in G it follows that this is zero. Hence $\mathrm{tr}_W(Nh) = 0$ for all $h \in G$. Because the h span $\mathrm{End}_K(W)$ over K and tr_W is K-linear, this means $\mathrm{tr}_W(NT) = 0$ for all operators T on W. But tr_W is non-degenerate (an easy exercise). Hence $N = 0$ and $g = I$. Thus each $g \in G$ is the identity on the non-zero subspace W. $\quad\square$

We remark that this corollary actually holds over any field k of characteristic zero simply by embedding k in its algebraic closure, K. Then unipotent elements of $\mathrm{GL}(V_K)$ are also unipotent in $\mathrm{GL}(V)$ and since $gv = v$, v is an eigenvector of g with eigenvalue 1. But since

$1 \in k$, we can find $v_K \in V_K$ so that $gv_K = v_K$. Moreover, by induction on the dimension of V_K we conclude G can be simultaneously put in unitriangular form. Hence,

Corollary 13.2.5. *Let G be a subgroup of* $\mathrm{GL}(V)$, *where V is a finite dimensional vector space over a field of characteristic zero consisting of* unipotent *elements. Then G can be put in simultaneous unitriangular form.*

13.3 Semi-simple Algebras and Wedderburn's Theorem

Here k denotes a field of characteristic zero and $\mathcal{A}$ a finite dimensional associative algebra over k.

Definition 13.3.1. An element $x \in \mathcal{A}$ is called *nilpotent* if $x^n = 0$ for some positive integer n. If every element of $\mathcal{A}$ is nilpotent we say $\mathcal{A}$ is nilpotent.

An example of a nilpotent associative algebra is the full strictly triangular algebra of matrices of order n over k. It has dimension $\frac{n(n-1)}{2}$. Notice that here the product of any n elements is always 0.

In fact, this situation is quite general.

Proposition 13.3.2. *Given a nilpotent associative algebra $\mathcal{A}$, there exists an integer n so that the product of any n elements of $\mathcal{A}$ is 0.*

Proof. Since λ is an associative algebra representation, by Lemma 13.3.3 $\lambda(\mathcal{A})$ is a nilpotent associative algebra of linear transformations. Hence there is some n so that $\lambda(x_1) \ldots \lambda(x_n) = \lambda(x_1 \ldots x_n) = 0$ for all $x_i \in \mathcal{A}$. Therefore, $x_1 \ldots x_n \in \mathrm{Ker}\,\lambda$ and so $(x_1 \ldots x_n)a = 0$ for all a and $x_i \in \mathcal{A}$. $\qquad\square$

The following lemma is one of the things that makes finite dimensional associative algebras easier than the related situation in Lie algebras. (The analogue of the Lemma and many of the succeeding statements are false in Lie algebras.) Concomitantly, the analogue of the definition just above would not be correct for Lie algebras. Another reason for the increased simplification here would seem to be that for a linear associative algebra over $\mathbb{R}$, the Lie bracket is $[A, B] = AB - BA$, where the associative algebra operation is AB. So some information is lost when passing from the associative algebra to the underlying Lie algebra.

Lemma 13.3.3. *Every subalgebra and every quotient algebra of a nilpotent associative algebra is nilpotent. If $\mathcal{I}$ is a nilpotent ideal in $\mathcal{A}$ and $\mathcal{A}/\mathcal{I}$ is nilpotent, then $\mathcal{A}$ itself is nilpotent.*

Proof. The first statement is certainly correct. As to the second, let $x \in \mathcal{A}$. Since $\mathcal{A}/\mathcal{I}$ is nilpotent, $(x + \mathcal{I})^n = x^n + \mathcal{I} = \mathcal{I}$ so $x^n \in \mathcal{I}$. Therefore because $\mathcal{I}$ is also nilpotent, $(x^n)^m = x^{mn} = 0$ for some m. $\square$

In order to define the radical of an associative algebra we need the following proposition.

Proposition 13.3.4. *If B and C are nilpotent ideals in the associative algebra $\mathcal{A}$ so is $B + C$.*

Proof. Clearly $B + C$ is an ideal in A. Since B is a nilpotent algebra so is $B/B \cap C = B + C/C$. Hence $B + C/C$ is nilpotent. Moreover, because C is a nilpotent ideal in $B + C$, it follows by 13.3.3 that $B + C$ itself is a nilpotent ideal in $\mathcal{A}$. $\square$

Since 0 is a nilpotent ideal in $\mathcal{A}$, the above proposition shows there is a *largest nilpotent ideal* in $\mathcal{A}$ which is called the *radical* of $\mathcal{A}$ and denote it by $\mathrm{Rad}(\mathcal{A})$. Actually, we have encountered this earlier and in greater generality for rings in Chapter 9, but we repeat ourselves here since we are in the category of finite dimensional associative k-algebras. The reader who is only interested in this will find the arguments here, may we say, refreshingly simpler than those in the more general context of Chapter 9.

Definition 13.3.5. An associative algebra will be called *semi-simple* if its radical is zero; alternatively, if it has no non-trivial nilpotent ideals.

Moreover, $\mathcal{A}$ is simple if it has no non-trivial ideals at all. Since the radical is an ideal, every simple algebra is semi-simple. In particular, $M(n, k) = \text{End}_k(V)$ is semi-simple.

The following exercise deals with these notions under field extensions.

Exercise 13.3.6. Show that if $\mathcal{A}$ is a simple k algebra and K is a field containing k, then as an algebra over K it is also simple.

Let $\mathcal{A}$ be an associative algebra. Using the left regular representation, λ, we now define a *bilinear form*, $\langle , \rangle$, on $\mathcal{A}$ by,

$$\langle x, y \rangle = \text{tr}(\lambda(x)\lambda(y)).$$

Notice that this form is indeed bilinear. It is also symmetric and is intrinsic to the associative algebra $\mathcal{A}$.

The following is a very useful criterion for an operator to be nilpotent.

Lemma 13.3.7. *Let V be a finite dimensional vector space over a field k of characteristic zero and T be a linear map of V. Then T is nilpotent if and only if $\text{tr}^i(V) = 0$ for all $i = 1, \ldots, n = \dim(V)$.*

Proof. By passing to the algebraic closure of k we may assume k is algebraically closed in both directions. Put T in triangular form. If T is nilpotent, then all the eigenvalues are zero while the conditions on the trace are

$$\lambda_1 + \lambda_2 + \cdots + \lambda_n = 0$$
$$\lambda_1^2 + \lambda_2^2 + \cdots + \lambda_n^2 = 0$$
$$\dotfill$$
$$\lambda_1^n + \lambda_2^n + \cdots + \lambda_n^n = 0$$

and are evidently satisfied.

Conversely, assume these conditions are satisfied. Our task is to show that the only simultaneous solution of this system of equations is

$\lambda_i = 0$ for all i. This we will do by induction on n. By the Cayley[1]-Hamilton[2] Theorem 6.5.11 of Volume I

$$T^n - \text{tr}(T)^{n-1} + \sigma_2(T)T^{n-2} - \ldots + (-1)^n \det(T)I = 0.$$

Taking the trace of both sides yields

$$\text{tr}(T^n) - \text{tr}(T)\text{tr}(T^{n-1}) + \sigma_2(T)\text{tr}(T^{n-2}) - \ldots + (-1)^n n\det(T) = 0.$$

Hence $\det(T)n = 0$ and since k has characteristic zero, $\det(T)$ itself is zero. Thus one of the eigenvalues, say $\lambda_n = 0$. This means after discarding the last equation, the system of equations reduces to the system involving $n - 1$ unknowns, which by inductive hypothesis has only the trivial solution.   □

Proposition 13.3.8.

1. *$\langle xy, z \rangle = \langle x, yz \rangle$.*

2. *If $\mathcal{I}$ is an ideal in $\mathcal{A}$, then so is $\mathcal{I}^\perp$.*

3. *If $x \in \mathcal{A}$ and for all positive integers n, $\langle x, x^n \rangle = 0$, then x is nilpotent.*

Proof.

1. The statement follows from the associative law.

2. For statement 2, let $x^\perp \in \mathcal{I}^\perp$ and $x \in \mathcal{A}$. We show $xx^\perp \in \mathcal{I}^\perp$. To do so let $y \in \mathcal{I}$. Then by 1 $\langle xx^\perp, y \rangle = \langle x^\perp, xy \rangle$ and since $\mathcal{I}$ is an ideal and $y \in \mathcal{I}$ so is xy. But then, since $x^\perp \in \mathcal{I}^\perp$, this is zero. Thus $xx^\perp \in \mathcal{I}^\perp$ so $I^\perp$ is a left ideal. A similar argument shows it is also a right ideal.

---

[1]Arthur Cayley (1821-1895) British mathematician, FRS and Professor at Cambridge University is well known for Cayley's theorem (embedding a group in its permutations), the Cayley multiplication table for groups, the Cayley graph of a group, the Cayley numbers (octonians), the Cayley transform and the Cayley-Hamilton theorem.

[2]William Hamilton (1805-1865) Irish mathematician is well known for the "Hamiltonian" in mechanics and quantum mechanics, the Cayley-Hamilton theorem and the invention of the quaternions.

3. Suppose $\text{tr}(\lambda(x)\lambda(x^n)) = 0$ for all n. Then $\text{tr}(\lambda(x)^{n+1}) = 0$ for all n. By Lemma 13.3.7 $\lambda(x)$ is nilpotent. That is, $\lambda(x)^j = \lambda(x^j) = 0$. But then, $x^j x = x^{j+1} = 0$.

$\square$

Corollary 13.3.9. *The associative algebra $\mathcal{A}$ is semi-simple if and only if the form $\langle,\rangle$ is non-degenerate.*

Proof. Suppose $\langle\,,\,\rangle$ is non-degenerate and $\mathcal{I}$ is a nilpotent ideal. Let $x \in \mathcal{I}$ and $a \in \mathcal{A}$. Then $ax \in \mathcal{I}$ and so ax is nilpotent. Hence $\text{tr}(ax) = 0$. Therefore since x is arbitrary in $\mathcal{I}$, $a \in \mathcal{I}^{\perp}$. Thus, $\mathcal{A} \subseteq \mathcal{I}^{\perp}$ and hence $\mathcal{A} = \mathcal{I}^{\perp}$. But because $\langle\,,\,\rangle$ is non-degenerate $\mathcal{I} = (0)$. In particular, $\text{Rad}(\mathcal{A}) = (0)$.

Now suppose $\mathcal{A}$ is semi-simple. Let $x \in \mathcal{A}$. We want to show that if $\text{tr}(\lambda(x)\lambda(y)) = 0$ for all $y \in \mathcal{A}$, then $x = 0$. Take $y = x^n$. Then, as we saw, x is nilpotent. Thus $\mathcal{A}^{\perp}$ consists of nilpotent elements. It is also an ideal since $\mathcal{A}$ is. But $\mathcal{A}$ is semi-simple so it has no such ideals except (0). Hence $\mathcal{A}^{\perp} = (0)$ and the form is non-degenerate. $\square$

Corollary 13.3.10. *The group algebra $\mathcal{A} = k(G)$ of a finite group over a field of characteristic zero is semi-simple.*

Proof. We show the form $\langle x, y \rangle = \text{tr}(\lambda(x)\lambda(y))$ is non-degenerate. Notice that the left regular representation of G on $\mathcal{A}$ extends to λ and so the form is completely determined by G as a basis. Let $g \in G$ and L_g be the linear transformation on $k(G)$ gotten by left translation by g. Then L_g is a permutation matrix which is non-trivial unless $g = 1$. Since any non-trivial permutation is a finite product of independent cycles, the corresponding matrix is block diagonal with blocks of the form (in the 4×4 case)

$$\begin{pmatrix} 0 & 1 & 0 & 0 \\ 0 & 0 & 1 & 0 \\ 0 & 0 & 0 & 1 \\ 1 & 0 & 0 & 0 \end{pmatrix}.$$

In particular, its trace is zero. As a sum of block diagonal matrices of trace zero, $\text{tr}(L_g) = 0$. (Since the trace is independent of basis it does not matter in what order we take the cycles.) On the other hand if $g = 1$, then $L_g = I$ and so here $\text{tr}(L_g) = n$. Hence for $g, h \in G$, $\langle g, h \rangle = \text{tr}(L_g L_h) = \text{tr}(L_{gh})$ which is 0 if $g \neq h$ and n if $g = h$. Hence the matrix of our form calculated with respect to the basis G is nI. Since $\text{char}(k) = 0$ and $n \neq 0$ the matrix is invertible and the form is non-degenerate. □

Exercise 13.3.11. Let G be a finite group and K be an algebraically closed field of $\text{char}(K) = 0$. What is the decomposition of $\mathcal{A} = k(G)$ into the direct sum of simple algebras in this case? What are the multiplicities n_i? (Notice that since the left regular representation of G on $\mathcal{A}$ extends to and determines λ, we have the same invariant subspaces and intertwining operators for L as for λ.) Here the reader should refer to Chapter 12 on Representations of Finite Groups.

Corollary 13.3.12. *If an associative algebra $\mathcal{A}$ over k is semi-simple and $K \supset k$ is an extension field, then $\mathcal{A}_K$ remains semi-simple.*

Proof. Consider the symmetric matrix that represents the form $\langle , \rangle$ for the algebra $\mathcal{A}$ with respect to a basis of $\mathcal{A}$. Since the form is non-degenerate this matrix is non-singular. But with respect to $\mathcal{A}_K$ this is the same matrix, hence also non-singular. Thus the form on $\mathcal{A}_K$ is also non-degenerate and $\mathcal{A}_K$ is semi-simple. □

Corollary 13.3.13. *A semi-simple associative algebra $\mathcal{A}$ over k is isomorphic with the direct sum of a finite number of simple algebras. Also, the two sided ideals of $\mathcal{A}$ are just subsums of these simple algebras. Finally, the direct sum of a finite number of simple algebras is semi-simple.*

Proof. If $\mathcal{A}$ is simple there is nothing to prove. Let $\mathcal{B}$ be a non-trivial ideal and $\mathcal{A}_1$ be a minimal non-trivial ideal. Then $\mathcal{A}_1^\perp$ is a non-trivial ideal and $\mathcal{A} = \mathcal{A}_1 \oplus \mathcal{A}_1^\perp$. Now if $\mathcal{I}$ is an ideal in either of these summands, since it commutes with the other it is actually an ideal in $\mathcal{A}$ itself. In particular, $\mathcal{A}_1^\perp$ is semi-simple since $\mathcal{A}$ was and $\mathcal{A}_1$ is simple. By induction on the lower dimensional $\mathcal{A}_1^\perp$, we reach our conclusion.

Now suppose $\mathcal{I}$ is an ideal in $\mathcal{A}$. Projecting onto the i-th factor (which is an algebra homomorphism) we see that $\pi_i(\mathcal{I})$ is an ideal. Since this factor is simple either $\pi_i(\mathcal{I}) = \mathcal{I}_i = (0)$ or $\mathcal{A}_i$. In the latter case $\mathcal{A}_i = \mathcal{A}_i^2$ because $\mathcal{A}_i$ has an identity. Hence $\mathcal{A}_i^2 = \mathcal{A}_i \mathcal{I}_i$ since $\mathcal{I}_i$ is an ideal. Finally, $\mathcal{A}_i \mathcal{I}_i = \mathcal{A}_i \mathcal{I} \subseteq \mathcal{I}$. Hence $\mathcal{I}$ contains all such $\mathcal{A}_i$ and therefore also their sum. Since it projects onto (0) in all other factors, $\mathcal{I}$ is this sum.   □

13.3.1   The Big Wedderburn Theorem

We now begin the proof of Wedderburn's[3] structure theorem for semi-simple algebras. For this we shall have to assume not only that K is of characteristic zero, but that it is also *algebraically closed*. Since this result depends on Burnside's theorem the same counter-example given there shows that if K were not algebraically closed, i.e. $K = \mathbb{R}$, these results are not valid. For example, the quaternions are a simple algebra over $\mathbb{R}$.

We begin with a special case.

Proposition 13.3.14. *Let $\mathcal{A}$ be a non-trivial simple finite dimensional associative algebra with identity over K. Then $\mathcal{A}$ is isomorphic to $\mathrm{End}_K(V)$ for some finite dimensional V over K, and in particular has dimension n^2.*

Proof. In Chapter 5 Volume I we proved that even if K is not algebraically closed, $\mathrm{End}_K(V)$ is simple. Now consider the restriction ρ of λ to a minimal invariant subspace, V. As an invariant subspace V is a left ideal of $\mathcal{A}$ and ρ is an irreducible representation of $\mathcal{A}$ on V. The dimension of V can not be 1 because $\mathcal{A}$ is simple. g K is algebraically closed Burnside's theorem tells us ρ is surjective, that is, $\rho(\mathcal{A}) = \mathrm{End}_K(V)$.

---

[3]Joseph Wedderburn (1882-1948) was a Scottish mathematician who spent most of his career at Princeton University. He proved the big and little Wedderburn theorems in this book. He also worked in group theory and matrix algebra. He was elected a Fellow of the Royal Society of Edinburgh, edited the Proceedings of the Edinburgh Mathematical Society and then the Annals of Mathematics. He received the MacDougall-Brisbane Gold Medal and Prize from the Royal Society of Edinburgh in 1921, and was elected to the Royal Society of London.

Moreover, $\text{Ker}\,\rho$ is a two sided ideal in $\mathcal{A}$. Since the latter is simple $\text{Ker}\,\rho$ is either (0) or $\mathcal{A}$. But because ρ is onto it can not be $\mathcal{A}$. Thus ρ is faithful and hence gives an associative K-algebra isomorphism of $\mathcal{A}$ with $\text{End}_K(V)$. $\square$

Theorem 13.3.15. *(The Big Wedderburn Theorem) A non-trivial semi-simple associative algebra $\mathcal{A}$ with unit over an algebraically closed field K of characteristic zero is isomorphic to a direct sum of $\text{End}_K(V_i)$. In particular, it has dimension $\sum n_i^2$.*

Proof. By Corollary 13.3.13 $\mathcal{A}$ is a direct sum of simple algebras, and by Proposition 13.3.14 each of these is of the form $\text{End}_K(V_i)$ for some finite dimensional V_i over K. $\square$

Corollary 13.3.16. *A semi-simple associative and commutative algebra $\mathcal{A}$ over an algebraically closed field k is isomorphic to the direct sum of a finite number of copies of K. If it is simple it is just K.*

Since $\text{End}_k(V)$ is simple even if k is not algebraically closed, the next proposition also holds generally.

Proposition 13.3.17. *Every automorphism α of the k-algebra $\text{End}_k(V)$ is inner. That is, there is some $g \in GL(V)$ such that $\alpha(T) = gTg^{-1}$ for all $T \in \text{End}_k(V)$.*

Proof. Let $P_{i,j} = \alpha(E_{i,j})$, where the $E_{i,j}$ are the matrix units. Then as is easily checked, each $P_{i,i} \neq 0$, $P_{i,i}^2 = P_{i,i}$, $P_{i,i}P_{j,j} = \delta_{i,j}P_{j,j}$ and $\sum_{i=1}^{n} P_{i,i} = I$. For each $i = 1 \ldots n$ choose a non-zero $w_i \in P_{i,i}(V)$. If $\sum c_i w_i = 0$, then

$$0 = P_{j,j}(0) = P_{j,j}\left(\sum c_i w_i\right) = \sum P_{j,j}(c_i w_i) = c_j w_j,$$

so that each $c_j = 0$. Thus the w_i are linearly independent and therefore are a basis of V. Hence there is an invertible operator g_0 satisfying $g_0(e_i) = w_i$, where the e_i is the standard basis of V. For $T \in \text{End}_k(V)$ let $\beta(T) = g_0^{-1}\alpha(T)g_0$. Then the reader can easily check that β is also an automorphism of the algebra $\text{End}_k(V)$. Moreover, $\beta(E_{i,i}) = E_{i,i}$

so the effect of replacing α by β is that we can assume α fixes each $E_{i,i}$. Hence $\alpha(E_{i,j}) = \alpha(E_{i,i}E_{i,j}E_{j,j}) = E_{i,i}\alpha(E_{i,j})E_{j,j}$. It follows that $\alpha(E_{i,j})(e_k) = 0$ if $j \neq k$ and $\alpha(E_{i,j})(e_j) \in ke_i$. Hence $\alpha(E_{i,j}) = a_{i,j}E_{i,j}$ for some non-zero scalar $a_{i,j}$. But also $\alpha(E_{i,j}E_{j,k}) = \alpha(E_{i,k})$. Therefore the scalars $a_{i,j}$ must satisfy the relations $a_{i,j}a_{j,k} = a_{i,k}$ and since each all the $a_{i,i} = 1$, $a_{i,j}^{-1} = a_{j,i}$ for all i, j. Letting $b_i = a_{i,1}$ we see $a_{i,j} = a_{i,1}a_{1,i} = b_i b_j^{-1}$. Then $g = \text{diag}(b_1, \ldots b_n)$ is invertible and $gE_{i,j}g^{-1} = b_i b_j^{-1} E_{i,j} = a_{i,j}E_{i,j}$. By the above for all i, j, $g^{-1}E_{i,j}g = E_{i,j}$. Thus, $\alpha(T) = gTg^{-1}$ for all T. □

Finally, we see that simple k-algebras over an algebraically closed field have only a single irreducible representation.

Corollary 13.3.18. *If K is algebraically closed, then up to equivalence the only irreducible representation of $\text{End}_k(V)$ is the identity representation.*

Proof. Suppose ρ is such an irreducible representation of $\text{End}_K(V)$ on say W. Then $\rho(\text{End}_K(V))$ is a subalgebra of $\text{End}_K(W)$. Since ρ is irreducible and an algebra representation, Burnside's theorem tells us $\rho(\text{End}_K(V)) = \text{End}_K(W)$. On the other hand, ρ is non-trivial and $\text{End}_K(V)$ is simple. Hence ρ is an isomorphism. Therefore $\text{End}_K(V))$ is isomorphic to $\text{End}_K(W)$ as K algebras. In particular, $\dim_K V^2 = \dim_K W^2$ so that $\dim_K V = \dim_K W$. Let T_0 be a linear isomorphism $T : V \to V$ and for $T \in \text{End}_K(V)$ let $\alpha(T) = T_0^{-1}\rho(T)T_0$. Then a direct calculation shows α is an automorphism of $\text{End}_K(V)$. By Proposition 13.3.17 we know there is a $g \in \text{GL}(V)$ which implements α. Thus $T_0^{-1}\rho(T)T_0 = gTg^{-1}$ for all T. Let $S_0 = (T_0 g)^{-1}$. Then $S_0 : W \to V$ is a linear isomorphism and

$$S_0\rho(T) = S_0 T_0 \alpha(T) T_0^{-1} = S T_0 g T g^{-1} T_0^{-1} = T S_0,$$

and so S_0 intertwines ρ with the identity representation. □

An important general result in this subject is the Jacobson Density Theorem [67] which can be used to give an alternative proof of both Burnside's and Wedderburn's theorems. The statement of the density theorem, which we invite the reader to prove, is given below.

Theorem 13.3.19. *Let M be a simple module over a ring R and $\mathcal{A} =$ $\mathrm{End}_R(M)$. Then given $f \in \mathrm{End}_{\mathcal{A}}(M)$ and $x_1, \ldots, x_n \in M$, there exists an $r \in R$ so that $rx_i = f(x_i)$ for all $i = 1, \ldots, n$.*

13.4   Division Algebras

Here we study the special case of finite dimensional associative algebras with unit $\mathcal{D}$ over k in which every non-zero element has a multiplicative inverse. As we saw earlier, such a $\mathcal{D}$ is always a simple k-algebra.

We can consider $\mathbb{C}$ and $\mathbb{H}$ to be division algebras over the rational field, $\mathbb{Q}$, or indeed any ordered field, k, with the property that a sum of squares is always positive unless all elements concerned are zero. Such fields are called *formally real*. Of course, this includes any subfield of $\mathbb{R}$ including $\mathbb{R}$ itself. Notice that such a field is necessarily of characteristic zero. For if $1 + \ldots + 1 = 0$, then writing this as $1^2 + \ldots + 1^2$ we get a contradiction. We do this by generators and relations. In the case of $\mathbb{C}$ we consider $\{a + bi : a, b \in K\}$ and $i^2 = -1$ and the usual (distributive) multiplication. For $\mathbb{H}$ we consider $\{a + bi + cj + dk : a, b, c, d \in K\}$ and $i^2 = j^2 = k^2 = -1$ and also the usual (distributive) multiplication. Then, as we have seen in Chapter 5 of Volume I, $\mathbb{C}$ and $\mathbb{H}$ are division algebras over $\mathbb{R}$ with $\mathbb{C}$ commutative. $\mathbb{H}$ can also be considered a division algebra over $\mathbb{C}$. Of course, any field k is a division algebra over itself.

As further important examples we now introduce *quaternion algebras,*

Example 13.4.1. Let k be a field of characteristic zero and α and $\beta \in k^\times$. We define $\mathbb{H}(\alpha, \beta)$ to be the algebra generated by $1, i, j$ and k with $k = ij$ and satisfying the additional relations $i^2 = \alpha 1$, $j^2 = \beta 1$, $ij = -ji$. Then,

$$k^2 = ijij = -i^2 j^2 = -\alpha\beta 1.$$

Here we also assume the relations $ik + ki = 0$ and $jk + kj = 0$. This gives a 4-dimensional associative algebra $\mathbb{H}(\alpha, \beta)$ over the field k. (Here we ask the readers indulgence as we are forced to use the symbol k in two different ways in the same paragraph!) The question is, when is $\mathbb{H}(\alpha, \beta)$

a division algebra?. We proceed just as with the usual quaternions. Consider $\mathbb{H}(-1, -1)$. By making use of the norm $N(q) = q\bar{q}$, where if $q = x1 + yi + zj + uk$, $\bar{q} = x1 - (yi + zj + uk)$. Just as in the standard case, this conjugation is an involution and an anti automorphism.

Exercise 13.4.2. Check this last assertion.

Clearly $N(q) \in k$ and so if it is non-zero, $q^{-1} = \frac{\bar{q}}{N(q)}$. One checks easily that

$$N(q) = x^2 - \alpha y^2 - \beta z^2 + \alpha\beta u^2.$$

In particular, if $k \subseteq \mathbb{R}$ and α and β are negative, then we get a positive definite form in 4 variables. Now even though, in general, this form is not positive definite the only trouble is *it might represent zero where* $q \neq 0$. Thus,

Proposition 13.4.3. $\mathbb{H}(\alpha, \beta)$ *is a division algebra if and only if* $N(q)$ *never represents zero for* $q \neq 0$, *that is, if and only if the form* $N(q)$ *is non-degenerate.*

Exercise 13.4.4.   1. Show $\mathbb{H}(1, 1)$ is not a division algebra by proving that it is isomorphic to $M(2, k)$ (which has many zero divisors). Calculate $N(\cdot)$ and show it is $\det(\cdot)$. What is the involution?

2. More generally, show $\mathbb{H}(\alpha, 1)$ is isomorphic to $M(2, k)$.

3. Show if a and $b \in k^{\times}$, then $\mathbb{H}(a\alpha, b\beta)$ is isomorphic as an algebra to $\mathbb{H}(\alpha, \beta)$.

Now let $\mathcal{D}$ be a division algebra. We consider $k \subseteq \mathcal{D}$ by identifying k with the subset K_1 of $\mathcal{D}$. Then k is a 1-dimensional subalgebra of $\mathcal{D}$. We now show when k is algebraically closed the possibilities are severely limited. We first deal with the situation when $\mathcal{D}$ is actually commutative.

Lemma 13.4.5. *If* $\mathcal{D}$ *is a finite dimensional commutative division algebra over an algebraically closed field* K, *then* $\mathcal{D} = K$.

Proof. First assume $\mathcal{D}$ has a single generator, x. Since $1, x, x^2, \ldots$ can not be linearly independent over K, there must be a linear relation $c_0 1 + c_1 x + \ldots = 0$. Since K is algebraically closed, $x \in K$ and $\mathcal{D} = K$. Now let $x_1, \ldots x_n$ be a basis of $\mathcal{D}$ over K. By the single generator case, for each i, $K[x_i]$ being a finite extension of K must coincide with K. Hence $K[x_i] = K$ and so $x_i \in K$. Therefore again $\mathcal{D} = K$.   □

We now prove our result in general.

Proposition 13.4.6. *If $\mathcal{D}$ is a finite dimensional division algebra over an algebraically closed field K, then $\mathcal{D} = K$.*

Proof. If $x \in \mathcal{D}$, then $K[x]$, the subalgebra of $\mathcal{D}$ generated by x, is commutative since it is a quotient ring of the polynomials in one variable with coefficients in K. Now if $x \neq 0$, $K[x]$ is also a division algebra in its own right. This is because $K[x] \subseteq \mathcal{D}$ and if $y \in K[x]$ and $yz = 1$, taking x as the first basis element of $\mathcal{D}$ over K and making use of the finite dimensionality, one sees easily that z is also a polynomial in x of degree perhaps one more than that of y. The coefficients of z are $a_i \alpha_0$, where the a_i are the coefficients of the polynomial y and $z = \alpha_0 x + \ldots$ when we write z in terms of this basis of $\mathcal{D}$. Since K is algebraically closed and $K[x]$ is a finite dimensional commutative division algebra over K as the lemma above shows, it coincides with K. Hence $x \in K$ and $\mathcal{D} = K$.   □

As a corollary of Proposition 13.4.6 we get an extension of Wedderburn's little theorem.

Corollary 13.4.7. *Let $\mathcal{D}$ be a finite dimensional associative algebra with unit. If $\mathcal{D}$ has no zero divisors, then $\mathcal{D}$ is a division algebra.*

Proof. Let $x \neq 0 \in \mathcal{D}$. Then, as we just saw in Proposition 13.4.6, $K[x]$ is a finite dimensional *commutative* algebra over K. Since it has no zero divisors it must be a field. Hence x has an inverse in $K[x]$ and in particular in $\mathcal{D}$. But x was an arbitrary element $\neq 0$ so $\mathcal{D}$ is a division algebra.   □

This concludes our study of finite dimensional division algebras over an algebraically closed field. We now wish to see what can be said for non-algebraically closed fields such as $k = \mathbb{R}$. Of course, here k is still a rather large field; that is, it is one small algebraic extension from being algebraically closed. The smaller k gets the weaker is our hypothesis on $\mathcal{D}$ and hence the more numerous will be the finite dimensional division algebras over k. So we can expect there will be many such algebras over $k = \mathbb{Q}$.

When dealing division algebras over an non-algebraically closed field k we will pass to the "complexification", namely, we will tensor with $\overline{k}$, the algebraic closure. When $k = \mathbb{R}$, this is the complexification. In fact, let $L \supset k$ be any field extension of k and $\mathcal{D}_L = L \otimes_k \mathcal{D}$. Then $\mathcal{D}_L$ can be made into an associative algebra over L by defining,

$$(l_1 \otimes d_1)(l_2 \otimes d_2) = l_1 l_2 \otimes d_1 d_2$$

and identifying each element $d \in \mathcal{D}$ with $1 \otimes d$ of $\mathcal{D}_L$. Then if $e_1, \ldots e_n$ is a basis for $\mathcal{D}$ over k with structure constants $c_{i,j}^s \in k$, then $e_i e_j = \sum_s c_{i,j}^s e_s$ and since $k \subset L$, these same formulas define multiplication in $\mathcal{D}_L$ with respect to $e_1, \ldots e_n$ which is a basis of $\mathcal{D}_L$ over L. However, to be of any use this procedure must preserve important properties of $\mathcal{D}$ when "complexifying". For example, the property of an associative algebra being semi-simple (see Corollary 13.3.12.)

Denote by $\mathcal{Z}(\mathcal{D})$ the center of $\mathcal{D}$. That is, $\mathcal{Z}(\mathcal{D})$ consists of all elements which commute with everything in $\mathcal{D}$. Since 1 commutes with everything and we are in an algebra, $k \subseteq \mathcal{Z}(\mathcal{D})$. Clearly, $\mathcal{Z}(\mathcal{D})$ is a subalgebra of $\mathcal{D}$.

Exercise 13.4.8. Let $\mathcal{D}$ be a finite dimensional division algebra over k. Show $\mathcal{Z}(\mathcal{D})$ is a division algebra over k and in particular, $\mathcal{Z}(\mathcal{D})$ is a field. Show $\mathcal{D}$ is a division algebra over $\mathcal{Z}(\mathcal{D})$.

Definition 13.4.9. A division algebra $\mathcal{D}$ over k is called *central* if $\mathcal{Z}(\mathcal{D}) = k$.

Exercise 13.4.10. Show when $\mathbb{H}$ is regarded as a division algebra over $\mathbb{R}$ (or any formally real field), it is central.

We now sharpen the statement about semi-simple algebras and extending the field (since a division algebra is always simple).

Proposition 13.4.11. *Let $\mathcal{D}$ be a central division algebra over k and $L \supset k$ be a field extension. Then, the algebra $\mathcal{D}_L$ is a simple algebra over L.*

Proof. To see that $\mathcal{D}_L$ is simple, let I be a non-zero two sided ideal. Every element of I (indeed of $\mathcal{D}_L$) can be written $x = \sum_{i=1}^{s} l_i d_i$, where $d_i \in \mathcal{D}$ and $l_i \in L$. Choose an $x \in I$ for which this representation has shortest length, say $x = \sum_{i=1}^{s} l_i d_i$. Then the d_i are linearly independent over k as are the l_i. Multiplying x by d_1^{-1} (on either side) gets us another non-zero element of I. If $s > 1$, then since d_2 is not in $k = \mathcal{Z}(\mathcal{D})$, there exists a $d \in \mathcal{D}^\times$ so that $dd_2 \neq d_2 d$. Hence

$$x - dxd^{-1} = \sum_{i=2}^{s} l_i (d_i - dd_i d^{-1}).$$

Now the l_i are linearly independent over k and $d_i - dd_i d^{-1} \neq 0$. This contradicts the choice of shortest length representation and we conclude $s = 1$. This means there is an $x \neq 0$ for which $x = ld$, that is $x \in L$. Multiplying by x^{-1} shows $1 \in I$ and so $I = \mathcal{D}_L$. $\qquad\square$

In particular, if $\overline{k}$ is the algebraic closure of k, we get the following corollary.

Corollary 13.4.12. *If $\mathcal{D}$ is a central division algebra over k, then the algebra $\mathcal{D}_{\overline{k}}$ is isomorphic to the matrix algebra $\mathrm{End}_{\overline{k}}(V)$ of dimension n^2. Hence also $\dim_k \mathcal{D} = n^2$.*

Definition 13.4.13. The number n is called the *degree* of the central division algebra $\mathcal{D}$ over k and is denoted by $\deg \mathcal{D}$.

Notice that when we apply this result to $\mathbb{H}$, which is a central division algebra over $\mathbb{R}$, we get $\mathbb{H}_\mathbb{C}$ is a matrix algebra. So of dimension n^2. But the dimension of $\mathbb{H}$ over $\mathbb{C}$ is 2 so the dimension of $\mathbb{H}_\mathbb{C}$ over $\mathbb{C}$ is 4, making $\deg(\mathbb{H}) = 2$. This means while $\mathbb{H}_\mathbb{C}$ is the full 2×2 complex matrix algebra, when we regard $\mathbb{H}$ as an algebra over $\mathbb{C}$ and take the left regular representation, we get a subalgebra of $M_2(\mathbb{C})$ of complex dimension 2. If $q = (z, w)$, then the corresponding matrix is

$$\begin{pmatrix} z & w \\ -w^- & z^- \end{pmatrix}.$$

The results of the following exercise are quite important in Physics.

Exercise 13.4.14. Check that this correspondence is a faithful linear representation. (Giving yet another proof that quaternionic multiplication is associative). Check that $1, i, j, k$ correspond respectively to

$$\begin{pmatrix} 1 & 0 \\ 0 & 1 \end{pmatrix}, \quad \begin{pmatrix} i & 0 \\ 0 & -i \end{pmatrix}, \quad \begin{pmatrix} 0 & 1 \\ -1 & 0 \end{pmatrix}, \quad \begin{pmatrix} 0 & i \\ i & 0 \end{pmatrix}.$$

Show $\det q(z, w) = N(q)$. Hence $N(\cdot)$ is a real valued multiplicative function.

Proposition 13.4.15. *If $\mathcal{D}$ is a central division algebra over k, then every maximal commutative subfield F of $\mathcal{D}$ has $\dim_k F = \deg \mathcal{D}$ and any isomorphism of maximal subfields extends to an inner automorphism of $\mathcal{D}$.*

Proof. Let F be a maximal commutative subalgebra of $\mathcal{D}$ and $L \supset k$ be (for the moment) any extension field of k, then F_L is a maximal commutative subalgebra of $\mathcal{D}_L$. This is because the maximality condition means that $F = Z_\mathcal{D}(F)$ and $F_L = Z_{\mathcal{D}_L}(F_L)$. These conditions are equivalent respectively to $\dim_k(F) = \dim_k(Z_\mathcal{D}(F))$ and $\dim_L(F_L) = \dim_L(Z_{\mathcal{D}_L}(F_L))$. Expressing the conditions of lying in one of these centralizers in coordinates, one has a system of homogeneous linear equations with coefficients in k (respectively L). Hence the dimension of the space of solutions remains the same when one passes to an extension field.

Now take L to be the algebraic extension of k for which $\mathcal{D}_L = \mathrm{End}_L(V)$. Then as a commutative division algebra $F_L = L \oplus \ldots \oplus L$, where we have, say m summands. Thus F_L, as a vector space over L, has a basis $e_1, \ldots, e_m$ satisfying $e_i^2 = e_i$ and for $i \neq j$, $e_i e_j = 0$ (regarding F_L as a subset of $\mathcal{D}_L = \mathrm{End}_L(V)$). Hence, the e_i are pairwise commuting projections, thus having for eigenvalues only 0 and 1. As we know, these can be simultaneously diagonalized in some basis. This means there is a $p \in \mathcal{D}_L^\times$ so that $pF_L p^{-1}$ consists of diagonal matrices. On the other hand, being conjugate to F_L, $pF_L p^{-1}$ is also a maximal commutative subalgebra of $\mathcal{D}_L$. Hence it consists of all diagonal matrices of order n and therefore $m = n$. Remembering that $m = \dim_L F_L = \dim_k F$, this proves the first statement.

Now, let E and F be two maximal commutative subalgebras of $\mathcal{D}$ which are isomorphic by say ϕ. Then ϕ extends to a linear isomorphism $\overline{\phi} : E_L \to F_L$. As we just saw, there exists a $p \in \mathcal{D}_L^\times$ so that $pE_L p^{-1} = F_L$. Now every automorphism of the algebra of diagonal matrices is a permutation of the diagonal entries, and hence is induced by conjugation by a permutation matrix in $\mathcal{D}_L = \mathrm{End}_L(V)$. Thus $pxp^{-1} = \overline{\phi}(x)$ for all $x \in E_L$. What we are now supposed to prove is there is some non-zero $d \in \mathcal{D}$ so that $dxd^{-1} = \overline{\phi}(x)$ for all $x \in E$, that is $dx = \overline{\phi}(x)d$ for all $x \in E$. This is also a system of homogenous linear equations with coefficients in k. But $px = \overline{\phi}(x)p$ for all $x \in E_L$ is the same set of equations with coefficients in L and we know it has a non-trivial solution. Hence, over k there is also a non-trivial solution.   □

13.4.1   Jacobson's Theorem

The following theorem due to N. Jacobson, arose from the little Wedderburn theorem and gave rise to the whole theory of polynomial identities in associative algebras, of which Jacobson was one of the leading figures.

Theorem 13.4.16. *(Jacobson's Theorem) Suppose $\mathcal{D}$ is a division ring such that for every $x \in \mathcal{D}$ there exists an integer $n(x) > 1$ so that*

$$x^{n(x)} = x,$$

then $\mathcal{D}$ is commutative.

Proof. If $0 \neq a \in \mathcal{D}$, then by assumption, $a^n = a$ and $(2a)^m = 2a$. Setting $q = (n-1)(m-1) + 1$ we have $q > 1$, $a^q = a$ and $(2a)^q = 2a$, so that $(2^q - 2)a = 0$. Thus $\mathcal{D}$ has characteristic $p > 0$ for some prime p. Now, if P is the prime field with p elements contained in the center $\mathcal{Z}$, then $P(a)$ has p^h elements, so that $a^{p^h} = a$. Hence, if $a \notin \mathcal{Z}$, by the proof of Wedderburn's Little Theorem of Volume I, there is a $b \in \mathcal{D}$ such that

$$bab^{-1} = a^u \neq a. \qquad (1)$$

Suppose $b^{p^k} = b$ and consider the set

$$W = \left\{ x \in \mathcal{D} \,\middle|\, x = \sum_{i,j}^{p^h, p^k} p_{ij} a^k B^j, \; p_{ij} \in P \right\}.$$

Evidently, W is finite and closed under addition. By (1) it is also closed under multiplication. Thus W is a finite ring and being a subring of the division ring $\mathcal{D}$ the two cancelation laws hold. It follows that W is a finite division ring. Hence by Theorem 5.8.6 of Volume I it is commutative, and since a, b $\in W$, this forces $ab = ba$, contrary to $a^u b = ba$. □

We conclude this section by mentioning the role of the Jacobson radical, $\mathrm{Jac}(\mathcal{A})$. Let $\mathcal{A}$ be a finite dimensional associative algebra with 1 over a field k. Then

$$\mathrm{Jac}(\mathcal{A}) = \cap M,$$

where the intersection is over the set of all maximal two sided ideals M. The two sided ideal $\mathrm{Jac}(\mathcal{A})$ measures the deviation of $\mathcal{A}$ from being semi-simple.

Exercise 13.4.17. Prove that $\mathcal{A}$ is semi-simple if and only if $\mathrm{Jac}(\mathcal{A}) = (0)$.

13.4.2 Frobenius' Theorem

Here we deal with the well known theorem of Frobenius which classifies division algebras over $\mathbb{R}$. As it turns out there are only the usual ones of Chapter 5 Volume I! If we are willing to accept non-associative ones there is also the octonions (also see Chapter 5). All these structures, including the octonions, play a very important role in real rank-1 simple Lie groups. Frobenius' original proof was very computational and hard to remember or understand well. As the reader will see the proof offered here is more conceptual.

Theorem 13.4.18. *(Frobenius Theorem) A finite dimensional division algebra $\mathcal{D}$ over $\mathbb{R}$ is isomorphic to $\mathbb{R}$, $\mathbb{C}$ or $\mathbb{H}$.*

Proof. First consider the case where $\mathcal{D}$ is *Abelian*. Then $\mathcal{D}$ is isomorphic to $\mathbb{R}$ or $\mathbb{C}$. This is because $\mathcal{D}$ is a commutative finite algebraic field extension of $\mathbb{R}$ so these are the only two possibilities. Thus in general $\mathcal{Z}(\mathcal{D})$, being such an algebra over $\mathbb{R}$ is isomorphic to $\mathbb{R}$ or $\mathbb{C}$. In the latter case, since $\mathcal{D}$ can be viewed as an algebra over $\mathbb{C}$ and since $\mathbb{C}$ is algebraically closed, by Proposition 13.4.6, its dimension is 1. Hence $\mathcal{D} = \mathbb{C}$.

On the other hand, if $\mathcal{Z}(\mathcal{D})$ is isomorphic to $\mathbb{R}$, then $\mathcal{D}$ is a *central algebra* over $\mathbb{R}$. There are again two possibilities; either $\mathcal{D}$ is Abelian i.e. $\mathcal{D} = \mathcal{Z}(\mathcal{D}) = \mathbb{R}$, or $\mathcal{D}$ is a non Abelian central division algebra of degree 2. In this case choose a maximal subfield of $\mathcal{D}$ which is clearly isomorphic to $\mathbb{C}$. Since this field is commutative, conjugation is an automorphism. By Proposition 13.4.15 conjugation of $\mathbb{C}$ extends to an inner automorphism of $\mathcal{D}$. Hence there is a $j \in \mathcal{D}$ so that $jzj^{-1} = \bar{z}$ for all $z \in \mathbb{C}$. In particular, j is not in $\mathbb{C}$. Therefore, $\mathcal{D} = \mathbb{C} \oplus \mathbb{C}j$. Now j^2 commutes with j and since conjugation is of order 2, it also commutes with all elements of $\mathbb{C}$. This forces $j^2 \in \mathcal{Z}(\mathcal{D}) = \mathbb{R}$. Now we can normalize j so that $j^2 = \pm 1$. But $j^2 = 1$ is not possible because then $(j-1)(j+1) = 0$, which would mean $j \in \mathbb{C}$, a contradiction. Thus $j^2 = -1$ and also $i^2 = -1$. Moreover, $jij^{-1} = -i$ and so $ij = -ji$. Taking $k = ij$ we see that $k^2 = ijij = i(-ij)j = -1$, $jk = jij = -ij^2 = i$, and $ki = iji = -ji^2 = j$. Thus $\mathcal{D} = \mathbb{H}$. $\qquad\Box$

We now give an alternative proof of Frobenius' theorem via topology that is even more conceptual.

Proof. If $\mathcal{D}$ is of odd dimension, then by Proposition 5.9.16 of Volume I, $D \cong \mathbb{R}$. So we can assume that $\dim(\mathcal{D}) > 1$ and is even. Let us first examine the case in which D is commutative. Consider the map

$$\hat{f} : S^{n-1} \longrightarrow S^{n-1} \quad : \quad x \mapsto \hat{f} := \frac{x^2}{\|x^2\|},$$

where $\| \cdot \|$ is the usual Euclidean norm on $\mathbb{R}^n$. This map is well defined since $\mathcal{A}$ is a division algebra. Now, by the section on Real Projective Space of Volume I, this map induces a well defined continuous map

$$f : \mathbb{R}P^{n-1} \longrightarrow S^{n-1}$$

which is injective. Indeed, if $f(x) = f(y)$, then $x^2 = \|\frac{x^2}{y^2}\|y^2 = \lambda^2 y^2$, where $\lambda^2 = \|\frac{x^2}{y^2}\|$. Hence $x^2 - \lambda^2 y^2 = (x - \lambda y)(x + \lambda y) = 0$. It follows that $x = \pm \lambda y$, which implies $\lambda = \pm 1$ by taking the norm of both sides. Thus the map f is injective. As an injective continuous map between compact Hausdorff spaces, f is a homeomorphism onto its image. Since any imbedding of compact manifolds of the same dimension (here $n-1$) is surjective, we must have $\mathbb{R}P^{n-1} \cong S^{n-1}$ and this can happen only if $n = 2$, since only then do these two spaces have the same fundamental and homology groups. It remains to prove that this 2-dimensional real division algebra is isomorphic to $\mathbb{C}$. To see this, pick an element $j \in \mathcal{D} - \mathbb{R}$. Then $\{j, 1\}$ is a basis in $\mathcal{D}$, and so

$$j^2 = a + bj, \quad a, b \in \mathbb{R}.$$

Therefore,

$$\left(j - \frac{b}{2}\right)^2 = a + \frac{b^2}{4} \in \mathbb{R}.$$

Hence we can write $j^2 = \lambda$ for some $\lambda \in \mathbb{R}$. Now, if $\lambda \geq 0$, say $\lambda = \gamma^2$, then $j^2 = \gamma^2$, which gives us $(j - \gamma)(j + \gamma) = 0$, i.e, $j = \pm \gamma \in \mathbb{R}$, a contradiction. Therefore, $\gamma < 0$. Set $\gamma = -\nu^2$, and we get $j^2 + \nu^2 = 0$. By rescaling, $j^2 = -1$, which shows that $\mathcal{D} \cong \mathbb{C}$. (The part of the proof to this point is due to A. Hatcher see [59] p.173. The remaining part is due to R. Palais see [105].) For a general $\mathcal{D}$, pick a non-zero element $j \in \mathcal{D} - \mathbb{R}$ and consider $\mathbb{R}, \langle j \rangle := \mathbb{R} \oplus j\mathbb{R}$. This is a 2-dimensional commutative division algebra over $\mathbb{R}$ and hence is isomorphic to $\mathbb{C}$. Therefore we can pick $j = i$ with $i^2 = -1$. Now consider $\mathcal{D}$ as a left $\mathbb{C}$-vector space, and call T the linear transformation on $\mathcal{D}$ given by right multiplication by i. This operator has eigenvalues i and $-i$ with corresponding eigenspaces $\mathcal{D}^+$ and $\mathcal{D}^-$. Since

$$x = \frac{1}{2}(x - ixi) + \frac{1}{2}(x + ixi),$$

$\mathcal{D} = \mathcal{D}^+ \oplus \mathcal{D}^-$. Clearly, $\mathcal{D}^+ = \mathbb{C}$ since if $d \in \mathcal{D}$ and is in $\mathcal{D}^+$, then it commutes with $\mathbb{C}$ and the algebra generated by $\mathbb{C}$ and d would be of dimension strictly larger than 2 and yet commutative, which is impossible as we showed earlier. Now take a non-zero element $d \in \mathcal{D}^-$. Then right multiplication by d clearly interchanges $\mathcal{D}^+$ and $\mathcal{D}^-$. Hence $\mathcal{D}^-$ is a 1-dimensional complex vector space, and $\mathcal{D}$ is a 4-dimensional real division algebra. To finish, note that $d^2 \in \mathcal{D}^- d = \mathcal{D}^+$ and obviously $d^2 \in \mathbb{R}\langle d \rangle$. Therefore

$$d^2 \in \mathbb{R}\langle d \rangle \cap \mathcal{D}^+ = \mathbb{R}.$$

Furthermore, $d^2 < 0$ since otherwise it would have two square roots in $\mathbb{R}$, in addition to the square root d. This amounts to three different square roots in the field $\mathbb{R}\langle d \rangle$, which is impossible. Hence we can rescale d to an element j such that $j^2 = -1$. Defining $k = ij$ we verify that all the quaternionic identities are satisfied, and clearly 1, i, j and k generate $\mathcal{D}$. Hence, $\mathcal{D} \cong \mathbb{H}$. $\square$

As a first corollary of the above proof we obtain a proof of the *Fundamental Theorem of Algebra*.

Indeed, if α is a root of the polynomial $f \in \mathbb{C}[x]$ in some algebraic extension of $\mathbb{C}$, then $\mathbb{C}(\alpha)$ is a finite field extension of $\mathbb{R}$, which by the above (since $\mathbb{H}$ is not commutative) must have degree 1 or 2 over $\mathbb{R}$.

Corollary 13.4.19. *The only connected and locally compact division rings are $\mathbb{R}$, $\mathbb{C}$ or $\mathbb{H}$ (see [68]).*

We can ask what happens if in the Theorem above we consider complex division algebras. As we have seen the only possibility is $\mathbb{C}$. Here we offer another proof of this.

Proposition 13.4.20. *Every finite dimensional division algebra over an algebraically closed field k is isomorphic to k itself.*

Proof. Let $\mathcal{D}$ be such a division algebra. Then $\mathcal{D}$ must contain k. Let $d \in \mathcal{D} - k$. Then the set

$$\{d^n \; : \; n \in \mathbb{N}\}$$

can not be linearly independent since the dimension of $\mathcal{D}$ over k is finite. Hence, there is a polynomial $p(x) \in k[x]$ such that $p(d) = 0$. Since k is algebraically closed, this implies $d \in k$, contradiction. $\qquad\Box$

13.4.3   The 4-Square Theorem

Lagrange's 4-square theorem states that every natural number can be represented as the sum of four integer squares[4].

Lagrange's 4-square theorem strongly suggests a connection to the quaternions (just as the two square Theorem 5.6.8 of Volume I suggests a connection to the complex numbers). There are several proofs of this, we shall follow the argument of Herstein. Another proof using Minkowski's convex body theorem is found in Chapter 8 of [1].

Lemma 13.4.21. *(Four Square Identity) If two numbers can each be written as the sum of four squares, then so can their product.*

Proof. Let $q_1 = a_1 1 + b_1 \mathbf{i} + c_1 \mathbf{j} + d_1 \mathbf{k}$ and $q_2 = a_2 1 + b_2 \mathbf{i} + c_2 \mathbf{j} + d_2 \mathbf{k}$ be two quaternions. Then $N(q_1) = a_1^2 + b_1^2 + c_1^2 + d_1^2$ and $N(q_2) = a_2^2 + b_2^2 + c_2^2 + d_2^2$. By the multiplicative property of norms (see Exercise 5.9.15 of Volume I), $N(q_1)N(q_2) = N(q_1q_2)$. $\qquad\Box$

It follows that it is sufficient to prove the result for prime numbers, p. Since $2 = 1^2 + 1^2 + 0^2 + 0^2$, we may assume p is an odd prime.

---

[4] *Arithmetica* (Greek: $A\rho\iota\theta\mu\eta\tau\iota\kappa\eta$) an ancient Greek text on mathematics written by Diophantus in the 3rd century CE, is a collection of 130 algebraic problems. Arithmetica originally consisted of thirteen books, but the Greek manuscripts that survived to the present contain only six. This was translated into Latin in 1621 by Claude Gaspard Bachet de Méziriac, who stated the theorem in the notes of his translation. The theorem was not proved until 1770 by Lagrange using results of Euler. Adrien-Marie Legendre completed the picture in 1797 with his three-square theorem which states a positive integer can be expressed as the sum of three squares if and only if it is not of the form $4^k(8m+7)$ for integers k and m. Later, in 1834, Carl Gustav Jakob Jacobi discovered a simple formula for the number of representations of an integer as the sum of four squares with his own four-square theorem. In book 4, he finds rational powers between given numbers. He also noticed (as we know) that numbers of the form $4n + 3$ can not be the sum of two squares.

Let $\mathbb{Z}[\mathbf{i}, \mathbf{j}, \mathbf{k}] = \{a + b\mathbf{i} + c\mathbf{j} + d\mathbf{k} \mid a, b, c, d \in \mathbb{Z}\}$, i.e. the set of all quaternions with integer coefficients. Geometrically, we can regard these as vertices of a tiling of $\mathbb{R}^4$ by unit 4-dimensional cubes and the half integer quaternion

$$\xi = \frac{1 + \mathbf{i} + \mathbf{j} + \mathbf{k}}{2},$$

as the center of the 4-dimensional unit cube $[0, 1]^4$. Now, consider the set

$$\Lambda = \{a\xi + b\mathbf{i} + c\mathbf{j} + d\mathbf{k}, \mid a, b, c, d \in \mathbb{Z}\}.$$

Elements of Λ are called *Hürwitz integers*[5].

Geometrically Λ is the set of all vertices $\mathbb{Z}[\mathbf{i}, \mathbf{j}, \mathbf{k}]$ together with the centers of these unit cubes.

Proposition 13.4.22. Λ *is a ring, and is a left PID (i.e. each left ideal in* Λ *is of the form* $\Lambda\alpha$ *for some* $\alpha \in \Lambda$).

Proof. Considering both the integer and half integer cases, one sees that Λ is closed under addition, multiplication, and taking additive inverses. Hence Λ is a ring and, since $N(qr) = N(q)N(r)$, for all quaternions it has no zero divisors.

We will show that Λ is a left Euclidean domain. That is, given any non-zero elements $a, b \in \Lambda$, there are q and $r \in \Lambda$ such that $a = qb + r$, where $N(r) < N(b)$. If $q \in \Lambda$, then so is $r = a - qb$.

We do this by imitating the proof in Chapter 5 of Volume I that the Gaussian integers are a Euclidean ring. Divide a by b in $\mathbb{H}$, i.e. form $x = ab^{-1}$. Then we ask what is the distance of x to the nearest point

---

[5]Adolf Hürwitz (1859-1919) a German-Jewish mathematician worked in Algebra, Analysis, Geometry and Number Theory. He was a mentor to the younger Hilbert and Minkowski. The proof of the four square theorem presented here is essentially his and he singled out the special interest of the 24 cell. Among his many other accomplishments, he proved that $|Aut(S_g)|$, the order of the holomorphic automorphism group of a compact Riemann surface of genus $g \geq 2$ is $\leq 84(g-1)$, and in particular is finite. He also proved an important theorem about the number of zeros of a holomorphic function (see Corollary 4.3.6 of [88]). By studying the fundamental domain for $\mathrm{SL}(2, \mathbb{Z})$ in $\mathrm{SL}(2, \mathbb{R})$ (see p. 119 of [88]) proved the Diophantine approximation of irrational numbers by rationals.

q of Λ? Since in $\mathbb{R}^4$ the greatest distance from a lattice point occurs at the mid point of the 4-dimensional cube *which is in* Λ, that distance is zero. Otherwise q does not lie at the midpoint of a cube. Therefore, in this case, $N(q - x) < (1/2)^2 + (1/2)^2 + (1/2)^2 + (1/2)^2 = 1$. Thus, in any case $N(q - x) < 1$ for some $q \in \Lambda$. Hence $N(ab^{-1} - q)N(b) = N(a - qb) = N(r) < N(b)$. $\qquad\square$

Proposition 13.4.23. *If p is a prime, the ideal Λp is not maximal.*

Proof. If Λp were a maximal ideal, then $\Lambda/\Lambda p$ would be a division ring, and in particular has no zero divisors. Let $\tau = a + b\mathbf{i} + \mathbf{j}$, where a and b are zeros of the equation $a^2 + b^2 \equiv -1 \pmod{p}$. Then $\tau\bar{\tau} \in \Lambda p$. But $[\tau]$ is a zero divisor in $\Lambda/\Lambda p$, a contradiction. $\qquad\square$

We are now ready to prove Lagrange's Four Square Theorem.

Theorem 13.4.24. (*Lagrange's Four Square Theorem*) *Every positive integer n can be written as the sum of four squares of integers, i.e.*

$$n = a^2 + b^2 + c^2 + d^2.$$

Proof. Since Λp is not a maximal left ideal, there is some new left ideal I such that $\Lambda p \subset I \subset \Lambda$ and I is not equal to either Λp or Λ. Since I is a left ideal we know $I = \Lambda\eta$ for some $\eta \in \Lambda$. Hence $p = \zeta\eta$ for some $\zeta \in \Lambda$. We have

$$p^2 = N(p) = N(\zeta)N(\eta).$$

If $N(\zeta) = 1$ then ζ is a unit and $I = \Lambda p$. This is false. If $N(\eta) = 1$ then η is a unit and $I = \Lambda$. Hence $N(\eta) = p$. But then we can write $\eta = a + b\mathbf{i} + c\mathbf{j} + d\mathbf{k}$ and $p = N(\eta) = a^2 + b^2 + c^2 + d^2$. If these are integers, we are already done. If all of these are half integers, let

$$a_1 = \frac{a+b}{2}, \ b_1 = \frac{a-b}{2}, \ c_1 = \frac{c+d}{2}, \ d_1 = \frac{c-d}{2}.$$

Obviously, these are all integers, and $a_1^2 + b_1^2 + c_1^2 + d_1^2 = p$. $\qquad\square$

13.4.4   The 24-cell Polytope

Here we consider some geometric aspects of the Hürwitz integers, Λ. As we shall see Λ is a regular polyhedron in $\mathbb{R}^4$, but exceptional in many ways. Not only are the units of the Hürwitz integers its vertices (which is analogous to the cube in 3 dimensions), and like the tetrahedron in 3 space, this polyhedron, which we will call the 24-cell, is also self-dual. As we shall see, there is nothing analogous to the 24-cell in 3 dimensions. Indeed, all regular polyhedra in 4 dimensions are analogues of the 5 Platonic solids in 3 dimensions, *except* the 24-cell because it is the only one that is self-dual and at the same time tessellates $\mathbb{R}^4$. In 3 dimensions the cube is the only Platonic solid which tessellates $\mathbb{R}^3$, but is not self-dual since its dual is the octahedron.

In order to discuss these matters we remind the reader about some details concerning regular polyhedra, in general, and in lower dimensional spaces.

A *regular polyhedron* in $\mathbb{R}^n$ is the term used for a bounded solid made up of points, straight lines, planar faces etc. in which all the lengths, angles, planar surfaces, etc. are all the same. Thus in 2 dimensions, the *regular polygons* have all their edges of the same length, and all the angles are all equal. Obviously, here we have the equilateral triangle, the square, the regular pentagon, and so on. In short, there is an infinity of regular polygons; one with n sides for each $n \geq 3$. The symmetry groups of these are all finite and we have calculated some of them in Chapter 1 of Volume I.

In 3 dimensions things change. The regular polyhedra, also known as the *Platonic solids*[6], are *only five*. These are the *tetrahedron*, with 4 triangular faces, the *cube*, with 6 square faces, the *octahedron*, with 8 triangular faces, the *dodecahedron*, with 12 pentagonal faces and the *icosahedron*, with 20 triangular faces. The *finite group* of symmetries of the Platonic solids also arise in mathematics as well as theoretical physics, for instance in the context of gravitational singularities. There

---

[6]The Platonic solids have in fact been known for millennia, and were known at least a thousand years prior to Plato to the neolithic people in Scotland. The solids are named after Plato for his use of them in philosophy.

is also a certain duality here as well. If one interchanges the number of vertices for faces (and vice versa) and leaves the edges alone, then the cube and octahedron become interchanged, the dodecahedron and icosahedron become interchanged, and the tetrahedron remains unchanged. For that reason it is called self-dual[7].

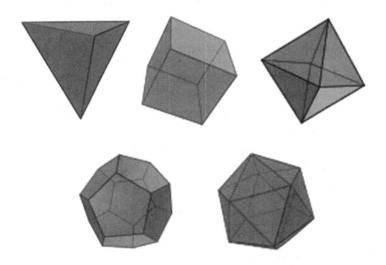

In 4 space all the 3 dimensional faces of a regular polyhedron must be lower-dimensional regular polytopes of the same size and shape, therefore these must be one of the 5 three dimensional ones just mentioned and all the vertices, edges, etc., have to be identical. Here there are exactly six regular polyhedra.

How can these be visualized? Just as in 3 space, where Platonic solids appropriately sized can be radially projected onto the circumscribing unit sphere S^2 in $\mathbb{R}^3$, Platonic solids in $\mathbb{R}^4$ project radially onto the unit sphere[8] S^3 of $\mathbb{R}^4$ and of course give a triangulation of S^3. Its surface consists of one of the regular polyhedra of 3 space. These are:

1. The *hypertetrahedron* with 5 tetrahedral faces.

---

[7]For the proof that there are only the 5 Platonic solids in $\mathbb{R}^3$ see e.g. [90].
[8]Therefore, in general, the Euler characteristic, $\chi(S^n) = 1 + (-1)^n$, plays a determining role. Here $\chi(S^3) = 0$.

2. The *hypercube* with 8 cubical faces.

3. The *hyperoctahedron* or 16-*cell*, with 16 tetrahedral faces.

4. The *hyperdodecahedron* or 120-*cell*, with 120 dodecahedral faces.

5. The *hypericosahedron* or 600-*cell*, with 600 tetrahedral faces.

6. The 24-*cell*, with 24 octahedral faces. This is a native of the 4-dimension with no analogue in any other dimension.

One might think that the higher one goes in dimension the more complicated the situation is. Not at all! In fact, 4-dimensional space has the greatest complexity concerning regular polytopes. After this, things become completely regular. This is one of many examples of how 4-dimensional geometry and topology are more complicated than geometry and topology in higher dimensions[9]. Indeed in 5 or more dimensions, there *are only three*! regular polytopes:

1. There is a kind of *hypertetrahedron*, called the n-simplex, having $n + 1$ faces, all of which are $(n - 1)$-simplices.

2. There is a kind of *hypercube*, called the n-cube, having $2n$ faces, all of which are $(n - 1)$-cubes.

3. There is a kind of *hyperoctahedron*, called the n-dimensional *cross-polytope*, having $2n$ faces, all of which are $(n - 1)$-simplices.

We now turn to the 24-cell polyhedron. To visualize it, imagine a hypercube with vertices $(\pm 1, \pm 1, \pm 1, \pm 1)$. Then imagine the 4-dimensional analogue of an octahedron, i.e. the *cross-polytope*, with vertices $(\pm 2, 0, 0, 0)$, $(0, \pm 2, 0, 0)$, $(0, 0, \pm 2, 0)$, $(0, 0, 0, \pm 2)$. The hypercube has 16 vertices and the cross-polytope has 8. We set things up so that all 24 of these points have the same distance from the origin. These are exactly the vertices of the 24-cell.

---

[9]This situation is eerily reminiscent of that of the Poincaré conjecture. There, in dimension 2 the situation is classical (and there are infinitely many solutions). In dimension ≥ 5, things are regular and the problem was solved by Smale, in the early 1960s. Then, in the early 1980s Freedman proved it for $n = 4$, and very recently Perelman did the same for $n = 3$.

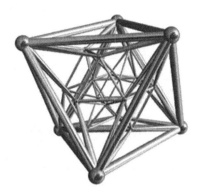

The 24-cell, denoted here by Π_{24}, is the *convex hull* of the 24 units in Λ. We need the following:

Lemma 13.4.25. *A quaternion q is a unit in Λ if and only if $N(q) = 1$.*

Proof. Let q be a unit. Then q^{-1} exists and $qq^{-1} = q^{-1}q = 1$. Thus $N(qq^{-1}) = N(q)N(q^{-1}) = N(1) = 1$. Hence, $N(q) = N(q^{-1}) = 1$ since $N(q) \in \mathbb{Z} \cup \{0\}$. Conversely, if $N(q) = 1$ then $q\bar{q} = 1$ and so $q = q^{-1}$, q is a unit and $\bar{q} \in \Lambda$.   □

Proposition 13.4.26. *The group of units $\mathcal{U}$ of Λ consists of the 24 points:*

$$\mathcal{U} = \left\{ \pm 1, \ \pm i, \ \pm j, \ \pm k, \ \frac{1}{2}(\pm 1 \pm i \pm j \pm k) \right\}.$$

Proof. Let $\zeta = \zeta_0 + \zeta_1 \mathbf{i} + \zeta_2 \mathbf{j} + \zeta_3 \mathbf{k}$ be a unit in Λ. Then, $N(\zeta) = \zeta_0^2 + \zeta_1^2 + \zeta_2^2 + \zeta_3^2 = 1$. We must show that $\zeta_k \in \mathbb{Z}$, or that $\zeta_k \in \frac{1}{2} + \mathbb{Z}$ for each $k = 0, 1, 2, 3$.

If $\zeta_0, \zeta_1, \zeta_2, \zeta_3 \in \mathbb{Z}$, the equation $\zeta_0^2 + \zeta_1^2 + \zeta_2^2 + \zeta_3^2 = 1$ implies that one of the ζ_k is 1 and the three others $= 0$. From this we get 8 units, i.e. $\pm 1, \pm \mathbf{i}, \pm \mathbf{j}, \pm \mathbf{k}$.

If $\zeta_0, \zeta_1, \zeta_2, \zeta_3 \in \frac{1}{2} + \mathbb{Z}$, then setting

$$\zeta_k = \frac{1}{2} + \mu_k, \ \text{where } \mu_k \in \mathbb{Z} \text{ for any } k = 0, 1, 2, 3$$

we get

$$\zeta_0^2 + \zeta_1^2 + \zeta_2^2 + \zeta_3^2 = \left(\frac{1}{2} + \mu_0\right)^2 + \left(\frac{1}{2} + \mu_1\right)^2 + \left(\frac{1}{2} + \mu_2\right)^2 + \left(\frac{1}{2} + \mu_3\right)^2 = 1.$$

This can be done only if

$$\left|\left(\frac{1}{2} + \mu_k\right)\right| \geq \frac{1}{2}, \quad k = 0, 1, 2, 3,$$

which implies (since the sum is 1) that each of them is $\pm\frac{1}{2}$. This gives us the other 16 units,

$$\frac{1}{2}(\pm 1 \pm \mathbf{i} \pm \mathbf{j} \pm \mathbf{k})\}.$$

$\square$

The 24-cell polytope has 24 octahedral faces, 96 triangles, 96 edges and 24 vertices, consistent with the Euler characterisic being zero, $\chi(\Pi_{24} = C - F + E - V = 0$. The same number of vertices and octahedral faces is because of the fact that the 24-cell is self-dual, and this is precisely because it is constructed from two polytopes that are dual to each other: the hypercube and the cross-polytope. The 24-cell is unique to 4-dimensional space *and it has no analogues in higher or lower dimensions.*

To see this let us try to do a similar construction in $\mathbb{R}^3$. Take a cube and an octahedron. Center them both at the origin, line them up, and rescale them so all their vertices are at the same distance from the origin, as for example $(\pm 1, \pm 1, \pm 1)$ for the cube, and $(\pm\sqrt{3}, 0, 0)$, $(0, \pm\sqrt{3}, 0)$, and $(0, 0, \pm\sqrt{3})$ for the octahedron. We get a 3-dimensional body with $8 + 6 = 14$ vertices. But *it is not* a Platonic solid, but rather a rhombic dodecahedron. This was first discovered by Kepler. Another way to see the uniqueness of the 24-cell is (see Baez [12]) to take two cubes, equal in size. Chop one up into 6 pyramids, each having one face of the cube as its base, and each having the cube's center as its apex. Now take these 6 pyramids and glue their bases onto the faces of the other cube. What do we get? A rhombic dodecahedron!

Let us do the same thing in 4 dimensions. Take two hypercubes of equal size. Chop one up into 8 *hyperpyramids*, each having one face of the hypercube as its base, and each having the hypercube's center as its apex. Now, take these hyperpyramids and glue their bases onto the faces of the other hypercube. What do we get? The 24-cell!

We represent the 24 units as points in $\mathbb{R}^4$ with coordinates $(\pm 1, 0, 0, 0)$, $(0, \pm 1, 0, 0)$, $(0, 0, \pm 1, 0)$, $(0, 0, 0, \pm 1)$ and $(\pm\frac{1}{2}, \pm\frac{1}{2}, \pm\frac{1}{2}, \pm\frac{1}{2})$. One of the octahedral faces of Π_{24} has vertices

$$1 = (1, 0, 0, 0),\ \mathbf{i} = (0, 1, 0, 0),\ \xi = \left(\frac{1}{2}, \frac{1}{2}, \frac{1}{2}, \frac{1}{2}\right),$$

$$\xi - \mathbf{k} = \left(\frac{1}{2}, \frac{1}{2}, \frac{1}{2}, -\frac{1}{2}\right),\ \ \xi - \mathbf{j} - \mathbf{k} = \left(\frac{1}{2}, \frac{1}{2}, -\frac{1}{2}, \frac{1}{2}\right),$$

and

$$\xi - \mathbf{j} = \left(\frac{1}{2}, \frac{1}{2}, -\frac{1}{2}, \frac{1}{2}\right).$$

This octahedral face lies in 3-space or a hyperplane in $\mathbb{R}^4$ generated by the orthogonal vectors $(1, -1, 0, 0)$, $(0, 0, 1, 1)$ and $(0, 0, 1, -1)$, and is orthogonal to the vector $(1, 1, 0, 0)$ in $\mathbb{R}^4$. We note that all units lie on the unit hypersphere centered at the origin, but $\frac{1+\mathbf{i}}{2}$ lies on a hypersphere centered at 0 and of radius $\frac{1}{\sqrt{2}}$ in $\mathbb{R}^4$.

Let $I = \{z \in \Lambda : N(z) \equiv 0 \ (\mathrm{mod}\ 2)\}$. That is, I is the set of Hürwitz integers with even norms. Obviously, the units do not belong to I.

Proposition 13.4.27. *I is the two sided ideal in $\mathbb{H}$ contained in Λ and generated by $1 + i$.*

Proof. An easy calculation shows that if z_1 and z_2 are in I, then both $N(z_1)$ and $N(z_2)$ are even integers and so are $N(z_1 + z_2)$ and $N(z_1 z_2)$. This implies that I is a subring of Λ. Let $\eta \in \Lambda$ and $z \in I$. Since the norm is multiplicative,

$$N(z\eta) = N(\eta z) = N(z)N(\eta) \equiv 0 \quad (\mathrm{mod}\ 2)$$

showing that $z\eta$ and $\eta z \in I$. Thus I is a two sided ideal. Now, we show that $I = (1 + i)$. By the proof of Proposition 13.4.22 we know that Λ is a Euclidean domain and so for any $\eta \in \Lambda$, we can find q and r in Λ such that

$$\eta = (1 + i)q + r, \ \ \text{with } N(r) < N(1 + i) = 2 \text{ or } r = 0.$$

From this we get $r = \eta - (1 + i)q$, and so

$$N(r) = N(\eta - (1 + i)q) \equiv 0 \quad (\mathrm{mod}\ 2).$$

Hence $N(r) \neq 1$ and therefore $r = 0$. It follows that

$$\eta = (1 + i)q,$$

i.e. I is generated by $1 + i$. $\square$

Since I is a two sided ideal in Λ, Λ/I is an $\mathbb{R}$-algebra and

$$\Lambda = \bigcup_{\eta \in \Gamma} (\eta + I),$$

where Γ is a transversal (cross section) consisting of exactly one element from each coset modulo I. If $\Gamma = \{0, 1\}$, then $\Lambda = I \cup (1 + I)$.

Theorem 13.4.28. $\mathbb{R}^4$ *is tessellated*[10] *by the translates* $\Pi_{24} + q$, $q \in I$.

Proof. First observe that two cosets, $\eta_1 + I$ and $\eta_2 + I$, are equal if $N(\eta_1 - \eta_2)$ is an even integer. One calculates

$$N(\eta_1 - \eta_2) = N(\eta_1) + N(\eta_2) - 2\Re(\eta_1 \overline{\eta_2})$$

i.e. the two cosets are equal if $N(\eta_1)$ and $N(\eta_2)$ are both even or both odd. Then

$$N(\eta - 1) = 1 + 1 + 2\Re(\eta) \equiv 0 \pmod 2$$

and therefore $(\eta - 1) \in I$.

For each point p of the convex hull Π_{24} we can find a unit η such that $p + \eta \in \Pi_{24}$ and $\eta + I = 1 + I$ since $\eta - 1 \in I$ and $N(\eta - 1)$ is even.

Let q be any quaternion (in $\mathbb{R}^4$). Since Λ is a lattice in $\mathbb{R}^4$, we can find a point $\lambda \in \Lambda$ nearest to q. Let $q^{'} = q - \lambda$. Then,

$$|q^{'}| = |q - \lambda| \le |q - \lambda_1| \text{ for all } \lambda_1 \in \Lambda.$$

Therefore, $q^{'} \in \Pi_{24}$, i.e. $q = q^{'} + \lambda$ and $q \in \Pi_{24} + \lambda$. But

$$\Lambda = I \cup (1 + I)$$

and so

$$q \in \{\Pi_{24} + \lambda, \ \lambda \in I\} \text{ or } q \in \{\Pi_{24} + \lambda, \ \lambda \in 1 + I\}.$$

This is equivalent to

$$q \in \{\Pi_{24} + \lambda, \ N(\lambda) \equiv 0 \pmod 2\} \text{ or } q \in \{\Pi_{24} + \lambda, \ N(\lambda) \equiv 1 \pmod 2\}.$$

It remains to show that $\lambda \in I$. Indeed, if $\lambda \in 1 + I$, then, since $q^{'} \in \Pi_{24}$ we can find a unit $\eta_0 \in \mathcal{U}$ such that $q^{''} = q^{'} + \eta_0 \in \Pi_{24}$. Hence, $q^{''} = q - \lambda + \eta_0 = q - (\lambda - \eta_0)$ and

$$N(\lambda - \eta_0) = N(\lambda) + N(\eta_0) + 2\Re(\lambda \overline{\eta_0}) \equiv 0 \pmod 2.$$

Thus $\{q + \Pi_{24}, \ q \in I\}$ is a tessellation of $\mathbb{R}^4$. $\qquad\square$

---

[10]In Latin, tesella is a small cubic piece of clay, stone or glass, used to make mosaics. the word *tesella* means small square (from the Greek word $\tau\acute{\epsilon}\sigma\sigma\epsilon\rho\alpha$ tessera for four).

13.4.5   Exercises

Exercise 13.4.29. If k is a field of characteristic 0 and t is an indeterminate, then the ring $\mathrm{End}_k(k[t])$ contains the operators

$$x : f(t) \to t f(t)$$

and

$$y : f(t) \to \frac{d}{dt} f(t).$$

Let A be the subalgebra of $\mathrm{End}_k(k[t])$ generated by x and y. This is the *first Weyl algebra*.

1. Prove that $yx - xy = 1$.

2. Prove that A has a basis $\{x^m y^n \mid m, n \geq 0\}$. (It looks like the polynomial ring $k[x, y]$ but here x and y do not commute).

3. Prove that $A \cong k\langle x, y \rangle / I$ where I is the two-sided ideal generated by the element $yx - xy - 1$.

4. Prove that A is simple, i.e. it has no two-sided ideals other than (0) and A itself.

Exercise 13.4.30. Suppose that $k \subset K$ are two fields and A is a finite dimensional k-algebra. Note then that $K \otimes_k A$ is a finite dimensional K-algebra and for any A-module M, $K \otimes_k M$ is naturally a $K \otimes_k A$-module.

1. For any finite dimensional A-module M, show that

$$\mathrm{End}_{K \otimes_k A}(K \otimes_k M) \cong K \otimes_k \mathrm{End}_A(M).$$

2. A finite dimensional irreducible A-module M is called *absolutely irreducible* if the $K \otimes_k A$-module $K \otimes_k M$ is irreducible for *every* field extension K of k. If M is absolutely irreducible, prove that $\mathrm{End}_k(M) \cong k$. (Hint: take K to be algebraically closed.)

3. If M is a finite dimensional irreducible A-module such that $\text{End}_k(M) \cong k$, prove that the corresponding representation $\rho : A \to \text{End}_k(M)$ is surjective. Therefore, show that M is absolutely irreducible.

Exercise 13.4.31. Let $\mathbb{Z}_n$ denote the cyclic group of order n and let $k(\mathbb{Z}_n)$ denote its group algebra over a field k.

1. Prove that $k(\mathbb{Z}_n) \cong k[x]/(x^n - 1)$.

2. How many inequivalent irreducible representations over $\mathbb{C}$ does $\mathbb{Z}_n$ have? What are their dimensions? How is this consistent with Wedderburn's structure theorem for the semi-simple algebra $\mathbb{C}(\mathbb{Z}_n)$?

3. Which of these irreducible modules are absolutely irreducible?

13.4.6 The Symmetry Groups of the Platonic Solids

Here we lead the reader through a series of exercises which results in finally calculating the symmetry groups of the 5 Platonic Solids.

Definition 13.4.32. Let X be a solid body in $\mathbb{R}^3$. The *symmetry axes* ℓ of X are straight lines in $\mathbb{R}^3$ about which if X is rotated through an angle $\theta \in [0, 2\pi)$ to a new orientation, the new object appears identical to X. The *symmetry planes* Π of X are imaginary mirrors in which X can be reflected while appearing unchanged.

If G is a group of symmetries of a platonic solid, the reflections of G correspond to planes through the origin that divide the solid into mirror images. We will call such planes, *reflection planes*. It is clear that a reflection plane must do at least one of the following two things:

1. contain an edge of the solid, or

2. split a face of the solid into mirror images,

and that every plane through the origin that does (1) or (2) corresponds to a reflection.

Definition 13.4.33. (*Duality*) For each regular polyhedron, the *dual* polyhedron is defined to be the polyhedron constructed by,

1. Placing a point in the center of each face of the original polyhedron,

2. Connecting each new point with the new points of its neighboring faces, and

3. Eliminating the original polyhedron.

One sees that the tetrahedron is self dual, the cube is dual to the octahedron, and the icosahedron is dual to the dodecahedron.

Duality is important because the symmetry groups of dual solids are isomorphic. This is because since the dual of a solid is created from the midpoints, it is perfectly symmetrical inside of the original solid. Therefore, when the original solid is rotated (in any way), its dual is rotated in exactly same way.

Definition 13.4.34. The *direct symmetry group* of the solid body X, is denoted by $\Sigma_d(X)$, is the symmetry group of X if only rotations are permitted. The *full symmetry* group of X, denoted by $\Sigma(X)$, is the symmetries of X if both rotations and reflections are allowed.

Exercise 13.4.35. Show that dual polyhedra have isomorphic symmetry groups.

Exercise 13.4.36. If X is any Platonic solid except for the tetrahedron, then
$$\Sigma(X) = \Sigma_d(X) \times \mathbb{Z}_2.$$

Solution 13.4.37. Let X be a Platonic solid other than the tetrahedron. Then $\Sigma(X)$ a subgroup of $O_3(\mathbb{R})$. If $\Sigma_d(X) = \Sigma(X) \cap SO_3((\mathbb{R}))$, i.e. the subgroup of rotations. If
$$z : \mathbb{R}^3 \longrightarrow \mathbb{R}^3 \; : \; z(x) = -x, \text{ for all } x \in \mathbb{R}^3,$$

then z (which is a reflection), is a central element in $O_3(\mathbb{R})$, and
$$O_3(\mathbb{R}) = SO_3((\mathbb{R})) \times \langle z \rangle = SO_3((\mathbb{R})) \times \mathbb{Z}_2.$$

Thus, $[O_3(\mathbb{R}) : SO_3((\mathbb{R}))] = 2$. This implies

$$[\Sigma(X) : \Sigma_d(X)] \le 2.$$

The element z is a symmetry of all X. Therefore, $[\Sigma(X) : \Sigma_d(X)] = 2$ and

$$\Sigma(X) = \Sigma_d(X) \times \mathbb{Z}_2.$$

Exercise 13.4.38. Direct and full symmetry groups of tetrahedron are isomorphic to A_4 and S_4, respectively.

Solution 13.4.39. We denote by $\Sigma_d(T)$ the full symmetry group of the tetrahedron. $\Sigma_d(T)$ acts on the set of vertices of the tetrahedron, and thus we have a map $\varphi : \Sigma_d(T) \to S_4$. We claim that φ is injective. If φ is not injective, then there is some element $\tau \in \Sigma_d(T)$ besides the identity matrix which fixes all the vertices. Pick any three of the vertices. These are three linearly independent vectors in $\mathbb{R}^3$, and since τ fixes them all it must be the identity matrix. Thus φ is injective. This map identifies $\Sigma_d(T)$ with a subgroup of S_4. We will now find the order of $\Sigma_d(T)$.

We now show $|\Sigma_d(T)| = 12$. To see this consider the action of $\Sigma_d(T)$ on the vertices of the tetrahedron. The action is transitive, so the orbit of any vertex has order 4 and the stabilizer of any vertex has order 3, since there are 3 rotations around that vertex that preserve the tetrahedron. Hence $|\Sigma_d(T)| = 12$. This implies that $\Sigma_d(T)$ is isomorphic to an index-2 subgroup of S_4. Such a subgroup is normal and (for any n) the only such subgroup of S_n is the alternating group A_n. This shows that $\Sigma_d(T)$ is isomorphic to the alternating group A_4. Therefore, by Exercise 13.4.36 we have $\Sigma(T) = S_4$.

(We remark that for the tetrahedron, every reflection plane that splits a face also contains an edge, and every reflection plane containing an edge contains exactly one edge. Thus the number of reflections is equal to the number of edges of the tetrahedron, namely 6).

Exercise 13.4.40. For the cube (and the octahedron) $\Sigma_d(X) = S_4$ and $\Sigma(X) = S_4 \times \mathbb{Z}_2$ respectively.

Exercise 13.4.41. Direct and full symmetry groups of the dodecahedron (and so for the icosahedron) are $\Sigma_d(X) = A_5$ and $\Sigma(X) = A_5 \times \mathbb{Z}_2$, respectively.

Appendix

A. Every Field has an Algebraic Closure

In this construction, due to Emil Artin [11], Zorn's lemma takes the form that any ideal can be imbedded in a maximal ideal which we proved earlier. For the reader's convenience, we recall the definition of an algebraically closed field and an algebraic extension.

Definition A.1. The field K is *algebraically closed* if any non-constant polynomial with coefficients in K splits over K. A field L is called an *algebraic extension* of the field k if every element of L is a root of some polynomial of $k[x]$.

Definition A.2. Let k be a field. An extension K/k is called an *algebraic closure* of k if K is algebraically closed and K/k is an algebraic extension.

We need the following lemma in which the reader should understand the number of variables is infinite.

Lemma A.3. *Let k be a field and $k[\{x_i\}]$ be the polynomial ring in the variables x_i, where i ranges over some set J. Suppose for each i that $f_i(x)$ is a monic irreducible polynomial over k. Then for each i the ideal I of $k[\{x_i\}]$ generated by $f_i(x_i)$ is a proper ideal of $k[\{x_i\}]$.*

Proof. Suppose the contrary, i.e. $I = k[\{x_i\}]$. Then, there is an n and polynomials $g_1,\dots,g_n$ such that

$$1 = \sum_{j=1}^{n} f_j(x_{i_j}) g_j(x_{i_1},\dots,x_{i_n}).$$

For simplicity we shall write x_j instead of x_{i_j}. We can assume that all the g_j are polynomials in the variables $x_1,\dots,x_n$ by increasing the number of f_j if necessary. Also, assume that n is chosen to be minimal.

If $S = k[x_1,\dots,x_n]$ then $(f_1(x_1),\dots,f_n(x_n)) = S$. Let $R = k[x_1,\dots,x_{n-1}]$. By the minimality of n,

$$(f_1(x_1),\dots,f_{n-1}(x_{n-1})) \neq R.$$

Let us view the above equation as taking place in $S = R[x_n]$. If $c_m = f_m(x_m) \in R$, then $J = (c_1,\dots,c_{n-1},f_n(x_n)) = S$. Now set $I_0 = (c_1,\dots,c_{n-1}) \subseteq R$. So $J = (I, f_n(x_n))$. There are ring homomorphisms

$$R[x_n] \longrightarrow (R/I_0)[x_n] \longrightarrow (R/I_0)[x_n]/\overline{(f_n(x_n))},$$

where $\overline{(f_n(x_n))}$ is the image of $f_n(x_n)$ in $(R/I_0)[x]$. Since R/I_0 is a non-zero ring and $f_n(x_n)$ is not a unit since f_n is a monic polynomial of degree at least 1, we see that this last ring is non-zero. Hence the kernel of the composite homomorphism is a proper ideal of S. But J lies in this kernel, and so $J \neq S$. This contradiction proves the lemma. $\square$

We now construct an algebraic closure of k.

Theorem A.4. *Any field k has an algebraic closure.*

Proof. Suppose we have constructed extension fields $k = K_1 \subseteq K_2 \subseteq \cdots$ such that for all n the following hold.

1. K_{n+1} is algebraic over K_n.

2. Every non-constant $f(x) \in K_n[x]$ has a root in K_{n+1}.

Let

$$K = \bigcup_n K_n.$$

Since the $(K_n)_{n \in \mathbb{N}}$ form an ascending chain of fields, K is a field extension of k and since each K_n is algebraic over k, K is also algebraic over k. We will show that K is algebraically closed. To do this, take a non-constant polynomial $g(x) \in K[x]$. Since there are only finitely many coefficients of $g(x)$, it follows that $g(x) \in K_n[x]$ for some n sufficiently large. By assumption 2, $g(x)$ has a root in $K_{n+1} \subseteq K$. Hence K is algebraically closed.

We construct the fields K_n by induction. Suppose we have the fields $k \subseteq K_1 \subseteq K_2 \subseteq \cdots \subseteq K_n$. To construct K_{n+1} consider the polynomial ring $K_n[\{x_f\}]$ in the variables x_f, one variable for each monic irreducible polynomial $f(x) \in K_n[x]$. Let I be the ideal of $K_n[\{x_f\}]$ generated by $\{f(x_f)\}$ for each such f. Then, the lemma above implies that

$$I \neq K_n[\{x_f\}].$$

Set $R = K_n[\{x_f\}]/I$, which is a non-zero ring. Let $\mathfrak{m}$ be a maximal ideal of R, and finally set

$$K_{n+1} = R/\mathfrak{m}.$$

Then we have a sequence of ring homomorphisms

$$K_n \longrightarrow K_n[\{x_f\}] \longrightarrow K_n[\{x_f\}]/I = R \longrightarrow R/\mathfrak{m} = K_{n+1}.$$

Since the map $K_n \longrightarrow K_{n+1}$ is not the zero map (because $1 \mapsto 1$), this map is injective, so we may assume $K_n \subseteq K_{n+1}$. The ring $K_{n+1} = R/\mathfrak{m}$ is a field since $\mathfrak{m}$ is a maximal ideal of R. Furthermore, since the last two maps above are onto, if a_f is the image in K_{n+1} of x_f, then $K_{n+1} = K_n[\{a_f\}]$. Now $f(x_f) \mapsto 0$ in R since $f(x_f) \in I$. Thus $f(x_f) \mapsto 0$ in K_{n+1}. But $f(x_f) \mapsto f(a_f)$, and therefore $f(a_f) = 0$. Thus each a_f is algebraic over K_n. Hence since K_{n+1} is generated over K_n by the a_f, K_{n+1} is algebraic over K_n. Finally, if $g(x) \in K_n[x]$ and $f(x)$ is a monic irreducible factor of $g(x)$, $f(x)$ has a root in K_{n+1}, namely a_f. Thus, $g(x)$ has a root in K_{n+1}. $\qquad\square$

Remark A.5. We notice that $K = \bigcup K_n$, the algebraic closure of k, is nothing other than the *direct limit* of this sequence

$$k = K_1 \subseteq K_2 \subseteq \cdots$$

with addition and multiplication defined as follows: for a, $b \in K$, there exists an n such that a, $b \in K_n$, so we define $a + b$ and ab by using the operations of K_n. K is obviously a field containing K_n for each n.

We shall conclude our discussion of algebraic closures by dealing with the uniqueness question. Namely that up to an isomorphism, which is the identity on k, k has only one algebraic closure. In order to do this, we shall show that if k and k_1 are isomorphic fields and $\varphi : k \to k_1$ is an isomorphism, and if K and K_1 are the respective algebraic closures of k and k_1, then there is an isomorphism from K to K_1 that extends φ. Here again we use Zorn's lemma.

Theorem A.6. *Let k and k_1 be isomorphic fields and $\varphi : k \to k_1$ be an isomorphism. Further, suppose that K and K_1 are algebraic closures of k and k_1, respectively. Then, there is an isomorphism $\overline{\varphi} : K \to K_1$ which extends φ.*

Proof. We shall use the notation (k, φ, k_1) to mean that k and k_1 are isomorphic fields and $\varphi : k \to k_1$ is an isomorphism. Let

$$\mathcal{P} = \{(E, \sigma, E_1) \mid k \subseteq E \subseteq K,\ k_1 \subseteq E_1 \subseteq K_1,\ \sigma \text{ is an extension of } \varphi\}.$$

Note that $\mathcal{P} \neq \varnothing$ since $(k, \varphi, k_1) \in \mathcal{P}$. We have a natural partial order in $\mathcal{P}$ by inclusion:

$$(k, \varphi, k_1) \leq (L, \tau, L_1) \text{ if } k \subseteq L,\ k_1 \subseteq L_1, \text{ and } \varphi \subseteq \beta.$$

We will show every chain $(E_i, \tau_i, E_i')_{i \in I}$ has an upper bound. Let E and E' denote the direct limits

$$E = \varinjlim_{i \in I} E_i, \text{ and } E' = \varinjlim_{i \in I} E_i,$$

and let

$$\tau = \bigcup_{i \in I} \tau_i.$$

Then, E and E' are subfields of k and K, respectively, and

$$\tau : E \longrightarrow E'$$

is an isomorphism (the restriction of τ to E_i is τ_i, and for any $a, b \in E$ there exists an i such that $a, b \in E_i$). It follows that $(E, \tau, E') \in \mathcal{P}$ and it is an upper bound for the chain. By Zorn's lemma, $\mathcal{P}$ has a maximal element, say (Ξ, ξ, Ξ'). We claim $\Xi = K$. Suppose not. Then, there is an $a \in K - \Xi$. Since K is algebraic over k, a is algebraic over k. Let p be the minimal polynomial for a over k, and let p' be the image of p under the (natural) extension of φ to an isomorphism from $k[x]$ to $k_1[x]$. Since K_1 is an algebraic closure of k_1, p' splits in K_1. If all of the roots of p' were in Ξ', then p would have all its roots in Ξ. Thus p' has at least one root $b \in K_1 - \Xi'$. Now, ξ has a unique extension ξ' from $\Xi(a)$ to $\Xi'(b)$. But then $(\Xi(a), \xi', \Xi'(b)) \in \mathcal{P}$. Since

$$(\Xi, \xi, \Xi') < (\Xi(a), \xi', \Xi'(b)),$$

this contradicts the maximality of (Ξ, ξ, Ξ'). Thus $\Xi = K$. By a similar argument, we obtain that $\Xi' = K_1$. □

B. The Fundamental Theorem of Algebra

Here we shall give another proof (in addition to the one, due to Horowitz (see 11.6.5), in the chapter on Galois Theory) of the Fundamental Theorem of Algebra. This one due to Oliveira Santos [114] uses methods of Lie groups.

Theorem B.1. *The degree of an irreducible polynomial in $\mathbb{R}[x]$ is ≤ 2. Hence, any irreducible polynomial in $\mathbb{C}[x]$ has degree 1.*

Proof. Let $p(x) \in \mathbb{R}[x]$ be an irreducible polynomial of degree n. Consider $\langle p(x) \rangle$, the ideal generated by $p(x)$. Since $p(x)$ is irreducible, $\mathbb{R}[x]/\langle p(x) \rangle$ is a field.

We now consider the following map

$$\phi : \mathbb{R}^n \longrightarrow \mathbb{R}[x]/\langle p(x) \rangle,$$

defined by

$$(a_0, ..., a_{n-1}) \mapsto a_0 + a_1 x + \cdots + a_{n-1} x^{n-1} + \langle p(x) \rangle.$$

Then ϕ is a group isomorphism between these additive groups. Hence this isomorphism induces a field structure on $\mathbb{R}^n$, with the usual addition. We also consider the multiplication on the right side and transplant it to the left by ϕ^{-1} giving a continuous, commutative and associative multiplication on $\mathbb{R}^n$ (with all non-zero elements invertible). We denote the multiplicative identity element by 1. Choose a norm $|\cdot|$ on $\mathbb{R}^n$ (regarded as a real vector space) so that $|1| = 1$, and define for each $x \in \mathbb{R}^n$,

$$\| x \| = \sup_{|y|=1} |x \cdot y|.$$

Then, $\| 1 \| = 1$, and $\| x \cdot y \| \leq \| x \| \| y \|$ for all x, $y \in \mathbb{R}^n$. Now, both series

$$\sum_{n=0}^{\infty} \frac{x^n}{n!}, \quad \sum_{n=1}^{\infty} (-1)^{n+1} \frac{(x-1)^n}{n}$$

are absolutely and locally uniformly convergent with respect to this norm. The first of these we will denote by $\exp(x)$ in $\mathbb{R}^n$, and the second, denoted by $\log(x)$ defined on $\{x \in \mathbb{R}^n \mid \| x - 1 \| < 1\}$. Using the commmutativity of multiplication, one checks easily that

$$\exp(x + y) = \exp(x) \cdot \exp(y), \text{ for all } x, y \in \mathbb{R}^n.$$

Moreover, since $\exp(x)\exp(-x) = \exp(0) = 1$, we see $\exp(x) \neq 0$ for all $x \in \mathbb{R}^n$. Thus exp is a continuous homomorphism

$$\exp : (\mathbb{R}^n, +) \longrightarrow (\mathbb{R}^n - \{0\}, \times).$$

Moreover, using the argument in [1], Proposition 6.2.2, p. 265, one sees that

$$\exp(\log(x)) = x \text{ where } x \in \mathbb{R}^n, \ \| x - 1 \| < 1 \tag{1}$$

and

$$\log(\exp(x)) = x \text{ where } x \in \mathbb{R}^n, \ \| \exp(x) - 1 \| < 1. \tag{2}$$

Now, equation (1) shows that if U is a neighborhood of 0, then $\exp(U)$ is a neighborhood of 1, and since exp is a homomorphism, exp is an open map. This implies exp is surjective. This is because $\exp(\mathbb{R}^n)$ is an open subgroup of $(\mathbb{R}^n - \{0\}, \times)$. Since $\mathbb{R}^n - \{0\}$ is connected exp is surjective.

Now, equation (2) implies exp is injective on a neighborhood of zero. Hence Ker(exp) is discrete. Therefore if Ker(exp) $\neq \{0\}$, there must exist linearly independent vectors $v_1, \ldots, v_m$ $(m \geq 1)$ such that

$$\mathrm{Ker}(\exp) = \bigoplus_{k=1}^{m} \mathbb{Z}v_k.$$

Applying the first isomorphism theorem to the open surjective map exp we get a topological group isomorphism,

$$\mathbb{R}^n / \mathrm{Ker}(\exp) \longrightarrow \mathbb{R}^n - \{0\} \cong (S^1)^m \times \mathbb{R}^{n-m}.$$

We will prove $n \leq 2$. Suppose $n > 2$. Then the space $\mathbb{R}^n - \{0\}$ on the left is simply connected. However, the space on the right can not be simply connected unless $m = 0$. But then exp itself is injective and so $\mathbb{R}^n /$ is isomorphic to $\mathbb{R}^n - \{0\}$, which is impossible. Therefore $n \leq 2$. $\square$

C. Irrational and Transcendental Numbers

Irrationality

Theorem C.1. *Let α be a zero of a polynomial $x^n + a_1 x^{n-1} + \cdots + a_n \in \mathbb{Z}[x]$. Then α is either irrational or $\alpha \in \mathbb{Z}$. In the latter case we have $\alpha | a_n$.*

Proof. Suppose $x^n + a_1 x^{n-1} + \cdots + a_n$ has a rational zero p/q with $p, q \in \mathbb{Z}$, and $(p, q) = 1$, $q > 0$. Then $p^n + a_1 p^{n-1} q + \cdots + a_n q^n = 0$. Hence $q | p^n$ and since $(p, q) = 1$ this implies $q = 1$.

Now, we can see that $p^n + a_1 p^{n-1} + \cdots + a_n = 0$ implies $p | a_n$. $\square$

We recall the following corollary.

Corollary C.2. *Let $m \in \mathbb{N}$. Then any integer n is either the m-th power of an integer k, or $k = \sqrt[m]{n}$ is irrational.*

Proof. Notice that $k^m - n = 0$, i.e. k is a zero of the polynomial $x^m - n = 0$. Then, according to the theorem, either $k \in \mathbb{Z}$, and hence n is the m-th power of an integer or $k \notin \mathbb{Q}$. □

We recall that a complex number a is called *algebraic* if there exists a polynomial $p(x) \in \mathbb{Z}[x]$, $p(x) \neq 0$ with $p(a) = 0$. Otherwise a is called *transcendental*[11].

The most prominent examples of transcendental numbers[12] are π and[13] e. Though only a few classes of transcendental numbers are known (in part, because it can be extremely difficult to show that a given number is transcendental) transcendental numbers are not rare: indeed, almost all real and complex numbers are transcendental, since the algebraic numbers are countable while the sets of real and complex numbers are uncountable. All real transcendental numbers are irrational, since all rational numbers are algebraic. The converse is not true: not all irrational numbers are transcendental, e.g. $\sqrt{2}$ is irrational but is an algebraic number.

The name "transcendental" comes from Leibniz in his 1682 paper where he proved that $\sin x$ is not an algebraic function of x. Euler

---

[11]The existence of non-algebraic numbers was conjectured by Euler in 1744, in his Introduction to the analysis of the infinite. There, he comments, without a proof, that "...the logarithms of (rational) numbers which are not powers of the base are neither rational nor (algebraic) irrational, so they should be called transcendental...".

[12]The fact that the ratio of the circumference to the diameter of a circle is constant has been known for so long that it is quite untraceable. The earliest values of π include the "Biblical" value of 3 and the Egyptian Rhind Papyrus (1650 BCE) where the value for π equals $4 \times (8/9)^2 = 3.16$. The first theoretical calculation seems to have been done by Archimedes of Syracuse (287-212 BCE) who obtained the approximation $223/71 < \pi < 22/7$. Archimedes knew what many people to this day do not, that π does not equal $22/7$ and made no claim to have discovered the exact value. The first to use the symbol π with its present meaning was by a Welshman named Jones in 1706. Euler adopted it in 1737 and it became standard.

[13]It first appeared in 1618 in an appendix to Napier's work on logarithms (probably written by Oughred). Bernoulli first discovered e while looking at compound interest, and used the Binomial theorem to say that the limit of $(1+1/n)^n$ was between 2 and 3. Euler is credited with using the letter e. In 1731 Euler published *Introduction to Analysis* and showed that $e = \sum_{k=0}^{\infty} \frac{1}{k!}$, and that e is also equal to $\lim_{n \to \infty}(1 + \frac{1}{n})^n$. Euler also connected e to sin and cos using DeMoivre's formula.

was probably the first person to define transcendental numbers in the modern sense.

Joseph Liouville first proved the existence of transcendental numbers in 1844, [4], and in 1851 gave the first decimal examples such as the Liouville constant

$$\Sigma_{n=1}^{\infty} 10^{-n!} = 0.11000100000000000000000001\ldots\ldots$$

Proposition C.3. *The number e is irrational.*

Proof. We will prove this by contradiction. To proceed, recall that

$$e = \sum_{k=0}^{\infty} \frac{1}{k!} = 1 + 1 + \frac{1}{2!} + \cdots + \frac{1}{k!} + \cdots$$

and consider the partial sum

$$s_n = 1 + 1 + \frac{1}{2!} + \cdots + \frac{1}{n!}.$$

For any $n \geq 1$ we get

$$\begin{aligned} 0 < e - s_n &= \frac{1}{(n+1)!} + \frac{1}{(n+2)!} + \cdots \\ &= \frac{1}{(n+1)!}\left[1 + \frac{1}{n+2} + \frac{1}{(n+3)(n+2)} + \cdots\right] \\ &< \frac{1}{(n+1)!}\left[1 + \frac{1}{n+1} + \frac{1}{(n+1)^2} + \cdots\right] \\ &= \text{using the formula for a geometric series} \\ &= \frac{1}{(n+1)!}\left(\frac{1}{1 - \frac{1}{n+1}}\right) = \frac{1}{(n+1)!} \cdot \frac{n+1}{n} = \frac{1}{n!} \cdot n. \end{aligned}$$

We just proved that $0 < e - s_n < \frac{1}{n \cdot n!}$, or equivalently that

$$0 < n!(e - s_n) < \frac{1}{n} \tag{1}$$

and this for any n.

Now, assume that e is a rational number, say $e = \frac{a}{b}$, with a, $b \in \mathbb{N}$, and with no common factor. Pick any integer n such that $n > b$. Then, obviously, $n > 1$ and so $\frac{1}{n} < 1$. Then, by (1)

$$0 < n!(e - s_n) < 1,$$

which in particular, implies that $n!(e - s_n)$ is not an integer. But here we have

$$
\begin{aligned}
n!(e - s_n) &= n!\left(\frac{a}{b} - s_n\right) \\
&= n!\frac{a}{b} - n!\left[1 + 1 + \frac{1}{2!} + \frac{1}{3!} + \cdots + \frac{1}{n!}\right] \\
&= \frac{n!}{b} \cdot a - \left[n! + n! + \frac{n!}{2!} + \cdots + \frac{n!}{n!}\right].
\end{aligned}
$$

Therefore, since $n > b$, $\frac{n!}{b}$ is an integer, and so we have a sum of integers, i.e. $n!(e - s_n)$ is an integer, a contradiction.   □

Concerning the irrationality of the number π we present a simple proof[14] due to Niven ([98]).

Proposition C.4. *The number π is irrational.*

Proof. Suppose the contrary, namely that $\pi = \frac{a}{b}$, where a and b are positive integers. Consider the polynomial

$$p(x) = \frac{x^n}{n!}(a - bx)^n.$$

Since the polynomial $n!p(x)$ has degree $2n$, and each of its terms has an integral coefficient of degree $\geq n$, this means all its derivatives up to order $2n$ are polynomials all of whose terms also have integral coefficients. Hence $p(x)$ and all derivatives up to the order $2n$ take integral values at $x = 0$ and at $x = \frac{a}{b} = \pi$. On the other hand,

$$p\left(\frac{a}{b} - x\right) = p(x).$$

---

[14]The original proof that π is irrational is due to Johann Lambert, in 1761.

Now, set

$$P(x) = p(x) - \frac{d^2 p(x)}{dx^2} + \frac{d^4 p(x)}{dx^4} - \cdots + (-1)^n \frac{d^{2n} p(x)}{dx^{2n}}.$$

Thus,

$$\frac{d}{dx}\left[P'(x)\sin x - P(x)\cos x\right] = P''(x)\sin x + P'(x)\cos x - P'(x)\cos x$$

$$+ P(x)\sin x = P''(x)\sin x + P(x)\sin x$$

$$= \left[P''(x) + P(x)\right]\sin x = p(x)\sin x.$$

By the Fundamental Theorem of Calculus:

$$\int_0^\pi p(x)\sin x \, dx = \left[P'(x)\sin x - P(x)\cos x\right]_0^\pi = P(0) + P(\pi).$$

But, as we observed above, the derivatives $\frac{d^k p(x)}{d^k x}(0)$ and $\frac{d^k p(x)}{d^k x}(\pi)$ are integers for $k = 0, ..., 2n$. Therefore, $P(0) + P(\pi)$ must be an integer. But for $0 < x < \pi$,

$$0 < p(x) = \frac{x^n (a - bx)^n}{n!} < \frac{\pi^n a^n}{n!}$$

and since $|\sin x| \leq 1$, it follows that,

$$0 < p(x)\sin x < \frac{\pi^n a^n}{n!} \qquad \text{for any } 0 < x < \pi.$$

This means that the integral above is positive, and for n sufficiently large it is arbitrarily small, contradicting the fact that this integral is an integer. □

There are many proofs of the irrationality of π. An interesting one is based on the fact that trigonometric functions taken at rational points give irrational values.

Proposition C.5. *For any non-zero $r \in \mathbb{Q}$, $\cos r$ is irrational.*

Proof. Here we follow [107]. Without loss of generality (since $\cos(-r) = \cos r$) we will assume $r = \frac{a}{b}$, where $a, b \in \mathbb{N}$. Consider the function

$$f(x) = \frac{x^{p-1}(a-bx)^{2p}(2a-bx)^{p-1}}{(p-1)!} = \frac{(r-x)^{2p}[r^2 - (r-x)^2]^{p-1}b^{3p-1}}{(p-1)!}$$

where $p \neq 2$ is a prime number. For $0 < x < r$,

$$0 < f(x) < \frac{r^{2p}(r^2)^{p-1}b^{3p-1}}{(p-1)!} = \frac{r^{4p-2}b^{3p-1}}{(p-1)!}.$$

Now set

$$F(x) = f(x) - f^{(2)}(x) + f^{(4)}(x) - \cdots - f^{(4p-2)}(x).$$

We obtain

$$\frac{d}{dx}\left[F'(x)\sin x - F(x)\cos x\right] = F^{(2)}(x)\sin x + F'(x)\cos x$$
$$- \left(F'(x)\cos x - F(x)\sin x\right)$$
$$= F^{(2)}(x)\sin x + F(x)\sin x$$
$$= f(x)\sin x.$$

Integrating over $[0, r]$ we get,

$$\int_0^r f(x)\sin x\,dx = F'(x)\sin r - F(r)\cos r + F(0). \qquad (\dagger)$$

Notice that since $f(x)$ is a polynomial in $(r-x)^2$, $F'(r) = 0$. In addition, we observe the function $f(x)$ is of the form

$$\frac{x^{p-1}g(x)}{(p-1)!} \quad \text{with } g(x) \in \mathbb{Z}[x] \qquad (\star)$$

with $\frac{d^m f}{dx^m}(0) \in \mathbb{Z}$ for all $m = 0, 1, \dots$. Moreover, $p \mid \frac{d^m f}{dx^m}(0)$ for all m, with a possible exception of $m = p-1$. Looking at the $p-$1st derivative of $f(x)$, we see

$$\frac{d^{p-1}f}{dx^{p-1}}(0) = \frac{(p-1)!\, a^{2p}(2a)^{p-1}}{(p-1)!} = a^{2p}(2a)^{p-1}.$$

This shows that if we choose $p > a$, then $p \nmid f^{(p-1)}(0)$. But $p \mid f^m(0)$ for all other $m \neq p$. Hence $F(0) \in \mathbb{Z}$ and $p \nmid F(0)$. Therefore, by setting $F(0) = q$, we get $(p, q) = 1$.

Let us see what happens to $F(r)$. To do this, we first calculate

$$f(r - x) = \frac{x^{2p}(r^2 - p^2)^{p-1}b^{3p-1}}{(p-1)!}$$
$$= \frac{x^{2p}(a^2 - b^2x^2)^{p-1}b^{p+1}}{(p-1)!}.$$

Notice $f(r - x)$ is of the form $(\star)$, but this time,

$$g(x) = x^{p+1}(a^2 - b^2x^2)^{p-1}b^{p+1}.$$

This implies that $f^{(m)}(r) \in \mathbb{Z}$ and $p \mid f^{(m)}(r)$ for any m, since now $g(0) = 0$. Hence there is some $k \in \mathbb{Z}$ such that $F(r) = pk$. Replacing in (†), we get

$$\int_0^r f(x) \sin x \, dx = -pk \cos r + q.$$

Now, if $\cos r$ was in $\mathbb{Q}$, then $\cos r = \frac{\gamma}{\delta}$, and so

$$\int_0^r f(x) \sin x \, dx = -pk \frac{\gamma}{\delta} + q.$$

In other words,

$$\delta \int_0^r f(x) \sin x \, dx = -pk\gamma + \delta q.$$

Now, it is time to add another condition on p: we require $p > \delta$. This forces (since $(p, q) = 1$) $p \nmid \delta q$, so $p \nmid (-pk\gamma + \delta q)$ and hence $-pk\gamma + \delta q \neq 0$.

Now,

$$\left| \delta \int_0^r f(x) \sin x \, dx \right| < \delta r \frac{r^{4p-2}b^{3p-1}}{(p-1)!}$$
$$= \delta r^3 b^2 \frac{(r^4 b^3)^{p-1}}{(p-1)!}$$
$$= \frac{\varepsilon_1 \varepsilon_2^{p-1}}{(p-1)!},$$

where the constants $\varepsilon_1 = \delta r^3 b^2$ and $\varepsilon_2 = r^4 b^3$ are independent of p. Taking limits,

$$\lim_{p \to \infty} \frac{\varepsilon_1 \varepsilon_2^{p-1}}{(p-1)!} = 0.$$

This shows that we can choose p sufficiently large and obtain

$$-1 < \delta \int_0^r f(x) \sin x dx < 1.$$

But

$$\delta \int_0^r f(x) \sin x dx = (-pk\gamma + \delta q) \in \mathbb{Z} - \{0\},$$

which is a contradiction. $\qquad\square$

A consequence of this proposition is the following corollary.

Corollary C.6. π *is irrational.*

Proof. If π were rational, then by the proposition above $\cos \pi$ has to be irrational. But $\cos \pi = -1$. $\qquad\square$

Corollary C.7. *The trigonometric functions are irrational at non-zero rational values of the arguments.*

Proof. If $\sin x \in \mathbb{Q}$ for $x \in \mathbb{Q}$, $x \neq 0$, then $1 - 2\sin^2 x = \cos 2x \in \mathbb{Q}$. But this conclusion contradicts Proposition C.5 since $2x \in \mathbb{Q}$.
If $\tan x \in \mathbb{Q}$ for $x \in \mathbb{Q}$, $x \neq 0$, then

$$\cos 2x = \frac{1 - \tan^2 x}{1 + \tan^2 x} \in \mathbb{Q}.$$

So, again we have a contradiction.

Finally, $\csc x$, $\sec x$ and $\cot x$ are irrational numbers as reciprocals of irrationals. $\qquad\square$

Corollary C.8. *Any non-zero value of an inverse trigonometric function is irrational for rational arguments.*

Proof. Let $x \in \mathbb{Q}$. If $\arccos x = y \in \mathbb{Q}$, then $\cos y = x$, contrary to Proposition C.5. The same holds for all the other inverse trigonometric functions. □

On the other hand there are many *naturally* occurring numbers for which no irrationality results are known. For example, it is not known whether Euler's constant

$$\gamma = \lim_{n \to \infty} \left(\sum_{k=1}^{n} \frac{1}{k} - \log n \right)$$

or $e + \pi$ or $e\pi$ is irrational. Motivated by the standard series for e P. Erdos asked the following question. Is

$$\sum_{k=1}^{\infty} \frac{1}{k! + 1}$$

irrational? Surprisingly this seems to be difficult to answer.

Transcendence

The Transcendence of π
Here we present the proof of I. Niven, [97].

Theorem C.9. *The number π is transcendental.*

Proof. We assume that π is algebraic and we will arrive at a contradiction. Observe that if π were algebraic, then, by 11.8.5, $i\pi$ would be as well. So, let $\theta_1(x) \in \mathbb{Z}[x]$ be a polynomial of degree n with $i\pi$ as one root. Denote by $\varrho_1 = i\pi$, ϱ_2, ... , ϱ_n the roots of $\theta_1(x) = 0$. Using Euler's equation

$$e^{i\pi} + 1 = 0$$

we get

$$(e^{\varrho_1} + 1)(e^{\varrho_2} + 1) \cdots (e^{\varrho_n} + 1) = 0. \tag{1}$$

Now, we construct an algebraic equation with integer coefficients whose roots are $\varrho_1 = i\pi$, $\varrho_2,...,\varrho_n$. To do this, first we consider the following expressions:

$$\varrho_1 + \varrho_2, \ \varrho_1 + \varrho_3, \ \ldots, \ \varrho_{n-1} + \varrho_n. \tag{2}$$

Equation $\theta_1(x) = 0$ implies that the elementary symmetric functions of $\varrho_1, \ldots, \varrho_n$ are rational numbers. Hence, the elementary symmetric functions of the expressions (2) are rational numbers. It follows that the quantities in (2) are roots of

$$\theta_2(x) = 0, \tag{3}$$

an algebraic equation with integral coefficients. Similarly, the sums of the ϱ's taken three at a time are the $\binom{n}{3}$ roots of

$$\theta_3(x) = 0. \tag{4}$$

Proceeding in this way, we obtain

$$\theta_4(x) = 0, \quad \theta_5(x) = 0, \quad \ldots, \quad \theta_n(x) = 0. \tag{5}$$

These are algebraic equations with integral coefficients, whose roots are the sums of the ϱ's taken 4, 5,... , n at a time, respectively. The product equation

$$\theta_1(x)\theta_2(x)\cdots\theta_n(x) = 0, \tag{6}$$

has roots which are precisely the exponents in the expansion of (1). The deletion of zero roots (if any) from equation (6) gives

$$\theta(x) = cx^r + c_1 x^{r-l} + \cdots + c_r = 0, \tag{7}$$

whose roots $\beta_1, \beta_2, \ldots, \beta_r$ are the non-vanishing exponents in the expansion of (1), and whose coefficients are integers. Hence, (1) may be written in the form

$$e^{\beta_1} + e^{\beta_2} + \cdots + e^{\beta_r} + k = 0, \tag{8}$$

where k is a positive integer. We define

$$f(x) = \frac{c^s x^{p-1}\big(\theta(x)\big)^p}{(p-1)!}, \tag{9}$$

where $s = rp - 1$, and p is a prime to be specified. We also define

$$F(x) = f(x) + f^{(1)}(x) + f^{(2)}(x) + \cdots + f^{(s+p+l)}(x), \tag{10}$$

noting that the derivative

$$\frac{d}{dx}e^{-x}F(x) = -e^{-x}f(x).$$

Hence,

$$e^{-x}F(x) - e^0 F(0) = \int_0^x (-e^{-\xi})f(\xi)d\xi.$$

Using the substitution $\xi = \tau x$ we get

$$F(x) - e^x F(0) = -x\int_0^1 e^{(1-r)x}f(\tau x)d\tau.$$

Let x range over the values β_1, β_2, ..., β_r and add the resulting equations. Using (8) we obtain

$$\sum_{i=1}^r F(\beta_i) + kF(0) = -\sum_{i=1}^r \int_0^1 e^{(1-r)\beta_i}f(\tau\beta_i)d\tau. \qquad (11)$$

This result gives us the contradiction we desire. For we can choose the prime p to make the left side a non-zero integer, and the right side as small as we please.

By (9),

$$\sum_{i=1}^r f^{(t)}(\beta_i) = 0 \text{ for } 0 \le t < p.$$

Also, by (9), the polynomial obtained by multiplying $f(x)$ by $(p-1)!$ has integral coefficients. Since the product of p consecutive positive integers is divisible by $p!$, the p-th and higher derivatives of $(p-1)!f(x)$ are polynomials in x with integral coefficients divisible by $p!$. Hence, the p-th and higher derivatives of $f(x)$ are polynomials with integral coefficients each of which is divisible by p. That each of these coefficients is also divisible by c^s is obvious from definition (9). Thus, we have shown that, for $t \ge p$, the quantity $f^{(t)}(\beta_i)$ is a polynomial in β_i of degree at most s, each of whose coefficients is divisible by pc^s.

By (7), a symmetric function of β_1, β_2, ..., β_r with integral coefficients and of degree at most s is an integer provided each coefficient is divisible by c^s by the Fundamental Theorem of Symmetric Polynomials (see 5.7.32 of Volume I). Hence,

$$\sum_{i=1}^{r} f^{(t)}(\beta_i) = pz_t, \quad (t = p,\, p+1,\, \ldots,\, p+s),$$

where the z_t are integers.

It follows that

$$\sum_{i=1}^{r} F(\beta_i) = p \sum_{t=p}^{p+s} z_t.$$

In order to complete the proof that the left side of (11) is a non-zero integer, we now show that $zF(0)$ is an integer relatively prime to p. From (9) it is clear that

$$
\begin{aligned}
&f^{(t)}(0) = 0, \quad (t = 0,\, 1,\, \ldots,\, p-2),\\
&f^{(p-1)}(0) = c^s c_r^p,\\
&f^{(t)}(0) = pK_t, \quad (t = p,\, p+1,\, \ldots,\, p+s),
\end{aligned}
$$

where the K_t are integers. If p is chosen greater than each of z, c, c_r (which is possible since the number of primes is infinite), the desired result follows from (10). Finally, the right side of (11) equals

$$-\sum_{i=1}^{r} \frac{1}{c} \int_0^1 \frac{\left(c^r \beta_i \theta(\tau\beta_i)\right)^p}{(p-1)!} e^{(1-r)\beta_i} d\tau.$$

This is a finite sum, each term of which may be made as small as we wish by choosing p very large, because

$$\lim_{p\to\infty} \frac{\left(c^r \beta_i \theta(\tau\beta_i)\right)^p}{(p-1)!} = 0.$$

$\square$

Theorem C.10. (*The Gelfond-Schneider Theorem*[15]) *If α and β are algebraic numbers with $\alpha \neq 0, 1$ and β is not a rational number, then every value of α^β is transcendental.*

For a proof see [99], p. 142.

Corollary C.11. *The number e^π is transcendental. Indeed, we notice*

$$e^\pi = (e^{i\pi})^{-i} = (-1)^{-i},$$

and since $-i$ is algebraic but not rational, by the Gelfond-Schneider theorem, it follows that e^π is transcendental.

By the same theorem we can see that, for example, the numbers 2^i, $3^{\sqrt{5}}$, $\sqrt{2}^{\sqrt{5}}$ and i^{-2i} are all transcendental.

Also $i^i = 0.207879576....$ Since i is algebraic but irrational, the theorem applies. Note also: i^i is equal to $e^{-\pi/2}$ and several other values. Consider $i^i = e^{i \log i} = e^{i \cdot i\pi/2}$. Since $\log$ is multivalued, there are other possible values for i^i.

Remark C.12. Up to today we do not know if π^e is transcendental or not. Also, the question remains open as to whether $\pi + e, \pi - e, \pi e, \pi/e, \pi^e, \pi^{\sqrt{2}}, \pi^\pi, e^{\pi^2}, \ln \pi, 2^e, e^e$ are rational, algebraic irrational, or transcendental numbers.

Remark C.13. The numbers α and β in the theorem are not restricted to be real numbers; complex numbers are allowed (but never rational, even if both the real and imaginary parts are rational).

If the restriction that α and β are algebraic is removed, the statement does not remain true in general. For example,

$$\left(\sqrt{2}^{\sqrt{2}} \right)^{\sqrt{2}} = \sqrt{2}^{\sqrt{2} \cdot \sqrt{2}} = \sqrt{2}^2 = 2.$$

---

[15]This theorem was proved independently in 1934 by Aleksandr Gelfond ([47]) and Theodor Schneider (in his PhD thesis). The Gelfond-Schneider theorem answers affirmatively Hilbert's seventh problem.

A Geometric Curiosity

It is known that the volume of the n-dimensional sphere of radius r is given by:

$$V_n = \frac{\pi^{\frac{n}{2}} r^n}{\Gamma(\frac{n}{2} + 1)},$$

where Γ is the gamma function. Since $\Gamma(n) = (n-1)!$, any even-dimensional unit sphere has volume

$$V_{2n} = \frac{\pi^n}{n!}$$

and, summing up the unit-sphere volumes of all even-dimensional spaces we see that

$$\sum_{n=0}^{\infty} V_{2n} = e^{\pi}.$$

Bibliography

[1] H. Abbaspour, M. Moskowitz, *Basic Lie Theory*, World Scientific Publishing Co., River Edge, NJ, 2007.

[2] A. Abbondandolo. and R. Matveyev, *How large is the shadow of a symplectic ball?*, arXiv: 1202.3614v3 [math.SG], 2013.

[3] I. T. Adamson, *Introduction to Field Theory*, Cambridge University Press, Cambridge, 2nd ed. 1982.

[4] M. Aigner, G. Ziegler, *Proofs from the book*, 4th ed., Springer-Verlag, Berlin, New York, 2010.

[5] Y. Akizuki, *Teilerkettensatz und Vielfachensatz*, Proc. Phys.-Math. Soc. Japan, vol. **17**, 1935, pp. 337-345.

[6] D. B. Ames, *An Introduction to Abstract Algebra*, Inter. Textbook Company, Scranton, Pennsylvania, 1969.

[7] V. I. Arnold, *Mathematical Methods of Classical Mechanics*, Springer-Verlag, Berlin, 2nd edition, 1978.

[8] M. Aristidou, A. Demetre, *A Note on Quaternion Rings over $\mathbb{Z}_p$*, Inter. Journal of Algebra, vol. **3**, 2009, no. 15, pp. 725-728.

[9] E. Artin, *Über einen Satz von Herrn J. H. Maclagan Wedderburn*, Abhandlungen aus dem Mathematischen Seminar der Hamburgischen Universität, **5** (1927), pp. 245-250.

[10] E. Artin, *Galois Theory*, University of Notre Dame, 1946.

[11] E. Artin, *Geometric Algebra*, Interscience Publishers, Inc., New York, NY., 1957.

[12] J. Baez, *Platonic Solids in all Dimensions*, math.ucr.edu/home/baez/platonic.html, 2006.

[13] J. C. Baez, *The Octonions*, Bull. Amer. Math. Soc., vol. **39**, 2002, pp. 145-205.

[14] J. Barnes, *Gems of Geometry*, Springer-Verlag Berlin Heidelberg, 2009.

[15] F. Beukers, E. Calabi, J. Kolk, *Sums of generalized harmonic series and volumes*, Nieuw Archief voor Wiskunde vol. **11** (1993), pp. 217-224.

[16] J. Barshey, *Topics in Ring Theory*, W.A. Benjamin, Inc., 1969.

[17] D. Birmajer, J. B. Gil, *Arithmetic in the ring of the formal power series with integer coefficients*, Amer. Math. Monthly, vol. **115**, 2008, pp. 541-549.

[18] D. Birmajer, J. B. Gil and M. D. Weiner, *Factoring polynomials in the ring of formal power series over* $\mathbb{Z}$, Intr. J. Number Theory, vol. **8**, no. 7, 2012, pp. 1763-1776.

[19] M. Boij and D. Laksov, *An Introduction to Algebra and Geometry via Matrix Groups*, Lecture Notes, 2008.

[20] A. Borel, *Compact Clifford-Klein forms of symmetric spaces*, Topology (2) 1963, 111-122.

[21] N. Bourbaki, *General Topology*, Chapters 5-10, Springer-Verlag, Berlin, 1998.

[22] P. Bürgisser, F. Cucker, *Condition. The Geometry of Numerical Algorithms*, Springer, Berlin, 2013.

[23] H. Cartan and S. Eilenberg, *Homological Algebra*, Princeton University Press, Princeton, N.J., 1956.

[24] K. Chandrasekharan, *Introduction to Analytic Number Theory*, Springer-Verlag, New York, 1968, p. 10.

[25] C. Chevalley, *On the theory of local rings*, Annals of Math., **44**, 1943, pp. 690-708.

[26] A. H. Clifford, *Representations Induced In An Invariant Subgroup*, Ann. of Math. v. 38, no. 3 (1937).

[27] R. Courant, H. Robbins, *What is Mathematics? An elementary Approach to Ideas and Methods*, 2nd Edition (revised), Oxford University Press, 1996.

[28] C. Crompton, *Some Geometry of the p-adic Rationals*, 2006.

[29] M. L. Curtis, *Matrix Groups*, 2nd Ed. Springer-Verlag, Berlin, 1987.

[30] C. Curtis and I. Reiner, *Representations of Groups and Associative Algebras*, John Wiley Sons. 1962.

[31] J. Dieudonné, *Topics in Local algebra.*, University of Notre Dame Press, 1967.

[32] J. Dieudonné, *Sur les générateurs des groupes classiques*, Summa Brad. Math., vol. **3**, 1955, pp. 149-180.

[33] L. E. Dickson, *On finite algebras*, Nachrichten der Gesellschaft der Wissenschaften zu Göttingen, (1905), pp. 358-393.

[34] L. E. Dickson, *Linear Algebras.* Cambridge University Press, Cambridge, 1930.

[35] J. Draisma, D. Gijswijt, *Invariant Theory with Applications*, Lecture Notes, 2009.

[36] S. Eilenberg, S. MacLane, *Group extensions and homology*, Annals of Math. vol. **43**, 1942, pp. 757-831.

[37] D. Eisenbud, *Commutative Algebra, with a View Toward Algebraic Geometry.*, Springer, Berlin, 2004.

[38] P. Erdös, *Über die Reihe* $\sum \frac{1}{p}$, Mathematica, Zutphen B **7** (1938), pp. 1-2.

[39] L. Euler, *Introductio in Analysin Infinitorum*, Tomus Primus, Lausanne (1748), Opera Omnia, Ser. 1, vol. **90**.

[40] W. Feit and J. G. Thompson, *Solvability of groups of odd order*, Pacific J. Math., vol. **13** (1963), pp. 775-1029.

[41] L. Fuchs, *Infinite Abelian Groups.* vol. I, Academic Press, 1970.

[42] H. Fürstenberg, *On the infinitude of primes*, Amer. Math. Monthly **62** (1955), p. 353.

[43] J. Gallier, *Logarithms and Square Roots of Real Matrices*, in arXiv:0805.0245v1, 2008.

[44] C. F. Gauss, *Disquisitiones Aritmeticae*, (translated by Arthur A. Clarke), Yale University Press, 1965.

[45] B. R. Gelbaum. J. M. H. Olmsted, *Counterexamples in Analysis*, Dover Publications, Inc. (1964).

[46] B. R. Gelbaum. J. M. H. Olmsted, *Theorems and Counterexamples in Mathematics*, Springer Verlag, Berlin, 1990.

[47] A. Gelfond, *Sur le septième Problème de Hilbert* Bulletin de l'Acadèmie des Sciences de l'URSS. Classe des sciences mathématiques et naturelles. VII (4): 1934, pp. 623-634.

[48] J. W. Gibbs, *Elements of Vector Analysis Arranged for the Use of Students in Physics*, Tuttle, Morehouse and Taylor, New Haven, 1884.

[49] J. W. Gibbs, *On Multiple Algebra*, Proceedings of the American Association for the Advancement of Science, vol. **35**, 1886.

[50] O. Goldman, *Hilbert rings and the Hilbert Nullstellensatz*, Math. Z., vol. **54**, 1951, pp. 136-140.

[51] D. M. Goldschmidt, *A group theoretic proof of the pq theorem for odd primes*, Math. Z., vol. **113**, 1970, pp. 373-375.

[52] R. Goodman, N. Wallach, *Symmetry, Representations, and Invariants*, Springer, Berlin, 2009.

[53] M. A. de Gosson, *Introduction to Symplectic Mechanics: Lectures I-II-III*, Lecture Notes from a course at the University of St-Paulo, May-June 2006.

[54] M. A. de Gosson, *Symplectic Geometry, Wigner-Weyl-Moyal Calculus, and Quantum Mechanics in Phase Space.*

[55] M. Gray, *A radical approach to Algebra*, Addison-Wesley Publishing Company, 1970.

[56] M. Gromov, *Pseudo-holomorphic curves in symplectic manifolds*, Invent. Math., vol. **81**, 1985, pp. 307-347.

[57] M. Hall, *The Theory of Groups*, AMS, Second Edition, 1999.

[58] P. Halmos, *What Does the Spectral Theorem Say?*, American Mathematical Monthly, vol. **70**, number 3, 1963, pp. 241-247.

[59] A. Hatcher, *Algebraic Topology*, Cambridge University Press, 2002.

[60] T. Head, *Modules; A Primer of Structure Theorems.*, Brooks/Cole Publishing Company, 1974.

[61] I. N. Herstein, *Wedderburn's Theorem and a Theorem of Jacobson*, Amer. Math. Monthly, vol. **68**, No. 3 (Mar., 1961), pp. 249-251.

[62] C. Hopkins, *Rings with minimum condition for left ideals*, Ann. Math. vol. **40**, 1039, pp. 712-730.

[63] L. Horowitz, *A proof of the Fundamental theorem of algebra by means of Galois theory and 2-Sylow groups*, Nieuw Arch. Wisk. (3) 14, 1966, p. 95-96.

[64] W. Y. Hsiang, *Lectures on Lie Groups*, World Scientific Publishing Co., 2000.

[65] T. W. Hungerford, *Algebra*, Springer, Berlin, 1974.

[66] N. Jacobson, *Lie Algebras*. Wiley-Interscience, New York, 1962.

[67] N. Jacobson, *Structure theory of simple rings without finiteness assumptions*, Trans. Amer. Math. Soc., vol 57, 1945, pp. 228-245.

[68] N. Jacobson, *Basic algebra I, II*. Dover publications, NY, 1985.

[69] D. M. Kan, *Adjoint Functors*, Trans. Amer. Math. Soc., vol. 87, 1958, pp. 294-329.

[70] I. Kaplansky, *Infinite Abelian groups*, University of Michigan Press, Ann Arbor 1954.

[71] I. Kaplansky, *Commutative Rings*, Allyn and Bacon, Boston, MA, 1970.

[72] M. G. Katz, *Systolic Geometry and Topology*, Am. Math. Soc., 2007.

[73] W. Krull, *Dimensionstheorie in Stellenringen*, J. reine angew. Math. 179, 1938, pp. 204-226.

[74] W. Krull, *Jacobsonsches Radikal und Hilbertscher Nullstellensatz*, in Proceedings of the International Congress of Mathematicians, Cambridge, MA, 1950, vol. 2, Amer. Math. Soc., Providence, 1952 pp. 56-64.

[75] S. Lang, *Algebra*, Addison-Wesley Publishing Company, 1965.

[76] J. Levitzki, *On rings which satisfy the minimum condition for the right-hand ideals.*, Compositio Mathematica, vol. 7, 1939, pp. 214-222.

[77] J. H. McKay, *Another proof of Cauchy's group theorem*, Amer. Math. Monthly **66**, 1959, p. 119.

[78] I. G. Macdonald, *Symmetric Functions and Hall Polynomials*, Oxford University Press, 2nd edition, 1995.

[79] G. W. Mackey, *The Scope and History of Commutative and Noncommutative Harmonic Analysis*, The History of Mathematics, AMS. vol. **5**, 1992.

[80] S. Mac Lane, *Homology*, Springer, 1995.

[81] P. Malcolmson, Frank Okoh, *Rings Without Maximal Ideals*, Amer. Math. Monthly, vol. **107**, No. 1 (2000), pp. 60-61.

[82] H. Matsumura, *Commutative ring theory*, Cambridge University Press, Cambridge 1986.

[83] J. P. May, *Munchi's Proof of the Nullstellenzatz*, The Amer. Math. Monthly, vol. **110**, No. 2, 2003, pp. 133-140.

[84] E. Meinrenken, *Symplectic Geometry*, Lecture Notes, University of Toronto, 2000.

[85] C. M. Mennen, *The Algebra and Geometry of Continued Fractions with Integer Quaternion Coefficients*, University of the Witwatersrand, Faculty of Science, School of Mathematics, 2015.

[86] P. Monsky, *On Dividing a Square into Trianles*, Amer. Math. Monthly, vol. **77** (2), 1970, pp. 161-164.

[87] L. J. Mordell, *A Statement by Fermat*, Proceedings of the London Math. Soc., vol. **18**, 1920, pp. v-vi.

[88] M. Moskowitz, *A Course in Complex Analysis in One Variable*, World Scientific Publishing Co., Inc., River Edge, NJ, 2002.

[89] M. Moskowitz, *Homological Albegra in Locally Compact Abelian Groups*, Trans. A.M.S., vol. **127**, no. 3, 1967.

[90] M. Moskowitz, *Adventures in Mathematics*, World Scientific Publishing Co., Inc., River Edge, NJ, 2003.

[91] L. Nachbin, *The Haar integral*, D. Van Nostrand Co. Inc., Princeton, N.J.-Toronto-London 1965.

[92] M. Nagata, *Local Rings.*, Interscience Publishers, 1962.

[93] M. A. Naimark, *Normed Rings*, P. Noordoff N.V., The Netherlands, 1959.

[94] R. Narasimhan, *Analysis on Real and Complex Manifolds*, Advanced Studies in Pure Mathematics, Masson & Cie-Paris, 1973.

[95] J. Neukirch, *Algebraic Number Theory*, Springer, Berlin, 1999.

[96] M. Neusel, *Invariant Theory*, Am. Math. Soc., 2007.

[97] I. Niven, *The Transcendence of* π, Amer. Math. Monthly, vol. **46**, No. 8, 1939, pp. 469-471.

[98] I. Niven, *A simple proof that* π *is irrational*, Bull. Amer. Math. Soc., vol. **53**, 1947, p. 509.

[99] I. Niven, *Irrational numbers*, Carus Mathematical Monographs, no. **11**, A.M.S., Washington, D.C., 1985.

[100] E. Noether, *Ableitung der Elementarteilertheorie aus der Gruppentheorie*, Nachrichten der 27 Januar 1925, Jahresbericht Deutschen Math. Verein. (2. Abteilung) 34 (1926), 104.

[101] Yong-Geun Oh, *Uncertainty Principle, Non-Squeezing Theorem and the Symplectic Rigidity*, Lecture for the 1995 DAEWOO Workshop, Chungwon, S. Korea.

[102] O. T. O'Meara, *Introduction to Quadratic fields*, Springer, Berlin, 1973.

[103] O. T. O'Meara, *Symplectic groups*, Amer. Math. Soc. 1978.

[104] O. Ore, *Number theory and its History*, Dover Publications, 1976.

[105] R. S. Palais, *The Classification of Real Division Algebras*, vol. **75**, No. 4, 1968, pp. 366-368.

[106] J. Pardon, *The Hilbert-Smith conjecture for three-manifolds*, J. Amer. Math. Soc., vol. **26** (3), 2013, pp. 879-899.

[107] E. Patterson, V. Shchogolev, *Algebraic and Transcendental Numbers*, Summer Number Theory Seminar, 2001.

[108] V. Perić and M. Vuković, *Some Examples of Principal Ideal Domain Which Are Not Euclidean and Some Counterexamples*, Novi Sad J. Math. vol. **38**, No. 1, 2008, pp. 137-154.

[109] H. Poincaré, *Analysis situs*, J. Éc. Polytech., ser. 21, 1895, pp. 1-123.

[110] H. Poincaré, *Second complément à l'analysis situs*, Proc. London Math. Soc., vol. **32**, 1900, pp. 277-308.

[111] L. S. Pontryagin, *Topological groups*, Translated from the second Russian edition by Arlen Brown, Gordon and Breach Science Publishers Inc., New York-London-Paris 1966.

[112] M. Reid, *Undergraduate Algebraic Geometry*, Cambridge University Press, 1989.

[113] G. Ricci, T. Levi-Civita, *Méthodes de Calcul Différentiel Absolu et Leurs Applications*, Math. Annalen, vol. **54**, 1901, pp. 125-201.

[114] O. Santos, *Another Proof of the Fundamental Theorem of Algebra*, Amer. Math. Monthly, January 2005.

[115] A. R. Schep, *A Simple Complex Analysis and an Advanced Calculus Proof of the Fundamental Theorem of Algebra*, Amer. Math. Monthly, vol. **116**, Number 1, 2009, pp. 67-68.

[116] A. Selberg, *On discontinuous groups in higher-dimensional symmetric spaces*, Contributions to function theory (Internat. Colloq. Function Theory, Bombay, 1960) pp. 147-164 Tata Institute of Fundamental Research.

[117] I. Shafarevich, A. O. Remizov, *Linear Algebra and Geometry*, Springer, Berlin, 2012.

[118] E. Sperner, *Neuer Beweis für die Invarianz der Dimensionszaht und des Gebietes*, Abh. Math. Sem. Hamburg, vol. **6**, 1928, pp. 265-272.

[119] D. Sullivan, Geometric Topology. Localization, Periodicity, and Galois Symmetry (The 1970 MIT notes), 2005.

[120] D. A. Suprunenko, *Matrix Groups* Translations of Mathematical Monographs, vol. **45**, AMS 1976.

[121] J. Tits, *Free subgroups in linear groups*, J. Algebra **20** (1972), 250-270.

[122] M. Tărnăuceanu, *A characterization of the quaternion group*, Ann. St. Univ. Ovidius Constantsa, vol. **21**(1), 2013, pp. 209-214.

[123] W. Vasconcelos, *Injective Endomorphisms of Finitely Generated Modules*, Proc. Amer. Math. Soc., **25**, 1970, pp. 900-901.

[124] V. S. Vladimirov, *P-adic Analysis and Mathematical Physics*, World Scientific Publishing Co., 1994.

[125] W. Voigt, *Die fundamentalen physikalischen Eigenschaften der Krystalle in elementarer Darstellung*, Verlag von Veit and Comp., Leipzig, 1898.

[126] E. Waring, *Meditationes Algebraicae* Cambridge, England, 1770.

[127] L. Washington, *Introduction to Cyclotomic Fields*, Springer, Berlin, 2nd edition, 1997.

[128] J. H. Maclagan-Wedderburn, *A theorem on finite algebras*, Trans. of the Amer. Math. Soc., **6** (1905), pp. 349-352.

[129] C. A. Weibel, *An Introduction to Homological algebra*, Cambridge University Press, 1994.

[130] A. Weistein, *Symplectic Geometry*, Bulletin A.M.S., vol. **5**, no. 1, 1981.

[131] H. Weyl, *The Classical Groups, their invariants and representations*, Princeton University Press, Princeton, N.J., 1939, 1946.

[132] H. Whitney, *Tensor Products of Abelian Groups*, Duke Math. Journal, vol. **4**, 1938, pp. 495-528.

[133] H. Whitney, *Elementary structure of real algebraic varieties*, Ann. of Math. **66**, 1957, 545-556.

[134] E. Witt, *Über die Kommutativität endlicher Schiefkörper*, Abhandlungen aus dem Maithematischen Seminar der Hamburgischen Universität, **8** (1931), p. 413, see also Collected Papers-Gesammelte Abhandlungen, ed. by Ina Kersten, Springer 1998.

[135] O. Zariski and P. Samuel, *Commutative algebra*, Vol. 1. With the cooperation of I. S. Cohen. Corrected reprinting of the 1958 edition. Graduate Texts in Mathematics, No. **28**. Springer-Verlag, New York-Heidelberg-Berlin, 1975.

[136] O. Zariski, *Generalized semi-local rings*, Summa Brasiliensis Math. **1**, fasc. 8, 1946, pp. 169-195.

[137] O. Zariski, *A new proof of Hilbert's Nullstellensatz*, Bull. Amer. Math. Soc., vol. **53**, 1947, pp. 362-368.

[138] H. J. Zassenhaus, *A group-theoretic proof of a theorem of Maclagan-Wedderburn*, Proceedings of the Glasgow Mathematical Association, **1** (1952), pp. 53-63.

[139] G. M. Ziegler, *Lectures on Polytopes*, Springer-Verlag New York, Inc., 1995.

Index

Printed in the United States
By Bookmasters